Honey Bee Social Evolution

A solitary female corbiculate bee, singly mated, and her cohabiting daughters exemplify the cusp of eusociality.

Honey Bee Social Evolution

Group Formation, Behavior, and Preeminence

KEITH S. DELAPLANE

JOHNS HOPKINS UNIVERSITY PRESS BALTIMORE

Printed in the United States of America on acid-free paper
9 8 7 6 5 4 3 2 1

Johns Hopkins University Press
2715 North Charles Street
Baltimore, Maryland 21218
www.press.jhu.edu

Library of Congress Cataloging-in-Publication Data

Names: Delaplane, Keith S., author.
Title: Honey bee social evolution : group formation, behavior, and preeminence/ Keith S. Delaplane.
Description: Baltimore : Johns Hopkins University Press, 2024. | Includes bibliographical references and index.
Identifiers: LCCN 2024010346 | ISBN 9781421450032 (hardcover) | ISBN 9781421450049 (ebook)
Subjects: LCSH: Honeybee. | Social evolution in animals. | Honeybee—Behavior. | Honeybee—Evolution. | Beehives.
Classification: LCC QL568.A6 D45 2024 | DDC 595.79/9—dc23/eng/20240613
LC record available at https://lccn.loc.gov/2024010346

CONTENTS

PART III
Honey Bee Phylogeography

PART IV
Beyond the Honey Bee Superorganism

PREFACE

THIS BOOK DEVELOPED out of a monthly serial in the *American Bee Journal*, "the oldest English language beekeeping publication in the world," published since 1861 by Dadant & Sons of Hamilton, Illinois. My articles on honey bee social evolution ran continuously there between January 2015 and August 2018, and I am grateful for permission from the Dadant family to rewrite some of that material for this book. My initial purpose in writing those pieces was to give beekeepers an underlying biological and evolutionary explanation for phenomena they see every time they open a bee hive. I wanted them to know the evolutionary reasons why honey bees behave the way they do. I also wanted to encourage beekeepers to observe the strategies that honey bees and their ancestors have adopted over millions of years for handling diseases and parasites and to learn, where possible, how modern beekeeping management can partner with those processes to optimize bee health and productivity.

But as time went on, it became clear that the story was larger than the stage I was giving it. Much of what I was writing was fodder for wonder, not for optimizing profits. Those monthly articles began saying less about beekeeping and more about the structures of organismal evolution on

Earth—processes that are recapitulated in the bee colony and whose traces are observable to anyone equipped with an observation bee hive and appreciable by anybody with a curious mind. When it comes to hearing wonders of the natural world, there are many ready to listen, both inside beekeeping and without. This book is for all of them. In every chapter, I bring the latest peer-reviewed science to the question of honey bee social evolution; but being aware that my audience draws from all walks of life, I made it my goal to present complex scientific ideas in a way any reasonably informed person can follow. At the same time, I emphasize the primary literature throughout, so this book is a credible statement of the state of the knowledge.

So, who should read this book? Scientists, amateur naturalists, graduate students, biology majors, beekeepers—yes, but also home gardeners, conservationists, foresters, lovers of bees and insects, and anyone sensitive to the great themes of life on this planet. It's all here in the honey bee, who, in my most expansive moments, I call a window into everything.

Everything? Well yes, I know that sounds pretentious. But at numerous junctures as I reflected on this book, I strayed from my learned reserve and thought, No, it isn't. The honey bee really *is* an insight into everything—or at least that part of everything that deals with the evolution of organismal life, which is a lot of everything. That epic story sweeps us into its wake—we, that is, *Homo sapiens*, those big-brained, linguistic, encultured bipedal apes who in modern times have learned so effortlessly to dissociate ourselves from the rest of the natural world and live as if we are somehow outside-nature-looking-in. Studying honey bees can cure us of that delusion.

The titanic forces that drive species toward complex eusociality are addressed in the nine chapters of Part 1; these chapters are the theoretical center of the book and foundational to all that follows. The eleven chapters of Part 2 are shorter, each dealing with a topical outcome or emergent property of complex eusociality. The two chapters of Part 3 contextualize the eusocial honey bee in its phylogeny and biogeography, and Part 4 lengthens our topical horizon by exploring what the end of a eusocial lineage might look like as well as the extent to which the most intelligent and ecologically dominant animal on Earth meets the criteria for eusociality.

Setting down this story required an extended immersion in the scientific literature on honey bee social evolution, including my own contributions as a practicing scientist of over 40 years. But literature never exhausts what its authors have to say. For that extra insight, at numerous junctures in this work, I turned to colleagues and others engaged at the beachhead of knowledge's advance. The following individuals gave me helpful suggestions to drafts of this work, provided images, proofed images, or helped me understand points of their specialty. I thank Paul Arnold (Young Harris College), Silas Bossert (Washington State University), Michael Burgett, (emeritus, Oregon State University), Sydney Cameron (University of Illinois), James Cane (ARS Pollinating Insects Research Unit, Logan, Utah), James Carpenter (American Museum of Natural History), Bobby Chaisson (Georgia Bee Removal), Julie Cridland (University of California, Davis), Bryan Danforth (Cornell University), David De Jong (University of São Paulo), Kathleen Dogantzis (York University), Lonnie Funderburg (Oneonta, Alabama), Andy Gardner (University of St. Andrews), Aleš Gregorc (University of Maribor), Brock Harpur (Purdue University), Greg Hunt (Purdue University), Holli Kircher (Hartwell, Georgia), Eugene Kritsky (Mt. St. Joseph University), Wyatt Mangum (University of Mary Washington), Alice Marzocchi (National Oceanography Center, Southampton), Joe McHugh (University of Georgia), Cristiano Menezes (Embrapa Amazônia Oriental), Jeffrey C. Miller (emeritus, Oregon State University), Ben Oldroyd (University of Sydney), David Queller (Washington University in St. Louis), Francis Ratnieks (University of Sussex), Benjamin Rouse (Athens, Georgia), Tom Seeley (Cornell University), Steve Sheppard (Washington State University), Joan Strassmann (Washington University in St. Louis), Anthony Vaudo (US Forest Service Rocky Mountain Station, Fort Collins, Colorado), Stuart West (University of Oxford), Michael Young (Hillsborough, Northern Ireland), and Georgia P. Zumwalt (Baci d'Fiori Honey Company).

Special thanks are extended to Robin Moritz (Martin Luther University, Halle-Wittenberg) and Kenneth Ross (University of Georgia), who read and commented on Part I and again to Robin Moritz for his metaphor of the flood and bee-eater in chapter 2; to Lewis Bartlett (University of Georgia) who reviewed chapter 19; to Jack Gilbert (University of California, San Diego) who reviewed chapter 10; to Olav Rueppell (University of Alberta) who reviewed and commented on my treatment of life history theory in chapter 3; to Deborah Smith (University of Kansas) who reviewed and commented on Part III, and to numerous anonymous reviewers. Any remaining errors of fact, interpretation, or omission are strictly my own. Kelsey Forester (University of Georgia Science Library) gave tireless assistance with bibliographic software.

All images and illustrations by the author and Sonja Delaplane unless otherwise indicated.

PART I

Structures of Honey Bee Social Evolution

CHAPTER 1

Honey Bee Social Evolution and Why It Matters

THIS BOOK TELLS THE STORY of a specially charismatic animal, the western honey bee, *Apis mellifera*. This is an animal that almost everyone knows something about. Foremost, they sting. They also make honey and pollinate crops and live in white stacked boxes, the interior architecture and workings of which, for most people, might as well be a magician's chest of mysteries. What their equally mysterious white-coveralled and anonymously veiled keepers do when they're working them could as easily be the rites of an extinct druid sect. There is general public awareness that "bees are dying," with bad implications for our food supply, an awareness indebted to an interested press and a consistent message from environmental educators on the necessary connectedness of humans to non-human nature.

But I'll wager that right now, before you read another word, you have another, more fundamental, insight into the honey bee whether you've ever thought about it, and it's this: the honey bee can be considered in two different ways, or levels. There's the honey bee that visits a clover blossom in your lawn. And if you've ever had a course in biology, you might remember that this flower visitor has three body segments, four wings, and six legs like all the other 5.5 million insect species on Earth (Stork, 2018).

But you probably also know that this flower visitor flies home to, and belongs to, a much larger, more complex entity, the colony—a nest of many, many others very similar to herself (for it is a she) who live together and interact with one another at a pitch of integration that even casual observers know to be qualitatively different from other kinds of social contact.

A family maybe? Well yes, but not quite. You've probably also heard the terms "queen" and "worker" applied to the relationships in this colony—words that, whatever else they mean, connote something more than "parent" and "offspring."

And while we're at it, what about the word "colony" itself? What biological realism is this anthropocentrism trying to attach to? An outpost of empire? What empire? Alas, here we find the limits of metaphor because I think "colony" is simply what lexicographers came up with when they intuited the insufficiency of "family" and had nothing better to suggest.[1] We know better now, and this book is all about that.

Thus, we have a honey bee as an individual, and a honey bee as a member of a colony. This is one difference between a solitary individual and a social one, or as you'll come to know, a *eusocial* one. The vast majority of organisms on Earth are solitary. For a solitary, social encounters within one's own species are accidental or ephemeral—things like territorial fighting, sharing a water hole, or sib juveniles dispersing from a common egg cluster. Mating is really the only essential exception. Solitary individuals are all about foraging, defending, and reproducing. The honey bees are too, but with a difference—they do it as a colony, and that's what makes them different than a family whose members leave the natal home to strike out on their own. Family members are reproductively autonomous. Colony members aren't.

This is all interesting in its own right, but the latest scholarship is hinting at bigger things—that the evolution of the honey bee recapitulates processes that are common to the evolution of all organismal life, including our own. Organismal evolution *is* social evolution.

How so? Well, for years now, evolutionary biologists, for whom honey bees have been an ever-ready model, have been showing, along with their colleagues

1. Some languages retain the familial emphasis. In Slovene, for example, "honey bee colony" translates to "družina medonosne čebele," or "honey bee family."

in ecology and quantum physics, what the mystics in all great spiritual traditions have intuited for centuries: the foundational, necessary, integral, and literal extent to which all of nature is connected. The science of ecology is little else than the dedicated study of relationships among organisms, species, populations, and ecosystems—how interacting relationships of parts create wholes that behave qualitatively differently than the sum action of their members.

Quantum physics is showing the almost eerie entity construction that arises out of interaction. Carlo Rovelli (2016) described this in a discussion on one of the discipline's early twentieth-century pioneers, German physicist Werner Heisenberg:

> Heisenberg imagined that electrons do not *always* exist. They only exist when someone or something watches them, or better, when they are interacting with something else. They materialize in a place, with a calculable probability, when colliding with something else. The "quantum leaps" from one orbit to another are the only means they have of being "real:" an electron is a set of jumps from one interaction to another. When nothing disturbs it, it is not in any precise place. It is not in a "place" at all.

This led Rovelli to conclude rhetorically:

> What does this mean? That the essential reality of a system is indescribable? Or does it mean that we lack only a piece of the puzzle? Or does it mean, as it seems to me, that we must accept the idea that reality is only interaction.

Reality is only interaction. Okay; if that's true, then it should not surprise us when the evolutionary biologists tell us that the same kinds of dynamics apply to the evolution of organisms: interactions between lower levels of biological organization lead to higher levels of biological organization. In other words, organisms—those packages of tissues and organs and gonads that describe you and me and pretty much all biological life on this planet—derive from the accretion of lower-level entities. Emergent properties—complex organization arising out of the actions of independent actors—bubble up from the cells or worker bees comprising those new entities, adding yet more information as scaffolding for natural selection to act upon.

Moreover, the lower entities do not become higher levels by being swallowed up or annihilated—they do not lose their contiguity like so many drops of water in an ocean—but rather they are integrated, incorporated, re-engineered, re-lifed, and graduated into newer, greater wholes. There's plenty here for the spiritually minded to think about. But for now, let's settle on the fact that in the honey bee colony, we see a recapitulation of the evolutionary phenomena that gave rise to organisms—us. This theme is woven throughout this book.

My reasons for thinking honey bees are a window into everything have increased and matured. In my early years as a professor in an agricultural college, it was all about beekeeping. I would recite a litany in class: "Beekeeping speaks to biology, evolution, genetics, behavior, ecology, farming, conservation, botany, horticulture, integrated pest management, food science," and then I'd diverge a little, "ag engineering, mechanical engineering, woodworking, manufacturing systems, artificial intelligence, marketing, business administration," and then diverge a lot, "feeling my oats" as my farmer father would say, adding the humanities "literature, poetry, mythology, and religion," being quick to cite credible anthropological sources such as Eva Crane's (1990) venerable *Bees and Beekeeping: Science, Practice and World Resources* and Hilda Ransome's (1937) timeless *The Sacred Bee*.

None of these reasons has gone away. I still believe beekeeping is distinctive for its wide purchase on all these spheres of human thought and enterprise. But apprehending, now as I approach my mature years, that the honey bee colony images in a microcosm the trajectory of evolution in a macrocosm—a trajectory in which each major evolutionary transition involves subsummation of the levels below it—I am eager to show this greatest superlative of all to others who are bee scientists, beekeepers, entomologists, or anyone whose heart is sensitive to the beauty and connectedness of the natural world.

Evolution

It can be inconvenient when new information comes along and shakes things up. A few years ago, this happened to me in a small way, and it forced me to revise the exams I give entomology students and Master Beekeeper candidates. The original question and answer went like this:

> (True or False) Honey bees are native to the North American continent. False

But now it looks like this:

> (True or False) Honey bees are native to the North American continent. True

All this came about because of a new paper announcing a fossil honey bee, member of the genus *Apis*, discovered in a shale deposit in Nevada dating from 14 to 14.5 million years ago. It was named *Apis nearctica* by its discoverers (Engel et al., 2009) (figure 1.1) and represents members of the genus that moved east from Asia, crossed the Bering land bridge during the early to mid-Miocene epoch, were cut off from retreat by rising seas, and settled down as residents of western North America for a few hundreds of

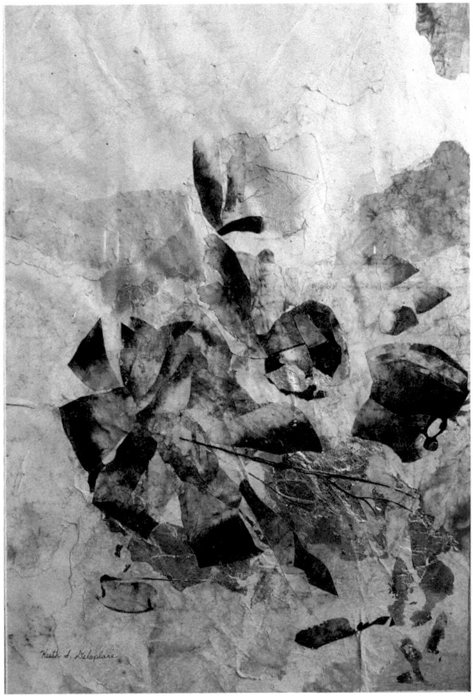

FIG. 1.1. A fossil specimen of North America's only known native honey bee, *Apis nearctica*, discovered from a shale deposit in Nevada dating from 14 to 14.5 million years ago. Original watercolor montage by the author redrawn from figure 2 in Engel et al. (2009). PHOTO: MICHAEL YOUNG.

thousands of years. Absence of additional fossils leaves us ignorant whether *A. nearctica* was the sole North American representative of *Apis*, but it seems unlikely that others didn't make the trip or that the group didn't spin off new species during its long tenure here.

In any case, its history is a story of the power of contingent events, in this case weather. Just as climate-induced low seas allowed its arrival, climate-induced high seas cut off its retreat, and climate shifts from warm-wet to cool-dry spelled its end. Bottom line—this native American honey bee went extinct and left no living descendants, leaving North America empty of honey bees until European humans reintroduced a much later *Apis* in historic times, the western honey bee, *Apis mellifera*, the domesticable maker of honey and beloved ward of beekeepers everywhere (figure 1.2).

This story reminds us that every species we know on Earth today is the outcome of a similar string of contingent events in the history of our planet, a history in which organisms and their genes are constantly being challenged and reshuffled by conditions or accidents of nature. Sometimes members of a population possess a "survivable" set of *phenotypes*—gene-encoded behaviors, morphologies, and physiologies—that lets their possessors reproduce and pass on those phenotypes to the next generation. In this manner, successful gene combinations accumulate in a population and over time literally shape the species, behaviorally and morphologically, into a finely-integrated member of its ecosystem (Darwin, 1859). A negative formulation of that process must also be named and credited: evolution by gene loss. Sometimes called the "less is more" principle, the loss of genes from environmentally-driven changes in selection pressure can be shown to exact adaptive population changes across phyla (Albalat and Cañestro, 2016).

By either mechanism, natural selection operates in an extravagance of time, which means that a species becomes adapted to all of the extremes its habitat can throw at it. This is what makes natural selection different from artificial selection, i.e., human-directed plant and animal breeding: in the window of our short lifetimes, we cannot see all the ramifications of our selections that may work well for our narrow purposes but fall short given a turn in the climate or other natural event that to us appear anomalous but are quite within the range of normal for that environment in geologic time.

FIG. 1.2. The moveable frame hive is the industry standard for modern beekeeping. The nineteenth-century American Congregationalist minister L. L. Langstroth popularized the idea of "bee space"—a gap that bees naturally leave around their combs and walls to permit their movement in the nest. Langstroth reproduced this gap around the walls and wood-reinforced comb frames of his hive design, introducing the first hive that permitted beekeepers to move and remove combs without damaging the colony structure. Standardization and mechanized handling followed shortly thereafter. Most subspecies of *A. mellifera* are highly tolerant of human management under which they can achieve colony populations and productivity far in excess of wild colonies. Honey bees are also responsive to human selection, intentional or inadvertent, for gentleness. Nowadays, it is not unusual for beekeepers to receive few or no stings if they use a modicum of protective clothing and a *smoker*—a device that directs smoke into the hive interior, disrupting the pheromones that normally synergize a group stinging response.

Thus, outcomes of natural selection are comparatively stable and intimately connected to geography of origin, whether at the scale of an African savannah if you're a honey bee or pool of water if you're a snail (Colgan and Ponder, 2000). This is why organisms aren't randomly dispersed around the world: zebras in America or elephants in Europe.

The rough and direct elegance of natural selection is its unbending bias for *what works*: phenotypes that survive and phenotypes that reproduce. If "what works" changes with time or circumstance, the lability of the genetic system underpinning inheritance means that natural selection can favor new sets of phenotypes if warranted. New species can evolve from existing ones in a process called *speciation*.

This brings us to *phylogeography*—the study of how relationships (phylogeny) between sister species or between populations of a species are structured by geography. The field makes explicit a point I made above: natural variation is integrally tied to *place*.

There are a handful of postulates that describe how geographic speciation—or more generally, genetic variation structured by geography—happens (Avise et al., 1987), two of the most meaningful for us in this book I discuss here.

The first form, called *allopatric* speciation, happens when a subset of a species's population is cut off from others. The barrier can be geographic or climatic—something like rising or falling seas or advancing or retreating glaciers, anything that separates one population from another. Left to their new life conditions and deprived of the ability to interbreed, the two populations begin separate histories and separate selection pressures until enough time lapses that the two can no longer interbreed and make fertile offspring, at which point the two are no longer one species but two. A *speciation event* has occurred.

A second form, *sympatric* speciation, happens when a region is colonized by a highly variable founder population, followed by subsequent loss, for idiosyncratic reasons, of local variants. What remains are discontinuous populations of wide genetic divergence.

Whenever we talk about *subspecies* or *races* or *biotypes*, variants within species, we are probably talking about early stages of speciation. But for now, the point is that species evolve from existing species, and this opens up the possibility of shared ancestries, shared DNA, and shared characters, some ancient and conserved and others more recent and derived. Indeed, the disciplines of *taxonomy* and *systematics*, biological nomenclature and classification, are the search for these natural relationships, groups nested within groups; and because we humans are categorizing animals we have created hierarchies of biological organization that we hope correspond to real lineages. These categories descend from older, more inclusive levels to increasingly modern and exclusive. In the case of the honey bee, I show these hierarchies, or *taxa* (singular *taxon*) in table 1.1.

TABLE 1.1. Taxonomy of the western honey bee, *Apis mellifera.*

DIAGNOSTICS BELOW ACULEATA BY MICHENER (2000)

TIME	TAXON	TAXON NAME	DIAGNOSTICS
More ancient and inclusive	Domain	Eukarya	Cells with membrane-bound nucleus and organelles; cells formed by symbiogenesis; taxon contains all animals, plants, fungi, and protists.
↑	Kingdom	Animalia	Multicellular motile organisms, acquire nutrients by digesting organic matter
	Subkingdom	Eumetazoa	Develop from embryo with three tissue layers, possess digestive tract, excludes sponges
	Phylum	Arthropoda	Invertebrate, exoskeletons, segmentation specialized by function; paired appendages
	Class	Insecta	Three body regions or tagmata, one pair antennae, three pair legs
	Subclass	Pterygota	Winged insects
	Superorder	Holometabola	Immatures undergo complete metamorphosis: egg, larva, pupa, adult
	Order	Hymenoptera	Hind wings linked to forewings by hamuli; haplodiploid sex determination
	Suborder	Apocrita	The "wasp-waisted" Hymenopterans
	Infraorder	Aculeata	Ovipositor modified into sting
	Superfamily	Apoidea	Pronotal lobe distinct; pronotum encircles thorax
	Form	Anthophila	The "true bees;" branched hairs, hind basitarsi broader than succeeding tarsomeres
	Family	Apidae	Four ovarioles per ovary
	Subfamily	Apinae	Includes corbiculate Apidae and others
	Tribe	Apini	"True" honey bees, mandibles lack teeth, hind tibial spurs absent
	Genus	*Apis*	Only modern members of Apini
	Subgenus	*Apis*	Medium-sized, cavity-nesting *Apis*
↓	Species	*mellifera*	"Honey making"
More recent and exclusive	Subspecies (races)	*ligustica*, *carnica*, etc.	Italian honey bee, Carniolan honey bee, many others

In technical usage, an ancestral trait is *plesiomorphic* and a derived trait *apomorphic*. These are relative terms, that is, always used in reference to two or more states of a character. The proper identification and chronological placement of these characters is essential in reconstructing phylogenies. A primitive trait common to two or more groups in shared ancestry is a *symplesiomorphy*, whereas a shared derived trait that unites a younger offshoot group and all its descendants is a *synapomorphy*. A derived character that is unique to a group is an *autapomorphy*. Among bees, a constricted waist is a symplesiomorphy they share with wasps, whereas vegetarianism by larvae is a synapomorphy that unites all bees and marks them as an evolutionary lineage distinct from wasps, whose larvae are carnivorous.

A synapomorphy shared by an earliest common ancestor and all its descendants defines a *monophyletic* group. A *paraphyletic* group is created if systematists name a clade with a common ancestor but for various reasons fail to include all its descendants; this is always considered a provisional situation holding until better data and understanding can enable revision of the group into one or more monophyletic taxa.[2] But for now, all bees together are a monophyletic group branching from the sphecoid wasps and defined by the synapomorphies of, among others, plumose hairs, pollen-feeding larvae, and the hind basitarsus (first foot segment) broader than subsequent segments (Michener, 2000) (figure 1.3).

The concept that unifies all this is of course evolution, one of the great narratives of our time—the recognition that life is malleable and adaptive, that organisms are vehicles for transmitting those famously self-replicating molecules DNA and RNA rich in information for building the proteins and other macromolecules that govern morphology and behavior, vehicles whose propagation every generation is subject to the unforgiving test of natural selection, the outcome of which explains the bewildering variants and unanticipatable specializations we find today in organisms across every habitat type and niche on Earth. Evolution is fitful and opportunistic, rife with false starts and surprising successes. It co-opts successful adaptations and ap-

2. We will consider an example of paraphyly in chapter 21, where I discuss the corbiculate ancestors of honey bees.

plies them to new uses. If an evolved trait works in one lineage, we may find something like it popping up in another, making it difficult for modern observers to tell the difference between a product of shared ancestry or multiple independent solutions to the same problem. Most of all, evolution operates randomly and impartially, ready to turn a blind eye to catastrophe—such as the extraterrestrial impact of 66 million years ago that wiped out the dinosaurs (Schulte et al., 2010)—or dole out benefits—such as to those lineages including the mammals that were primed and ready to exploit the new niches opened up by the apocalypse (Cockell and Bland, 2005).

Contemporaneous Relationships between *Apis mellifera* and *Homo sapiens*

To understand honey bee evolution and why it matters, let us turn our attention away from *Apis nearctica*, who for our purposes represents an evolutionary dead-end, and concentrate on the phylogeographic history and social evolution of the western honey bee *Apis mellifera*. The latest thinking is that cavity-nesting *Apis* species emerged in Southeast Asia from about 7–8 million years ago (Dogantzis et al., 2021; Ji, 2021) to 20 million years ago (Smith, 2021) and began expanding north and west into more temperate latitudes. By about 6.5 million years ago, *Apis mellifera* had appeared in western Asia (Dogantzis et al., 2021) and began dispatching various lineal diasporas into Europe, Africa, and the Middle East,[3] a radiation culminating in no fewer than 25 subspecies between the cape of southern Africa and the Arctic circle (Ruttner, 1992).

The diversity of this swath of terra firma represents virtually every terrestrial habitat type on Earth, and *Apis mellifera* was versatile enough to colonize all of it. Among these 25 subspecies, there are some adapted to tropical habitats with constant nectar flows, others adapted to temperate seasonality, and others to gradations in between. This versatility is part of the reason why *Apis mellifera* has become successfully established wherever

3. Alternative scenarios abound. One places the origin of *A. mellifera* in Africa (Whitfield et al., 2006) from which its radiation and diaspora into numerous lineages commenced much later at a scale of hundreds of thousands of years ago (Wallberg et al., 2014).

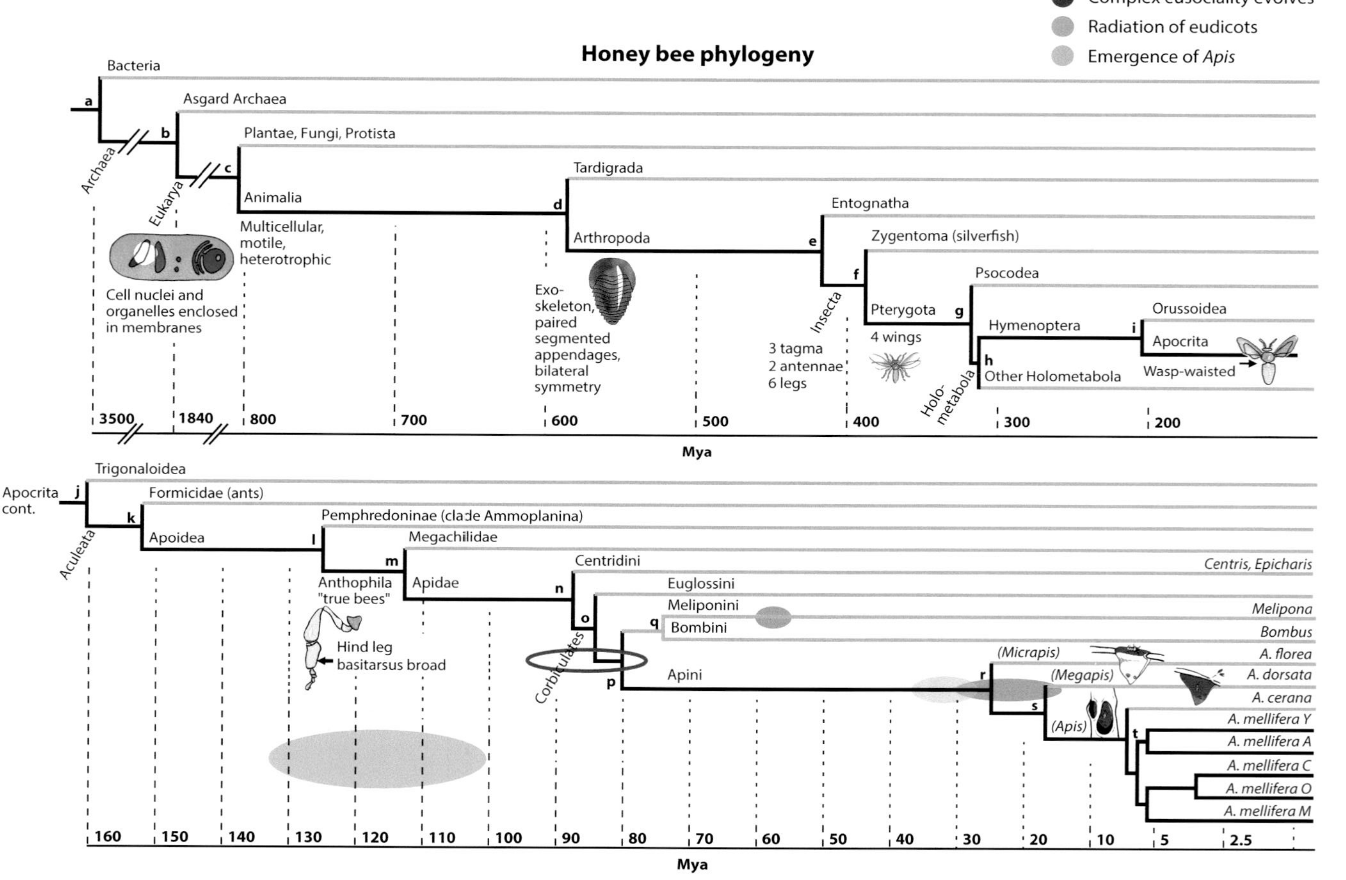

Honey bee phylogeny
Simple eusociality evolves
Complex eusociality evolves
Radiation of eudicots
Emergence of Apis
Bacteria
Asgard Archaea
Plantae, Fungi, Protista
Tardigrada
Animalia
Entognatha
Arthropoda
Zygentoma (silverfish)
Psocodea
Pterygota
Hymenoptera
Orussoidea
Apocrita
Wasp-waisted
Other Holometabola
Archaea
Eukarya
Insecta
Holo-metabola
Cell nuclei and organelles enclosed in membranes
Multicellular, motile, heterotrophic
Exo-skeleton, paired segmented appendages, bilateral symmetry
3 tagma
2 antennae
6 legs
4 wings
a b c d e f g h i
3500 1840 800 700 600 500 400 300 200
Mya
Trigonaloidea
Apocrita cont.
Formicidae (ants)
Apoidea
Pemphredoninae (clade Ammoplanina)
Megachilidae
Apidae
Anthophila "true bees"
Hind leg basitarsus broad
Centridini
Euglossini
Meliponini
Bombini
Apini
Aculeata
Corbiculates
(Micrapis)
(Megapis)
(Apis)
Centris, Epicharis
Melipona
Bombus
A. florea
A. dorsata
A. cerana
A. mellifera Y
A. mellifera A
A. mellifera C
A. mellifera O
A. mellifera M
j k l m n o p q r s t
160 150 140 130 120 110 100 90 80 70 60 50 40 30 20 10 5 2.5
Mya

its European human colonizers moved it. It also didn't hurt that the bee is a prodigious honey producer and tolerant of human management.

If this story sounds familiar, it's because we've heard something like it before—another highly adaptable and resourceful species, our own, *Homo sapiens*, that migrated out of its region of origin and eventually colonized every continent. Rapid evolution in our genus was happening on the African continent contemporaneous to the arrival and spread of *Apis mellifera*. It's unthinkable that these early humans were not keenly aware of these calorie-rich jackpots of honey in hollow trees and rock crevices, as no doubt they were of their defenders! I make no formal attempt to establish evidence for coevolution between the two other than to point out what for me is the best candidate—the honey bee's coordinated and ferocious colony defense reaction, which to this day is explosively evident in the African subspecies *A. m. scutellata*. No lesser reaction would do, it seems, against the world's most dangerous terrestrial predator. Similarly, I make note of the intense fear response in many people today against anything that remotely looks like a bee. Each of these is, I suggest, evidence of our mutual evolutionary influence on the other.

But I think the most fascinating possibility is the question whether honey bees contributed to that most important of all developments that rendered humans, well, human—and that's our oversized brains. It's been established

FIG. 1.3. (FACING PAGE) An abbreviated phylogeny of the western honey bee, *Apis mellifera*. Time scale varies. Lines of direct descent are shown in bold black. The true bees, Anthophila, are comprised of seven families, but only two are shown here—the Apidae, which contains *Apis mellifera*, and its sister family Megachilidae. The genus *Apis* is further divided into three subgenera: the single small-comb open nesters *Micrapis*, the large single-comb open nesters *Megapis*, and the multiple-comb cavity nesters *Apis*. Subspecies of *A. mellifera* fall within at least five evolutionarily distinct lineages: the M lineage of Eurasia, A lineage of Africa, C lineage of Europe, and O and Y lineages of western Asia. Letters at nodes indicate authority cited for divergence date and sister group: [a](Betts et al., 2018); [b](Betts et al., 2018; Williams et al., 2020); [c](Erwin et al., 2011); [d,e,f](Grimaldi and Engel, 2005); [g](Misof et al., 2014; Nel et al., 2013); [h](Ronquist et al., 2012); [i](Branstetter et al., 2017; Ronquist et al., 2012); [j,k](Branstetter et al., 2017); [l](Almeida et al., 2023; Danforth et al., 2019); [m](Almeida et al., 2023); [n](Almeida et al., 2023; Martins et al., 2014); [o–s](Almeida et al., 2023); [t–rest](Dogantzis et al., 2021). Emergence dates for apid eusociality by Cardinal and Danforth (2011). Emergence dates for *Apis* by Kotthoff et al. (2013) and Smith (2021). Dates for eudicot radiation by Ramirez-Barahona et al. (2020). Some terminal branches labeled with crown taxa discussed elsewhere.

that the dramatic increase in brain volumes that distinguishes the genus *Homo* from other primates was fueled by selection favoring high-energy diets. Between 5 and 4.7 million years ago, a difference of only 300,000 years, average brain volumes jumped from 600 cubic centimeters in *H. habilis* to 900 cubic centimeters in *H. erectus* (Leonard, 2002). This time interval is synchronous with the arrival and diversification of *Apis* on the African continent.

Acquiring the energy to fuel rapid brain growth would have favored selection in *Homo* for foraging on the most energy-dense foods. Leonard (2002) pointed out that around 100 grams of meat provides 200 calories, whereas the same amount of fruit yields 50–100 calories, and foliage just 10–20 calories. Today, a single bee colony available to the Mbuti honey bee hunters of the Democratic Republic of the Congo yields 1,398 calories from bee brood (at 17% protein), 555 calories from fat in bee brood, and 6,239 calories from carbohydrate in honey and brood, for a total of 8,192 calories (Skinner, 1991). Such depositories of concentrated calories and protein would not have been overlooked by early *Homo*, hungry for fuel to sustain their energy-demanding brains.

The antiquity of African bee hunting is supported by the existence of modern honey hunters such as the Mbuti people, who during rainy season obtain up to 80% of their calories from bee nests and can subsist for up to a month solely on honey and brood (Ichikawa, 1981). The Mbuti are efficient hunters: one searcher can discover two to three bee colonies during a one- to two-hour hunt within three to four kilometers of his camp (Ichikawa, 1981). Honey remains an important food among the three modern African hunter-gatherer human societies for which dietary data exist (Marlowe et al., 2014).

The earliest direct evidence of ancient bee hunting is a Neolithic cave painting in eastern Spain dating to around 3000 BCE (Kritsky, 2020) (figure 1.4). The image shows a human silhouette, very likely female, dangling before a bee nest on a cliff face with what appears to be a basket in her hand. One can easily imagine her extracting chunks of honey- and brood-laden comb and lowering it to her family below. The extension of this dietary habit to early humans is suggested by skeletal evidence of symptoms of hypervitaminosis A—Vitamin A toxicity—in a *H. erectus* skeleton from Kenya dating to 1.6 million years ago (Leakey and Walker, 1985). This disorder

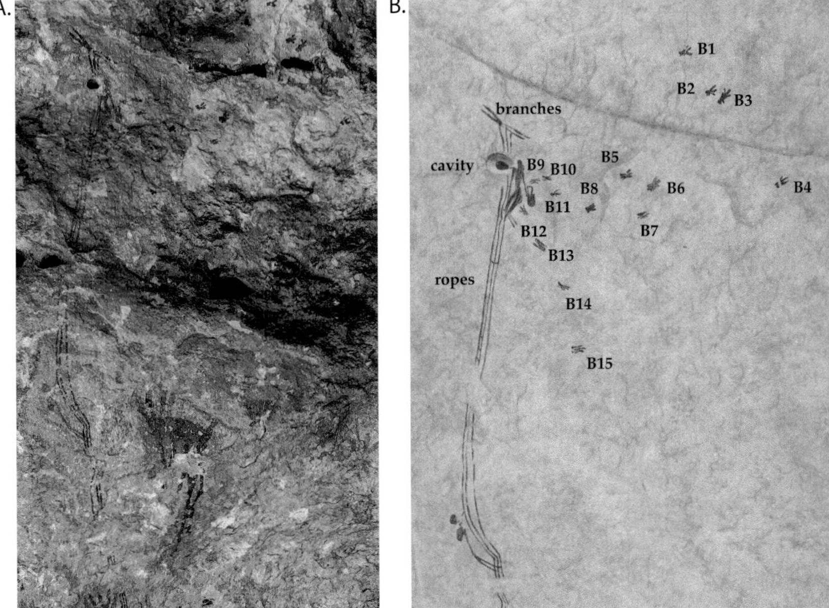

FIG. 1.4. (**A**) Photograph of a Neolithic rock painting from eastern Spain dating to 3,000 BCE showing the earliest known image of human interaction with honey bees. In 1924, Francisco Benitez Mellado painted the rock image as in (**B**). The bees are labeled B1–B15 by Eugene Kritsky over Mellado's painting. There are two honey hunters, the topmost, probably female, is hanging by ropes in front of the bee nest on a cliff face. The ancient artist used an existing cavity in the rock to represent the bee nest. The second human figure is rendered more crudely as a stick image and is positioned below the break in the line of ropes. PHOTO COPYRIGHT: EUGENE KRITSKY, USED BY PERMISSION. IMAGE BY MELLADO IS IN THE PUBLIC DOMAIN.

is normally attributed to excessive consumption of carnivore livers, but Skinner (1991), using modern bee colony densities in the Congo and values for Vitamin A in bee brood, showed that modern bee demographics could present levels of dietary Vitamin A within ranges known to induce human skeletal disorders associated with hypervitaminosis A. This line of evidence suggests that consumption of bee brood was habitual enough among early humans to hazard Vitamin A toxicity.

Marlowe et al. (2014) made the important point that chimpanzees, bonobos, and orangutans all eat honey. Chimpanzees have been recorded on video tolerating stings to acquire it, and they are known to insert sticks into bee nests, withdraw them, and lick the adhering honey—a practice still observed in the modern Hadza people. Hadza men have been observed using stone tools to break into trees to obtain honey. With the advent of fire control around 1.8 million years ago (Goren-Inbar et al., 2004), early humans would have used smoke to calm stinging bees and fire to penetrate and enter nests in hollow trees, greatly enlarging honey harvests.

Altogether, the known consumption of honey by humanity's near relatives, the range overlap between *Homo* and *Apis*, the patent density of protein and calories in time and place afforded by a bee nest, and the known practice of efficient honey bee hunting in modern hunter-gatherer human populations all point to honey bee nests as a nutrient asset for fueling brain development in the ancestors of *Homo sapiens*. In short, bee brood and honey helped fuel our transition to humanhood.

At some point, bee hunting transitioned to bee keeping, a transition which in its most minimal form requires humans to (1) provide bees a suitable cavity in which they can build their nest and the beekeeper maintain them, and (2) to abstain from harvesting all of the honey and brood, allowing the colony to survive the harvest and rebound. Fortunately for us (the jury's still out on the bees), *A. mellifera* responds well to human selection for honey production and gentleness (Collins et al., 1984), facts that no doubt synergized the emergence of beekeeping.

Archaeological evidence for this transition is equivocal. There is evidence of metal casting by the lost-wax method near the Dead Sea in Israel dating to at least 3500 BCE (Kritsky, 2015). With this craft, a model is made from beeswax and encased in clay. When the clay is fired, the wax melts and drains out, leaving behind a cavity mold into which molten metal can be poured to create the cast object. The existence of this industry implies a reliable source of beeswax but does not prove the existence of managed bee hives. The earliest historic indications of an association between humans and bees are the appearance of the Egyptian hieroglyph for "bee" (Crane, 1999) and titular references to an office called "Sealer of Honey" (Ransome, 1937), both dating from the First Dynasty of Egypt, *circa* 3150–2613 BCE.

The earliest known evidence of managed bee husbandry comes to us from the celebrated stone relief panel in the solar temple of Newoserre Any (2474–2444 BCE), sixth pharaoh of the Fifth Dynasty of the Old Kingdom of ancient Egypt (Kritsky, 2015). Here we see scenes depicting beekeepers "calling" a queen, pouring honey into jars, straining honey, and sealing and storing honey. Hives were horizontal hollow clay cylinders with ends covered except for a small flight entrance, a design still in use in Egypt to this day (Crane, 1999).

Thus, the relationship between honey bees and humans is ancient regardless of whether it is measured by the standards of biology or human culture. The relationship is peculiar among similar human cross-species relationships because honey bees straddle the line between wild and domesticated. In their company, we know that we are in touch with something more primal, less altered, than a dog, cow, pig, or goat. There is reason to believe that honey bee brood and honey contributed significantly to the nutrient demands of the evolving human brain. If these are not reasons enough for wonder, in the next section, I begin laying out a case that the processes of social evolution that gave us the honey bee are fundamentally no different than the processes that gave us organismal life, including our own. It is with good reason that honey bees have a permanent grip on the human imagination.

Honey Bees as Models of Organismal Evolution

Honey bees are a truly social, or *eusocial,* species. Most people have an intuitive understanding of this. It means more than the ephemeral societies that pop up in the life cycles of solitary species—things like mating aggregations, territorial aggression, hatchlings emerging from an egg cluster, or even a family of cockroaches living under a sink, even though the idea of parents and offspring gets us moving in the right direction. We will talk a lot in this book about eusociality, but for now, suffice it to say that eusocial life is focused on a stationary nest with a large population of individuals more or less related and exhibiting a high degree of behavioral coordination that promotes survival of the group. Honey bees are by no means the only such species, nor anywhere close to the most complex—ants and termites take that prize. But honey bees are the most well-known and have the distinction of being manageable. The same standardized equipment that is stock-in-trade for the beekeeper serves the bee scientist equally well when conducting experiments. So it's no surprise that honey bees have made an outsized contribution to our understanding of social evolution.

One of the biggest ideas to emerge out of that body of research is the realization that in eusocial insect colonies, we see a recapitulation of the same kinds of evolutionary gravity that pulled together organisms like you and me. A eusocial colony is a living demonstration that life on Earth happens as a sequence of nested hierarchies with lower levels of biological organization subsuming into higher units. Just as single-celled eukaryotes pooled together into colonies—one can imagine their organic mats floating on the primeval waves—with cells specializing into different functions, each cell subordinating its genetic interests to the whole, culminating in the *multicellular organism*, so do worker honey bees specialize and subordinate their genetic interests to the colony. Sociality is not just an odd quirk of nature, an ant farm to amuse a child (or grown-up); sociality is the way nature makes organisms, and organisms are the way life works.

When we feel that irresistible pull toward an ant farm or an observation bee hive, it's akin to those breath-holding moments in nature when we happen upon a mother deer and her fawn; we're seeing an *animal* as it really is. But imagine further that you can peer inside the deer and watch how its body works; and I'm not imagining X-ray glasses letting me watch things happen inside the body like blood circulating or food moving along the gut. Rather, I'm talking about the deer's cells that differentiate and perform the functions of circulation, digestion, and all other systems; I'm talking about the neural excitations that process sensory data and convince the deer to stay put or flee; I'm talking about the way the deer works at a cellular level. When we look at an observation bee hive, we're seeing into the colony at that level of detail, with the advantage that the "cells" going about their functions are themselves organisms and observable to the human eye.

With that simile of the deer, I have snuck in another concept that will color these pages—the idea that the parallels between organisms and eusocial colonies are not only remarkable but are also grounded in convergent evolutionary impulses. Subsummation and the compromise of interests have happened for both: cells into an organism for the deer; organisms into a *superorganism* for the bees. Something new has emerged; another level of biological organization has been achieved. The term superorganism captures this.

If a honey bee colony recapitulates the same kinds of biology that made organisms like us, then this is the most excellent reason for studying them.[4] In his book *The Meaning of Human Existence*, E. O. Wilson (2014) pointed out that the humanities study their subject—human beings—with the breadth, sophistication, technical acuity, and exhaustiveness one can expect of any mature scholarship. But if we bring to our understanding of human nature only the fruits of the humanities—human behavior and psychology; human cultural, political, and economic history; human philosophy, art, literature, and religion—while marginalizing, ignoring, or worse, denying the very concept of a human *natural* history, then we do so at the peril of misunderstanding Earth's apical species and visiting all the dangers on ourselves and the non-human biosphere that accompany that ignorance. If we hold human evolutionary biology somehow outside our mosaic of self-knowledge, we will be ignorant of the elemental dynamics that knit our bodies together and explain our deepest impulses, temptations, loves, and passions. We will not know how to live with ourselves nor the rest of the world.

But if we do peer back into that older history, we will find adaptive advantage rewarded for such traits as cooperation, compromise, altruism, and nuanced conflict resolution between self and group. If these sound like unlikely virtues to be attached to evolution, it may be because we are influenced by the caricature of it so often trafficked in common speech and for rhetorical advantage by its detractors, where slogans such as "survival of the fittest" paint evolution as manifest selfishness. There is indeed a thread of truth to that: natural selection is, to our experience, uncompromisingly partial to genes that are fit and unforgiving to those that aren't.[5] But it's also true that the way for a gene to be fit almost always means to cooperate and bundle

4. In this chapter, I have moved fluidly across levels of selection, seeing social forces at work at a scale as diverse as cooperating nucleic acids to cooperating tribe members. Although the social principles are general, each level demands its formal conditions and will be treated accordingly in later chapters.

5. Note that I have slipped anthropomorphisms into this sentence, describing evolution as "partial" and "unforgiving." This is bad form among purists, but I'm not that kind of purist. It does not demand volition of weather to say that a desertification process that lasts a hundred years is "partial" to drought resistant genes or "unforgiving" to drought susceptible genes. There is a kind of scientific writing that is precision-wise and pedagogic-foolish.

together with other genes. Evolving this cooperation is existentially necessary for the appearance and maintenance of organisms.

It is time that this fuller understanding of evolution become more general knowledge. Not only does it dull the edge of the caricature used in popular culture to repel people from taking our own natural history more seriously, but it gives warrant to our better impulses as members of human communities and societies and the greater biosphere.

And for those with eyes to see and ears to hear, it's all modeled in the honey bee superorganism. Let's see what she has to teach us.

CHAPTER 2

Organismality, Individuality, and the Preeminence of Group Formation

WE WILL TALK A LOT in this book about organisms and superorganisms. So let's first spend some time thinking about what an organism is. One would think that by the twenty-first century, this would be settled, but the question is still surprisingly open.

Definitions

In its traditional usage, *organism* applies to any life form, i.e., an organic package of one or more cells collectively capable of responding to external stimuli, procuring and metabolizing nutrients, maintaining homeostasis and development, and propagating its genes. Writers in the early twentieth century emphasized such observable characters as the organism's unity, integration, and separateness from other such entities [see for example Wheeler (1911) and Huxley (1912)].

When it comes to a single-celled free-living life form, one may refer to it as an organism, but it's more informative to refer to it as either a *prokaryote*—lacking a membrane-bound nucleus, or a single-celled *eukaryote*—possessing a membrane-bound nucleus and other organelles (figure 2.1). The prokaryotes are primitive and the eukaryotes advanced, or, to use value-neutral

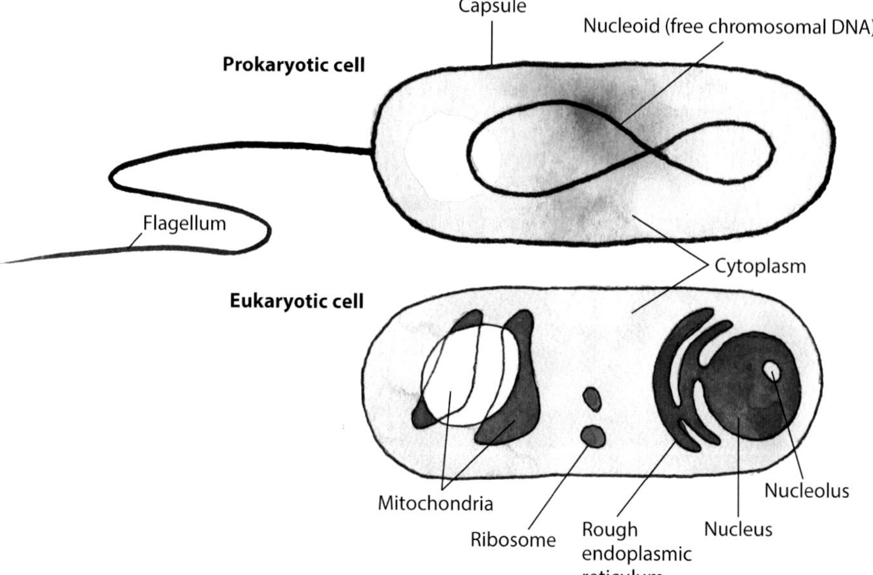

FIG. 2.1. The cell types of living things on Earth are divided into two groups—the prokaryotes and eukaryotes. The chromosomal DNA in a single-celled prokaryote (top) is not contained within a membrane-bound nucleus but rather a loosely organized region called the *nucleoid*; neither does a prokaryote possess membrane-bound organelles. Reproduction is by binary fission. In a eukaryotic cell (bottom), the chromosomal DNA is enclosed within a membrane-bound nucleus and there are numerous membrane-bound organelles with distinctive functions. Reproduction is either asexual or sexual. Prokaryotes today are represented by the Bacteria and Archaea, and together with the Eukaryota, these taxa make up the two domains of all cellular life. Only in the Eukaryota did single cells begin the cooperative actions that led to the evolution of complex multicellular organisms such as fungi, plants, and animals.

language, they are ancestral or derived, which establishes relative position in a lineage. As if to emphasize that "primitive" does not mean "inferior," the prokaryotes persist to this day in legion, including all bacteria as well as the archaebacteria, a group that includes beneficial human gut microbes (Bang et al., 2014) and a large fraction of the *extremophiles*—unicellular organisms adapted for harsh conditions of temperature, pH, and salinity (Woese et al., 1978). Invoking their proper names, the Bacteria (or Eubacteria), Archaea, and Eukaryota (or Eukarya) constitute the three *domains* of life (Woese et al., 1990), although in recent years, the traditional three-domain model has been challenged with a two-domain model where the Eukarya are nested inside the Archaea (Doolittle, 2020; Williams et al., 2020). I incorporated this newer model into figure 1.3.

It is now widely accepted that the first eukaryotic cells were interspecific partnerships, and eventually fusions, of formerly free-living prokaryotes. To this day, ghosts of those fusions persist in the cells of each of us—our mitochondria, those *organelles* (intracellular organs) responsible for the cell's energy metabolism who retain their ancient DNA more or less independently of the cell's, which resides in the nucleus (figure 2.1). Mitochondria are quite simply ancient aliens living in our cells. Among plants, the same can be said of chloroplasts—those organelles in the plant cell responsible for photosynthesizing sunlight into carbohydrates and, like the mitochondria, possessing their own DNA as evidence of their ancient origins as independent prokaryotes who were engulfed by others and never let go.

This idea of different species fusing together to form a new organism has a name—*symbiogenesis*, whose Greek translates as "creation of new life by the fusion of life forms already living in close association." Moreover, some authors consider symbiogenesis to be a defining feature of the Eukarya (Margulis and Chapman, 2010) (table 1.1). As a hint of things to come, this pattern of fusing lower entities into higher has been repeated over and over in the Eukarya; it is not limited to fusion of different kinds or species but happens in members of the same species as well, as the next level will show us.

That next level is the eukaryotic *multicellular organism*, all representatives of which reside in one of two clades, the Plantae + Fungi + Protista or our own and the honey bees', Animalia (figure 1.3). Called by Strassman and Queller (2010) the "paradigmatic organisms," these are the kinds of organisms most of us think about when we hear that word. This is the kind of organism Darwin had in mind when he wrote *On the Origin of Species*.

This important and most familiar type of organism was formed by the fusion of numerous like eukaryotic cells into an integral unit, with strong division of labor among its member cells and exhibiting near complete suppression of internal genetic conflicts. This non-conflict is brought about at some point in the life cycle by a single-cell bottleneck—a *zygote*—a fertilized 2N egg which subsequently divides over and over, resulting in a body with two cell types—*diploid* (2N) *somatic* or body cells making up the functional body tissues, and *haploid* (1N) *gametes* or *germ* cells comprising the female eggs and male sperm. Any gene of any somatic cell has, in theory, the same 0.5 chance of being represented in a gamete. Because all somatic cells derive by fission from one cell, somatic cells are clones, and by definition their genetic conflict is zero. However, this clonality is not absolute because time and enlarging body size, that is, the number of cell divisions distant from the original zygote, admit the possibility for gene mutations. For example, the somatic cells of an old elephant are more genetically diverse than those of a young mouse. Somatic clonality is nevertheless an important distinction between this and other levels of organsimality. Because their genomes are identical, somatic cells differentiate into different tissues and functions by *epigenesis*—the regulated expression, or suppression, of cellular genes during development.

This is a good place to insert the definition of an *animal*: a kind of multicellular organism that is motile for at least part of its life cycle and—in contrast to plants and algae that produce their own nutrients by photosynthesis—feeds on organic matter, digesting it to acquire its constituent nutrients.

A *superorganism* is an integrated biological entity whose fundamental units are themselves multicellular organisms. The idea that social insect colonies bear many resemblances to organisms was first expressed in 1889 by Weismann, but it was Wheeler who coined the term superorganism in its modern usage in 1928. In this book, we are primarily concerned with the honey bee superorganism—the colony whose tens of thousands of worker bees is each an organism, yet together make up a qualitatively different higher-level unity. In contrast to multicellular organisms, the constituent parts (worker bees) are not clones. Genetic conflict is therefore a greater hazard for superorganisms, but we will see that this level also succeeds at repressing conflict, albeit through different means.

A Brief Primer on Nucleotides, Nucleic Acids, and Chromosomes

Genes are sequences of *nucleotides*—the molecular building blocks of the nucleic acids DNA and RNA. A gene is a genomic sequence (DNA or RNA) directly encoding functional protein or RNA product molecules. DNA occurs as a double helix, a twisting ladder-like structure. Each step of the DNA ladder is made of two of the four nucleotides adenine (A), thymine (T), cytosine (C), or guanine (G). Nucleotides on opposing strands of the double helix always pair in the manner of AT or CG. The four DNA nucleotides constitute the alphabet of inheritance, the numbers and sequences of A, T, C, and G that occur along a strand of DNA literally encoding protein-building instructions in the same way that human alphabets encode words. In eukaryotes, during periods between cell divisions the DNA is copied in preparation for the next division, permitting an exact copy to be transmitted to the daughter cell. Enzymes unravel the DNA helix and attach appropriate nucleotides to the exposed sites, creating a new strand; thus, the parent DNA strand serves as a template for its new opposite (figure 2.2). Minor errors in DNA duplication occur from time to time; these *mutations* are an important source of genetic change, or variation, for natural selection to work upon.

In RNA, the thymine is replaced by uracil (U), which pairs with adenine. RNA usually occurs as a single strand folded back on itself (figure 2.3). This structure is architecturally less stable than the double helix of DNA. One of RNA's main jobs is protein synthesis—making a template of the sequence of nucleotides occurring along a strand of DNA in the nucleus and transporting that information to the ribosomes in the cell's cytoplasm where amino acids—protein precursors—are assembled in the appropriate order to construct the gene's protein. The fundamental unit of information in this system is the *codon*—a sequence of three successive nucleotides (i.e., TGC or UGA), any one of which codes for one amino acid. There are many types of RNA, some protein-encoding and others involved in regulating protein synthesis or activating or suppressing gene expression. Turning genes on and off is important in multicellular organisms for clonal somatic cells to specialize into different functions. Many viruses use RNA as the vehicle for transmitting their genome.

Each *chromosome* occurs as a single molecule of DNA folded densely into a single cylindrical package. In sexually produced organisms, there are pairs of homologous chromosomes in the nucleus of each somatic cell, one of the pair from each parent. The number of homologous chomosomes is called *ploidy* and varies by species: in honey bees, it is 32 (16 from each

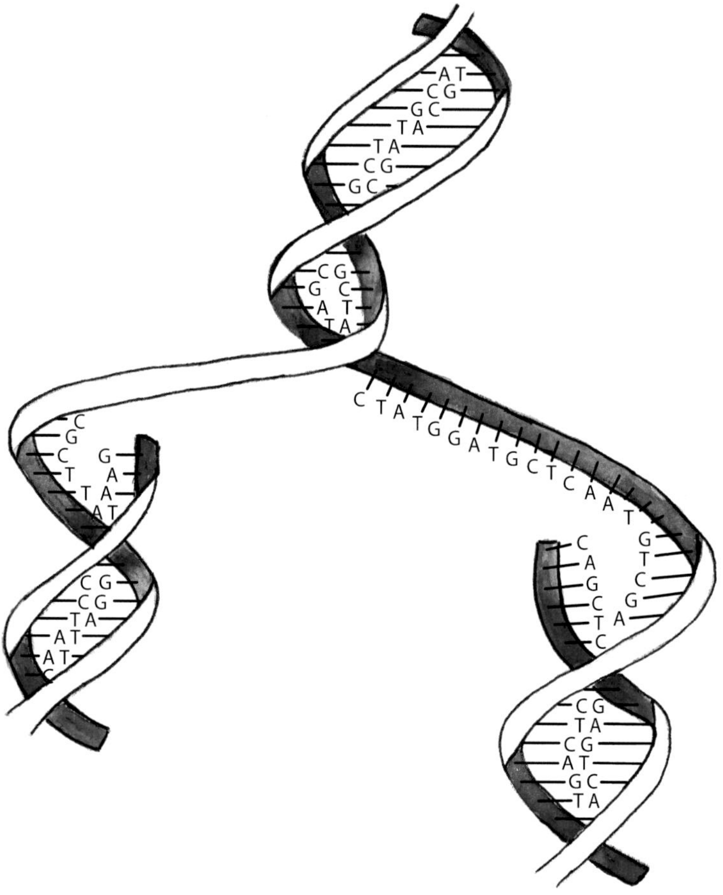

FIG. 2.2. DNA replication showing the parent molecule unraveling to create two identical copies of itself.

parent); in humans, it's 46. During DNA replication, each chromosome duplicates itself, each strand now called a *chromatid* and joined to the parent strand at a site called the *centromere* (figure 2.4). This is the moment at which chromosomes assume their well-known X or V shapes, depending on the location of the centromere. In the final stages of cell division (*mitosis*), each chromatid pulls away from its mate and joins one of the newly formed cells.

Meiosis is a type of cell division that happens in the gonads of sexual organisms to produce gametes—eggs or sperm. One diploid (2N) parent cell gives rise to four daughter cells, each with only one set of chromosomes (haploid, or 1N). Moreover, each haploid daughter cell is genetically unique

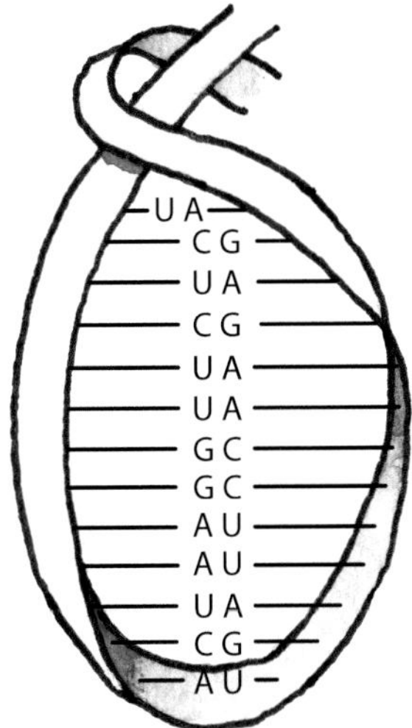

FIG. 2.3. The single-stranded RNA molecule.

from the parent cell. This happens by a process called *recombination* or *crossing over* (figure 2.5): homologous chromosomes pair up and exchange sections of DNA before separating to form gametes. In sexual union, the haploid gametes pair to restore full 2N ploidy to the embryo. Honey bees are known for their high rates of recombination, which contributes to genetic diversity within populations (Beye et al., 2006).

Evolutionary Transitions

So far we have focused on terms, but my goal is not to weary you with a list of definitions like so many lecture notes before an exam. Instead, my goal is to set the stage for something I think is eminently worth writing about, and that is the *tour de force* of social evolution—nothing less than the way life is organized: what things count as organisms and how they come to be.

To get there, first consider that there has been a kind of directionality to these descriptions, a vector toward complexification as lower levels are subsumed into higher ones. Are all of these emerging levels organisms? Or at least organisms in the same sense of the word? Were we even expecting a continuum when we asked the question? And if the answer is yes, then upon which of these levels is nature "selecting" when it practices natural selection?

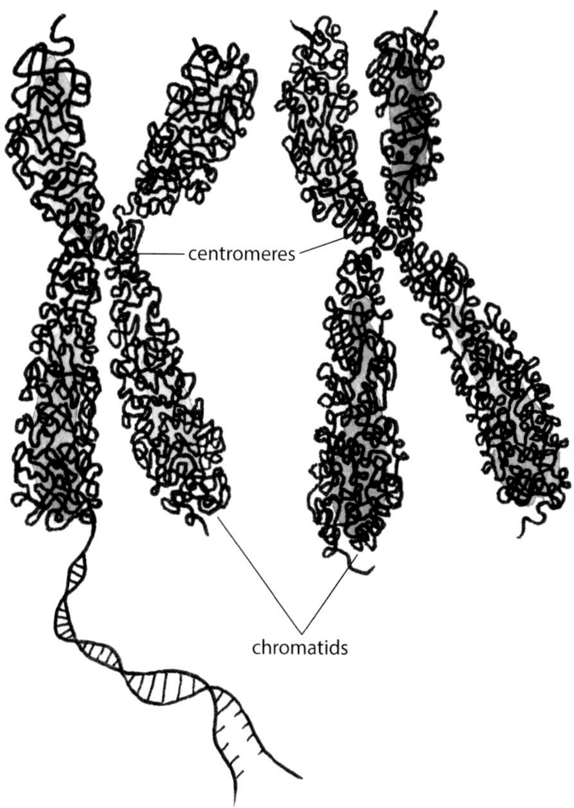

FIG. 2.4. A pair of homologous chromosomes. During DNA replication each chromosome duplicates itself. The resulting new strand, now called a chromatid, attaches with the parent at the centromere.

As classifying animals with a stake in the outcome, it seems important for us to know life's salient units of organization.

The idea of transition points along a time vector is not new in evolution. What twentieth-century school child can forget the imaginatively drawn text book illustrations of the Age of Fish transitioning to the Age of Amphibians transitioning to the Age of Reptiles transitioning to the Age of Mammals? As attractive as these nineteenth-century notions are, they lack any cohering framework for explaining these transitions. They neither seek, nor find, properties common to each level, treating each transition point as if it were a new problem for which evolution had to reinvent the wheel.

This leads us to a general starting point for questions about deep natural history: for every proposed transition, we should ask whether evolution

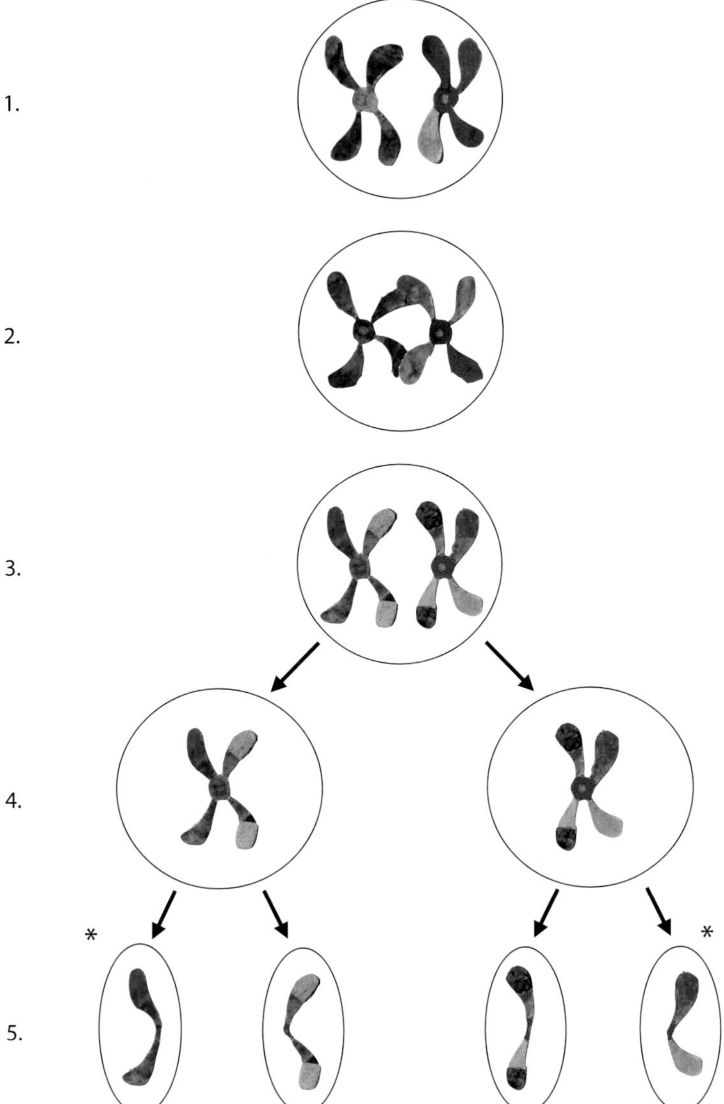

FIG. 2.5. Recombination, also called crossing over. During meiosis in the gonads of each parent, homologous chromosomes pair up in a parental cell (1) and exchange sections of DNA between non-sister chromatids (2–3) before separating into intermediary cells each containing one of the homologs (4). Individual chromatids then separate into one of four gametes (5). The two chromatids indicated with an asterisk retained their original configuration from the parental cell, but this would be very rare in honey bees, which are known to contain among the highest known rates of recombination.

encountered, and solved, a new and idiosyncratic set of problems, or was there instead a small number of recurring problems solvable with the same tool kit? Finding such a unifying problem-solver would satisfy science's bias for *parsimonious* explanations—ones that are most economical in cause and effect. But more importantly, finding such a principle would simplify our understanding of Earth's natural history, provide a theoretic base for understanding the drivers of major transitions, and help us know the important actors in the history of life.

A turning point in the conversation happened in 1976, when Richard Dawkins published the first edition of his watershed book *The Selfish Gene*, which foregrounded the fact that the *gene* is the fundamental and compelling unit of replication. Major evolutionary transitions could now be seen as radically *gene*-centered, with genes bundling with other genes into cells, then into multicellular organisms, and then into superorganisms—the somas of all of which being simply the vehicles for propagating the immortal germ line (figure 2.6). As genes threw their lot together with other genes, it was by extension an exercise in cooperation.

But the essential selfishness of genes is not to be understated. Each gene codes for a specific element of a phenotype, a particular amino acid sequence comprising a protein or a particular RNA molecule—one of a constellation of cellular products responsible for all structures, functions, and behaviors of the cell or organism. If the phenotype spreads by cooperating, then cooperation will be rewarded—but the minute the calculus tips in a different direction, natural selection can be counted on to reward or punish a gene strictly on the basis of its fitness for the present set of environmental circumstances. If selfish behavior optimizes replication for the long term, then natural selection will impartially—by definition, must—reward the selfish behavior. Cooperation may be a winning strategy on planet Earth, but genes are nevertheless always poised for mutiny.

The idea of gene-centered evolutionary transitions is attractive because it focuses on the pertinent players in the game: genes and the natural forces that act on them. However, to identify the actors is not to identify the principles. Again, what properties, if any, are common at each transition? Did evolution solve the same kinds of problems from eukaryotic cell to multicellular organism as it did from multicellular organism to superorganism?

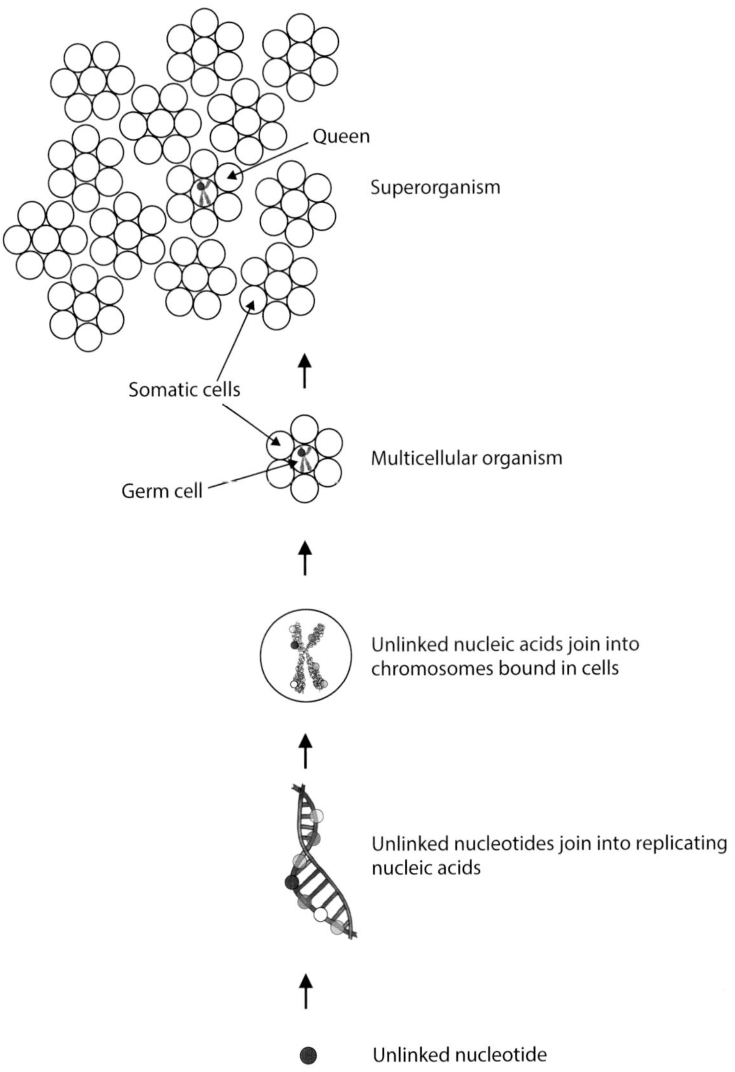

FIG. 2.6. The "immortal germ line" is a perspective on evolutionary transitions that emphasizes that nucleic acids (DNA and RNA) and the genes (sequence of nucleotides) on them are the only self-replicating molecules. From the beginning, there has been a pattern of lower units joining with higher units, each essentially throwing its reproductive fate into the hands of a group. Groups of nucleotides and nucleic acids were acted on by natural selection and responded by coding for proteins that enhanced the group's survival. These group products were at first the architecture of nucleic acids themselves, then chromosomes, then cells to contain the chromosomes. With the arrival of multicellular organisms arose the possibility of specialization among cells, with most cells diverging into "housekeeping" functions that supported propagation of the genome. Thus, there were now two types of cell—the *germ* cells that divide and propagate the genome and *somatic* cells that constitute the vehicle for the genome. Resource limitation resulted in programmed cell death for somatic cells; however, the genome in the germ cells—the self-replicating molecule DNA—is potentially immortal.

In a highly influential work published almost 20 years after Dawkins, Maynard Smith and Szathmáry (1995) offered a table of evolutionary transitions based on major changes in the way information is transmitted (table 2.1). These authors noted three recurring themes that run through these transitions: (1) entities that were reproductively independent before the transition can now replicate only as part of the larger whole to which they have been joined, (2) the joining of entities allowed for task specialization among its parts, division of labor, increasing the fitness of the whole, and (3) there have been fundamental changes in the way information is stored and transmitted: namely, the origin of genetic code, sexual reproduction, and human language.

These transitions defined by Maynard Smith and Szathmáry have been so influential in part because of the breadth of their ambit. Change in information transmittal is a fruitful criterion that captures evolution of human language alongside other transitions that we will in later chapters assign to social evolution. However, it's also possible that change in information transmittal is an *effect* of transition, not a cause. Maynard Smith and Szathmáry treaded near the same precipice as earlier modelers. What dynamic can explain in common the evolution of eukaryotes, the genetic code, sex, and human speech?

Now to be sure, science's bias for parsimony must bend to the real world and not the other way around. It is almost banal to say that evolutionary transitions in the broad sense are driven by more than one dynamic. But it's also true that *transitions that involve the emergence of new kinds of organisms are the work of social evolution.*

This realization began its hesitant emergence during the decades after Darwin (1859, pp. 352–354) aired his ruminations on the problems posed to his theory by the self-sacrificing "neuters" in ant and bee colonies. Generations of biologists carried his torch, working to reconcile the extreme altruism and cooperation expressed by the worker termites, ants, social wasps, and bees with Darwin's predictions that animals should favor their own direct gene transmittal. Social insect biologists, long considered troops dispatched to the enigmatic frontiers of entomology, were dismantling the objections to Darwin's theory throughout the twentieth century, naively uncovering along the way principles common to all of organismal evolution.

Today, a synthesis is happening between two formerly discrete domains of biology: social evolution and the evolution of organismality.[1] *Sociogenesis* is the study of the evolution and maintenance of groups, and we now know that the impulses of sociogenesis are general across evolutionary transitions and not peculiar to some subset of entomology. This has come about by recognizing that two conditions necessary for the emergence of integrated groups (figure 2.7) apply also to the emergence of new kinds of individuals (Bourke, 2011; Queller and Strassmann, 2009; West et al., 2015):

1. Formerly independent replicators join together and now reproduce only as part of the larger unit, exhibiting a high degree of cooperation among parts.
2. The new individual that results is contingent on the elimination of internal conflict.

Queller and Strassmann (2009) envisioned these conditions, cooperation and conflict, as two intersecting continua, leading to a sort of organizer for thinking about social possibilities in evolution (figure 2.8). In their analysis, an organism is a biological unit occupying space in the quadrant of high cooperation and low conflict.

David Queller (1997) nuanced our understanding of the emergence of new kinds of unions when he pointed out that some of them are *egalitarian*—the union of unlike partners, and others are *fraternal*—the union of like partners (completing his revolutionary metaphor by saying that *libertarian* organisms go it alone). Egalitarian groups are unions of unrelated cells, individuals, or even different species, whereas fraternal, i.e., kin-based, groups include cells in a multicellular organism or worker bees in a colony. When such a union happens and its constituent parts express high cooperation and low conflict, a new kind of individual has emerged.

It is this focus on socially driven emergence of new levels of *individuality* that has characterized the most recent discussions on evolutionary transitions. And so I introduce a new word for our attention.

---

1. Formalizing this alliance is a chief aim of Andrew Bourke in his 2011 book, *Principles of Social Evolution*.

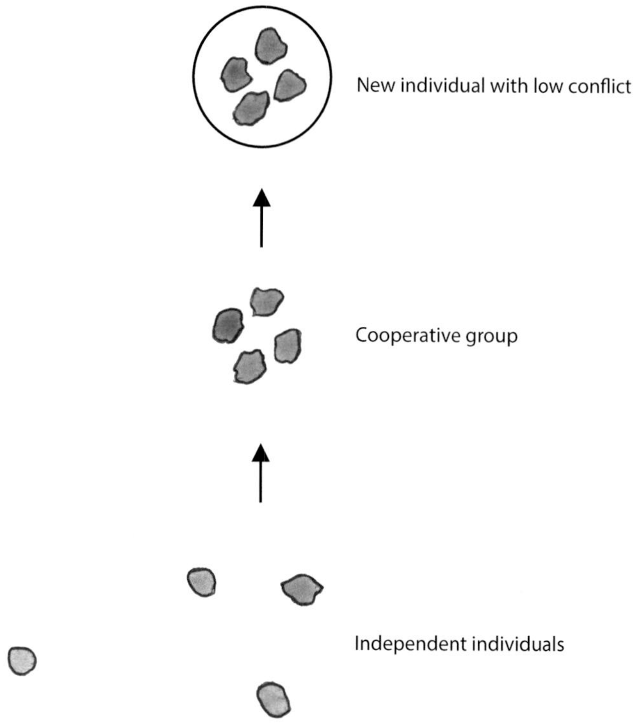

FIG. 2.7. Major evolutionary transitions are defined as the emergence of new types of individual, for example, the transition of independent eukaryotic cells into multicellular organisms. The multicellular design is a new type of *individual* resulting from the merger of lower entities that were themselves independent individuals. Two necessary steps must happen or the new individual will fail to emerge: (1) formerly independent replicators merge into one, exhibiting a high degree of cooperation among parts, and (2) the new individual that results exhibits near-total elimination of internal conflict among its parts. Each of these prerequisites is a pillar of social evolution learned largely from studying social insects and only recently applied formally to life in general.

I hope that by now, it has become apparent that "organism" is a word groping for clarification. It is perhaps irrecoverably freighted with common application to multicellular animals like ourselves. It is also the context for another word—superorganism—that is useful and owns no small traction in the social insect literature. But even though "organism" and "superorganism" possess contextual usefulness, they distract us when we're talking about major evolutionary transitions. For this, individuality comes to the rescue.

An *individual* is an entity composed of formerly independent parts who now share a high degree of integration and cooperation; whose genetic in-

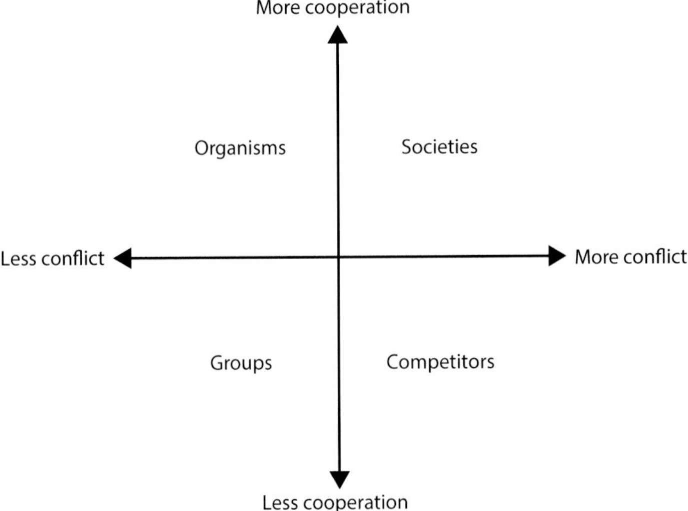

FIG. 2.8. An organism can be defined as a union of lower independent biological entities (cells or social insect workers) into a new higher entity characterized by extreme cooperation and absence of conflict among its parts. As such, the emergence of new forms of organism, or individual, is a product of evolutionary processes formerly thought to be particular to eusocial organisms like honey bees and termites. Queller and Strassman (2009) envisioned cooperation and conflict as two intersecting continua and use it here to organize the kinds of social constructs available in evolution. Organisms occupy the space of maximum cooperation and minimum conflict.

terests are aligned or at least subsumed to the interest of the group; whose genome is transmitted in common; and who express near complete eradication of within-group conflict. We can perhaps replace organismality with individuality, and we're back on track—and readers will be pleased to know that I use organism and superorganism in their traditional usage for the rest of this book.

Every major life transition that involves the emergence of a new kind of individuality involves the two social preconditions indicated above. "Kind of individuality" is synonymous with "level of biological organization." Any particular individual is a product of this social evolutionary process: an individual can be single-celled, multicelled, or superorganismal. An individual, as a reproductively autonomous packaged genome, is the product of adaptation and therefore a *unit of selection* in the classical Darwinian sense. By extension, a superorganism can be subject to natural selection at up to four levels—among colonies, among members of a colony, among cells within a member, and among genes within the genome of a focal cell (discussed in the "Multilevel Selection" section). And although they prefer to retain and apply "organism" to all biological levels, Strassman and Queller (2010) are willing to extend the label to symbiotic relationships or even non-contiguous entities[2] as long as they meet the criteria of cooperation and non-conflict.

But, back to that earlier caveat: it is neither necessary nor possible to lay all the glory at the feet of social evolution because quite simply, some transitions are important but don't involve the emergence of new kinds of individuality. However, when they do, then sociogenesis is what's doing it.[3]

Table 2.1 provides a hybrid of three important published lists of major evolutionary transitions. Some involve socially driven emergence of individuality; others don't. Each of them is a critical juncture in the history of life, but this synopsis highlights the preeminent role of social evolution in the history of organismal development.

Multilevel Selection

This discussion on evolutionary transitions compels me to jump ahead a little and address multilevel selection, something covered more directly in the chapters ahead on social group formation. *Multilevel selection theory* asks at which level does natural selection act? The individual or the group, sequentially or simultaneously? The question "Can natural selection act on the group?" can be asked at any point along a continuum from simple eusocial family to complex eusocial superorganism. The question becomes easier to answer in the affirmative as an incipient group moves toward a state of true individuality, that is, expressing the two criteria described in the previous section: high cooperation and elimination of within-group conflicts.

To illustrate this, consider a complex colony that chooses a bad nesting site so that the entire colony is washed away by a flood. It is the colony that is dead even if 10 bees survive. Nature (in this case the flooding stream) se-

2. These authors cite pathogenic microbes that may qualify as non-contiguous organisms bounded and connected only by their host's body.
3. In chapter 6, I argue in favor of the term "sociobiogenesis" for indicating socially-driven emergence of new biological levels.

TABLE 2.1. Major evolutionary transitions, some involving sociogenic emergence of new forms of individuality

CONSTITUENT STAGE	EMERGENT STAGE	SOCIOGENIC EMERGENCE OF NEW INDIVIDUAL?	EGALITARIAN OR FRATERNAL?	NOTES
Replicating molecules	Populations of molecules in compartments	yes	both	Nucleic acids clump into compartments. Natural selection acting on compartments promotes cooperation among molecules.
Independent replicators	Chromosomes	yes	both	Independent nucleic acids link together and now must replicate as a unit.
RNA as gene and enzyme	DNA and proteins	no	NA	DNA provides greater structural stability, and protein-based enzymes are more diverse and versatile.
Independent unicells (prokaryotes)	Symbiotic unicell (eukaryote)	yes	egalitarian	Animal mitochondria and plant chloroplasts, formerly free-living prokaryotes, transition from a symbiotic relationship with host cell to symbiogenesis.
Asexual clones	Sexual populations	no	both	Reproduction by meiotic production of haploid gametes and their fusion to form a diploid zygote.
Unicellular eukaryotes	Obligately multicellular organism	yes	fraternal	Unicellular eukaryotes link with others, specialize, cooperate, and now only reproduce as a unit. Resulting multicellular organism has a clonal soma.
Multicellular organisms	Eusocial colony (superorganism)	yes	fraternal	Multicellular organisms link with others, specialize, cooperate, and now only reproduce as a unit. Resulting superorganism has a non-clonal soma.
Primate societies	Human societies	no	NA	Language and culture
Different species	Obligate interspecific mutualism	yes	egalitarian	Different species co-dependent and physically conjoined

Note: From Maynard Smith and Szathmáry (1995), Bourke (2011), Fisher et al. (2013), and West et al. (2015). I include Queller's (1997) categories for socially-derived transitions: egalitarian—the merger of unlike units, and fraternal—the merger of like units.

lected for siting nests high up in branches, not near the roots. Conversely, let's say that 10 bees are consumed by a bee-eater bird. This too is selection (by the bee-eater), but it does not matter for the colony. So the individual level is typically less important than the group level unless large numbers of bees are affected.

Some authors have foregrounded group selection as the glue that holds the superorganism together (Nowak et al., 2010; Wilson and Sober, 1989; Wilson and Hölldobler, 2005), but this view has generally been marginalized to the one I emphasize in this book—that groups emerge from dynamics described by *inclusive fitness theory*: the cooperation and alignment of fitness interests of group members (discussed in chapter 6's "Hamilton's Rule and Inclusive Fitness Theory" section). Inclusive fitness theory does not reject group-level selection in principle but ultimately insists on a gene-centered view of natural selection so literal as to almost make questions about multilevel selection drop away. It was Dawkins (1989) who emphasized that the ultimate object of natural selection is the replicator—the gene. Dawkins sees the soma as the "vehicle" for genes and all the soma's permutations as the gene's "extended phenotype." However observers may demarcate "units of selection" along an organizational continuum, the ultimate target resolves to the gene.

Chapter 6 will discuss Hamilton's rule and demonstrate how it shows that cooperation—group formation and maintenance—is favored when the response to between-group selection exceeds response to within-group selection (West et al., 2007). Similarly, Hamilton's rule can predict the reverse—conditions for within-group selfishness and social dissolution. The lesson from inclusive fitness theory seems to be license for relaxing the boundaries of selection levels: there are indeed multiple levels of selection, but they can vary in their importance on a contextual basis (West et al., 2007).

It can also be helpful to distinguish levels of *selection* from levels of *adaptation*, as done by Gardner and Grafen (2009) in their formalization of the superorganism as a real unit of selection, or as they put it, "a purposeful being with its own agenda." One can appreciate this distinction if one considers the divergent forms of caste that occur in the complex eusocial colony: the group is the level that is *selected*, but it is the castes that are *adapted*. Gardner and Grafen see the group as a potential "maximizing agent" for genome transmittal and a legitimate target for selection, but only if its internal genetic conflicts are totally abolished. They seem to acknowledge that this strident criterion can be stretched in the case of the complex eusocial species in whom sufficient restraints against intra-group genetic conflict have evolved.

But it remains that the superorganism is inherently less stable than the organism, privileged as the organism is with the conflict-erasing feature of the single-celled zygote that renders all somatic cells clones (chapter 5). As an exercise in retaining factions with divergent interests, the superorganism falls short of a harmonized genome. This persistent risk of disruption is what prompted Ratnieks and Reeve (1992) to caution against applying the label "superorganism" to any group. Rather, they encourage us to consider that species can be "superorganism-like" in some contexts and not in others. Bourke (2011, p. 58) illustrates what this may look like in the honey bee: the celebrated dance recruitment language (chapter 14) appears to be a superorganismal-level adaptation to optimize group nutrient income—an outcome over which there should be no dispute, whereas the lethal fighting that occurs between rival virgin queens (Gilley and Tarpy, 2005) is an unresolved conflict between selfish parties in the group.

Operating as it does at the interface of organisms and superorganisms, sociobiology has long treated multilevel selection as an acceptable working hypothesis (Tarpy et al., 2004a), and rather than weary my readers with constant qualifiers, I will continue talking in this book about individual-, group-, and colony-level selection, trusting them to remember that these levels are ultimately provisional: the thing which is selected may not be that which is adapted, and the gene, forever, is natural selection's ultimate target.

Cooperation and Conflict Resolution as a Pedagogic for Evolution

So far I have tossed out "cooperation" and "conflict eradication" as the central drivers for the emergence of individuality. But entire libraries have been written on the conditions for *these* drivers to emerge and sustain; it is no overstatement that cooperation and conflict resolution are the central problems of sociobiology. To unpack them a little more: What ecological benefits are to be gained by cooperation? What forces sustain cooperation in the early stages when it is fragile? How are genetic interests of members aligned, sustained in the early stages, and maintained? How is group internal homeostasis selected for? Caste and task specialization? What mechanisms of nesting, integrated defense, thermoregulation, decision-making, and communication are rewarded and sustained at the level of group? And, mindful that evolution is neither necessarily progressive (Maynard Smith, 1988) nor static, we must also ask: what conditions are dissolutive to cooperation and conflict eradication, once attained?

In this book, I show how the honey bee has answered these questions and modeled the emergence of individuality. Along the way, we will enjoy the biology of this celebrity species and witness phenomena that attach to the grand actions of cooperative evolution, namely complexification and *emergent properties*—the global order that emerges when numerous independent actors respond to local stimuli. For although complexity is not a necessary end point of evolution, the subsummation of entities into higher entities entails a certain necessary complexification, and when that results in large populations of independent actors such as worker honey bees, their collective behaviors can create emergent order which itself provides scaffolding for evolution to act upon.

It's an interesting story in its own right but also hints at a Bigger Story, and that is that social evolution—the evolution of cooperation—has emerged as a general explanation for the evolution of life. To quote Bourke (2011, p. 5):

> Social evolution has grown outward from the study of the beehive and the baboon troop to embrace the entire sweep of biological organization. It claims as its subject matter not just the evolution of social systems narrowly defined, but the evolution of all forms of stable biological grouping, from genomes and eukaryotic unicells to multicellular organisms, animal societies, and interspecific mutualisms.

Beekeepers will be proud to learn that their charges are keepers of such important secrets; naturalists will appreciate the elegance and economy of honey bees as exemplars of such overarching principles; and avocational readers of biology will marvel anew at Charles Darwin's Great Idea and its seeming inexhaustible capacity to predict and explain.

Summary

The pattern for the evolution of life on Earth has been for formerly independent replicating units to join together into larger units, forming in the process a new, higher-level unit of organization. This pattern has recurred with the alliance of formerly free nucleic acids into a packaged compartment, the union of formerly independent prokaryotes into the eukaryotic cell, the clumping of formerly independent eukaryotes into the multicellular organism, and the union of formerly independent organisms into the superorganism. Each of these steps is considered one of life's major evolutionary transitions, and each involves the emergence of a new kind of *individuality*, defined here as an entity composed of formerly independent parts who now share a high degree of integration and cooperation; whose genetic interests are aligned or at least subsumed to the interest of the group; whose genome is transmitted in common; and who express near complete eradication of within-group conflict. The kind of individuality known at the multicellular level, including plants, fungi, and animals, also goes by the name *organism* or "paradigmatic organism" for authors who prefer to retain the term organism up and down levels of evolutionary transition.

Today, a synthesis is happening between two formerly discrete domains of biology: social evolution and the evolution of organismality. Sociogenesis is the study of the emergence and maintenance of groups, and we now know that the dynamics of sociogenesis are general across major evolutionary transitions. This has come about by recognizing that two conditions necessary for the emergence of integrated groups apply also to the emergence of new kinds of individuals: (1) formerly independent replicators join together and now reproduce only as part of the larger unit, exhibiting a high degree of cooperation among parts; and (2) the new individual that results is contingent on the elimination of internal conflict. It is neither necessary nor possible to credit sociogenesis with every major evolutionary transition because some transitions—such as the transition from RNA to DNA as Earth's dominant replicating molecule—are important but do not involve the emergence of a new kind of individual. But when they do, it is sociogenesis that is doing it.

The subsummation of entities into higher entities entails a certain necessary complexification, and when complexification results in large populations of independent actors such as worker honey bees, the actors' behaviors can create emergent properties that in turn provide scaffolding for evolution to act upon. The dependence of organismal evolution upon cooperation and conflict resolution is a pedagogic counterweight to the caricature of Darwinism sometimes painted as license for dystopian selfishness.

CHAPTER 3

The Honey Bee, Life at Two Levels

IN CHAPTER 2, I developed the idea that biological life occurs at multiple levels. Each emergence of a new kind of organism happened under the compulsions of social evolution, principles uncovered for the most part by social insect entomologists. And let's just say it, a lot of those entomologists were working with the honey bee.

The honey bee, *Apis mellifera*, exists as a superorganism (Moritz and Southwick, 1992). To refresh what that means, think about a multicellular organism and the eukaryotic cells that make it up. That's two levels. Each cell has a small degree of autonomy. It possesses its own set of chromosomes, simple organs called organelles, and a membrane that separates it from other cells. Cells are capable of dividing by mitosis, a form of reproduction. Each absorbs nutrients and metabolizes energy. Each reacts to local stimuli. In an animal, cells differentiate into different functions—liver cells, muscle cells, brain cells, and so forth. It's not difficult to see that each cell can do a lot of the same things as the organism of which it is a part, just at a much smaller and more restricted scale. The organism, therefore, is a collection of highly integrated yet differentiated cells working cooperatively to optimize the fitness of the organism. It should be no surprise, therefore, to say that biologists who study the evolution of organismality understand it as a process of increasing cooperation among formerly free-living eukaryotes.

Now translate everything I just said in the last paragraph to a honey bee colony and the individual bees that make it up. That's also two levels—but with a difference: this time the higher level (the colony or superorganism) is made up of lower levels (the individual bees) that are themselves organisms. We have moved up one step on the chain of biological organization: instead of moving from cell to organism, now we're moving from organism to superorganism. Shifts like these along an organizational continuum are among the major transitions in the history of life (table 2.1).

Each emergence of a new kind of individuality must pass through two checkpoints: (1) formerly independent parts come together in a high degree of cooperation, expressing adaptations that favor the reproduction of the group, and (2) the group survives because it eliminates genetic conflict within its parts. Honey bees have modeled these steps, not once but twice!—as organisms and as a superorganism.

So let's pause for an overview of the annual life cycle of this wonderful creature, the western honey bee *Apis mellifera*, living life on at least two levels. And readers now understand that the two levels we're talking about are the bees and the colony. First, we need an overview of the honey bee's modern life history. We do this to tell the story of *Apis mellifera*'s evolution as a *social* insect. And to continue the theme that social means something explicit and that sociality speaks to the evolution of organismality, let's quote Queller and Strassmann (2009):

> The evolution of organismality is a social process. All organisms originated from groups of simpler units that now show high cooperation among the parts and are nearly free of conflicts. We suggest that this near-unanimous cooperation be taken as the defining trait of organisms.

Some Things Common to All Honey Bees

Honey bees are winged insects and members of the superorder Holometabola, which share in common the characteristic that the young pass through very different morphological stages during development: the egg, larva, pupa, and adult. Also called "complete metamorphosis," holometabolism is a successful life strategy, in part because it spreads environmental risk across life stages who may occupy entirely different niches; compare, for example, mosquito larvae and their aquatic life and predatory diet of microorganisms versus the terrestrial life and diet of blood and nectar required of adults. All told, the Holometabola account for nearly 85% of all insect species (Grimaldi and Engel, 2005).

Honey bee eggs hatch into *larvae* (singular larva), which are dedicated to eating and rapid growth. True to all Holometabola, the larval cuticle has limited stretching ability, so growth is accomplished by a series of episodic molts in which the larva sheds its old skin, exposing a new one underneath. The molting insect expands its body volume with air to stretch the new cuticle before it hardens, thus achieving its next incremental stage of growth. The individual in between larval molts is referred to as an *instar* (figure 3.1). Thus, for example, an egg hatches into a first-instar larva; after its first larval molt, the individual is now a second-instar larva, and so on. The immature undergoes five larval instars, a complex transition period referred to as a prepupa, and an adult-like pupal stage before emerging as a fully winged adult.

Honey bees practice *progressive provisioning*, which means that adults feed larvae more or less continuously with repeated small quantities of *brood food*—a protein-rich product of nurse bee mandibular and hypopharyngeal glands metabolically derived from pollen. The opposite of this is *mass provisioning*, in which adults provide each larva all the food it needs at once for its entire immature development, typically by sealing an egg on top of a ball of pollen in its own cell (see figure 4.3). Across bees as a group, mass provisioning is considered the ancestral state, but in the honey bees (Apini) and their nearest social kin the tribes Meliponini and Bombini (see figure 1.3) there is no clear association between mode of larval feeding and social complexity. The complex social Apini and representatives of the simple social Bombini practice progressive provisioning whereas the complex social tropical Meliponini retain the ancestral state of mass provisioning (da Silva, 2021). In light of this apparent uncoupling of larval feeding mode from social drivers, Moritz and Crewe (2018, pp. 25–26) consider instead nutrient economics, suggesting that progressive provisioning affords colonies a more

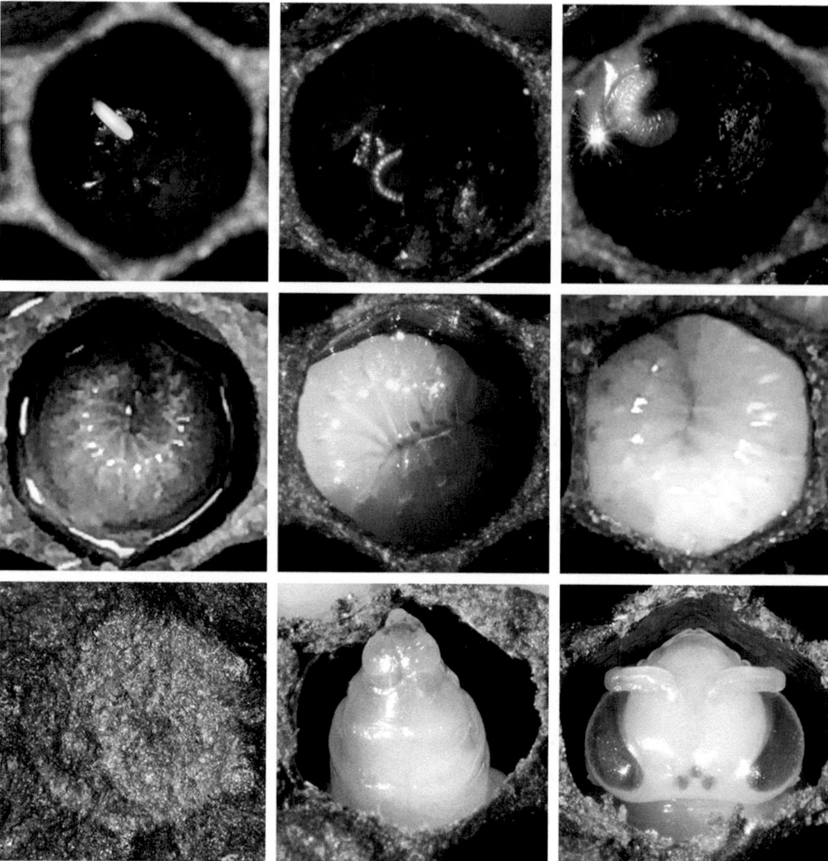

FIG. 3.1. An assembly of honey bee brood showing stages of the holometabolous immatures: (left to right; top, middle, bottom): egg, first instar larva, second instar larva, third instar larva, fourth instar larva, fifth instar larva open and unextended, fifth instar larva (not shown) under capped cell, fifth instar larva uncapped to show its extended prepupal stage, and pupa uncapped to reveal it in the purple-eyed stage. Molts post-capping produce the prepupa, pupa, and mature winged adult. It is possible to see the brood food as a glistening liquid in the cells of first, second, and third instar larvae. PHOTOS: THE AUTHOR AND BENJAMIN ROUSE.

nimble response to increasing foraging resources, leading to rapid individual development and colony growth. Such a dynamic could be highly beneficial for temperate-evolved Apini or Bombini colonies coming out of winter or after desert rainfalls.

Once the larva's feeding career is over, workers cap the cell with beeswax, and the larva's foregut and hindgut—formerly disconnected to prevent fecal contamination of the larva's food—are now joined, permitting the larva its first defecation. This is the reason entomologists use the prosaic adjectives pre- or post-defecating to describe larvae at either side of this benchmark. The larva, now called a *prepupa*, begins elongating with its head toward the cell's capping, spinning a cocoon as it does so, applying silk from its mouthparts to the cell walls and underside of the capping.

At the final larval molt, the individual transforms into a *pupa*, a quiet, non-feeding stage that superficially resembles the adult. During these days, the tissues and organs of the former larva are reorganized into adult tissues and organs. The pupa starts out completely white. Pigmentation begins with the eyes, starting as a pink-eyed pupa, maturing to purple-eyed with a gradual darkening of the whole body (see figure 3.1). At the final molt to adult, the fully formed wings appear for the first time. The so-called *teneral* adult stays in its cell for a few more hours, hardening and strengthening before it chews away its cell capping and emerges to begin participating in the colony's life.

Developmental times for the three bee types in the colony are different, yet relatively fixed within type, owing in part to the extreme environmental constancy made available by the eusocial nest. These fixed times are roughly 21 days for a worker, 24 days for a drone, and 16 days for a queen, but Winston (1987) reported a range for these values of 16–24, 20–28, and 14–17 days, respectively, values that vary according to genetics, latitude, and local micro-habitat.

Besides their common mode of holometabolous development, honey bees share with all members of the insect order Hymenoptera the property of *haplodiploidy*, which leads us to the discussion of sex determination.

First, a quick review: the state of diploidy applies to all mammals and means that in the nucleus of each of its somatic cells, each sex possesses two complete sets of chromosomes (2N), one set from each parent received at conception. Somatic cells reproduce by mitosis, a process of cell division that sustains full diploidy at each division. A different kind of cell division happens in the reproductive tissues, i.e., the ovaries and testes, that reduces the chromosome number by half. This type of cell division is called meiosis and results in gametes—eggs and sperm—each having only one set of chromosomes (1N), a condition called haploidy. When gametes fuse in sexual reproduction, full diploidy is restored for the embryo.

In diploid mammals, among one's complement of chromosomes are at least two, the sex chromosomes X and Y, called *allosomes*; all the rest of the chromosomes in one's genome are called *autosomes*. The presence of allosome Y results in a male embryo regardless of the number of autosomes or X chromosomes (Ohno, 1979).

Like in other haplodiploid taxa, males in Hymenoptera are derived from unfertilized eggs, and the genome of any male is derived wholly from its mother. This state of haploidy extends to somatic cells in haplodiploid insects like the thrips and whiteflies, but in most Hymenoptera, including the wasps and bees, a cellular doubling of DNA content occurs in muscle, essentially rendering the somatic tissue diploid, an adaptation, it is believed, permitting strong flight and dispersal ability (Aron et al., 2005). It is important to stress, however, that Hymenopteran males remain functionally haploid in terms of allele transmission dynamics.

Females, in contrast, are diploid in the classical sense with two parents, possessing, like mammals, a 2N complement in somatic cells and 1N in reproductive cells. The fact that drones come from unfertilized (haploid) eggs was recognized as early as 1845 by the Polish bee scientist Johann Dzierzon, who buried his insight in the prosaically entitled communication, *Opinion on the Issues Raised by Mr. Stöhr in the First and Second Chapters of the General Report*, leading his early successors to believe that sex determination in bees is simply a matter of fertilization. At one level this is true: the queen indeed has muscular control over releasing or withholding sperm and fertilizing an egg as it passes down her median oviduct. Those eggs from which she withholds sperm do indeed become males.

But it is more nuanced than that. It has since been established that sex in honey bees is controlled at a single sex-determining gene, or *locus*, on a chromosome. The number of these *alleles*, or gene variants, ranges from 19 to 53 locally (Adams et al., 1977; Cho et al., 2006; Lechner et al., 2014) to over 100 at the species level (Zareba et al., 2017). Diploid individuals who are *heterozygous* at this locus, possessing two variants of the gene, are female. Individuals who are *hemizygous* at this locus, possessing only one variant of the gene (which is automatically the case in a haploid individual), are male. But individuals who are *homozygous* at this locus, receiving the same allele from each parent, are also male; but these homozygous drones cannot produce fertile offspring. Moreover, these individuals are recognized by nurse bees and summarily eaten within hours after hatching into first-instar larvae (Woyke, 1963) (figure 3.9); thus, they have a Darwinian fitness value of zero.

The honey bee sex-determining locus appears to be ancient and conserved, appearing more than 100 million years ago (Miyakawa and Mikheyev, 2015) with evidence that its molecular elements reach back to the Aculeata or even the entire order Hymenoptera (Schmieder et al., 2012) (see figure 1.3).

Honey Bees as Individuals

A honey bee colony has three classes of individuals: workers (figure 3.2) from 10,000 to 50,000 in number depending on time of year; one queen (figure 3.3); and male bees, also called drones (figure 3.4), ranging in number from a few thousand in spring to zero over winter.

This mention of different classes of individuals is a handy moment to introduce the idea of *caste*—defined as a categorical variant based on morphology or behavior within sex. In the case of honey bees, there are two sexes, the females of which occur as two castes: workers and queens. Males occur in only one form, so it is improper to refer to male honey bees as a "third caste." The number of castes in a social species is one indicator of the species's evolutionary complexity. The honey bee, with only two castes limited to one sex, is rather non-remarkable on this count compared with the ants and termites.

WORKERS

For everything you've ever heard about honey bees, chances are it's the workers responsible for it (figure 3.2). The workers make honey; the workers sting; the workers pollinate plants. The workers make up the soma and productive outcomes of the superorganism. Through their local decisions, the workers, not the queen, arrive at functional group decisions that affect the whole colony. All this decision-making proceeds in a lateral *heterarchical*

FIG. 3.2. Honey bee workers not only comprise the largest fraction of the colony's population by far, but workers express the widest range of behaviors, cognitive aptitudes, and morphological specialization of any other bee type in the nest. Workers are task specialists, phasing in and out of different labor niches depending on genetics, age, and colony needs. These workers are "festooning," joining legs to form connecting chains across combs or hanging from cavity ceilings, a behavior associated with the secretion of beeswax.

FIG. 3.3. A honey bee queen. The worker and queen are the two female castes of the colony. Any female (fertilized diploid) egg has the potential to develop into either a worker or queen depending on the diet it receives as a larva. Genes coding for queen-like morphology and behavior are activated in response to a diet of royal jelly, a glandular brood food produced by nurse workers and reserved for larvae destined to be queens.

manner, in contrast to the top-down *hierarchical* commands from a sovereign queen imagined by naturalists of earlier centuries.

Workers and queens both come from the same kinds of eggs—fertilized, i.e., diploid, which in the Hymenoptera means female. The first 12 hours post-hatching of any female honey bee egg are a crucial window of opportunity. During this window, the individual's fate is determined directly by the kind of diet it receives from nurse bees: larvae fed *royal jelly* become queens; larvae fed *worker jelly* become workers. The overwhelming majority are fed worker jelly.

One of the most important features of worker honey bees, amounting to nothing less than a distinction of higher social insects in general, is their age-based division of labor, also called *age polyethism*. Division of labor is the great ecological advantage of social life. A solitary free-living female (and

FIG. 3.4. A male bee, or drone, is shown here surrounded by workers. Being produced from an unfertilized egg, a drone is entirely an expression of his mother's genome. One outcome of this is that recessive genes (in this photograph, the visible eye mutation *ivory*) show up in more often in drones than workers. A drone has no paternal genome to mask maternal recessive genes. All of a queen's sons propagate exclusively her genes, but because of chromosomal crossing over in the formation of her eggs (figure 2.5), these sons are not clones to one other, i.e., each possesses a different combination of their shared mother's genes. Drones perform no colony work and instead join local aggregations of drones from numerous colonies, flying in circuits along geographic landmarks called "drone congregation areas" (DCAs) that tend to be consistent from year to year. Virgin queens mate in or near these DCAs with *circa* 12 drones. PHOTO: GEORGIA P. ZUMWALT.

it usually *is* the female, burdened as she is with producing the next generation) must single-handedly solve all of life's problems: mating, wintering, emerging, nesting, foraging, ovipositing, feeding, and defending. As a unity, she can by extension only work *serially*—doing one task at a time. But in a group of cooperators, we have the synergizing alternative of tasks being performed in *parallel*, with numerous workers who are (1) specialists, and (2) perform different tasks simultaneously (Robinson, 1992; Wilson, 1985). It is impossible to overstate the importance of division of labor and its benefit of parallel task performance on the emergence of complex multicellular organisms (with specializing cells) and superorganisms (with age-based specialization in workers).

As but one example, consider the situation faced by a solitary bumble bee queen after she emerges from winter hibernation. Bumble bees are an annual eusocial species, cycling in and out of solitary and social phases. Males and new daughter queens are produced and mate during late summer, after which the colony begins a slow decline, ending in the deaths of the old queen, workers, and males. Only the newly-mated daughters survive to enter a solitary phase, overwintering in solitary shelters or *hibernacula*. In spring, each new queen emerges and single-handedly forages for energy, digs a nest, stocks it with a lump of pollen, deposits an egg on it, and proceeds to incubate the egg with her own body heat, precisely analogous to a bird incubating an egg. Insect development is tightly regulated by temperature, which means that development is slow and erratic for the first clutch of workers, ramping up while their mother incubates them and cooling down while she's away foraging. Only after the first cohort of workers emerges (sometimes the second) is the queen relieved of foraging duties and able to concentrate on tending her brood. With additional broods, the queen is further assisted in brood tending as workers take up feeding and incubating larvae, and only after this point does the colony appreciably and rapidly grow to its mature state.

In other words, the maximum state of colony development only emerges once the shift from serial to parallel task performance takes place. Looked at another way, the relative social simplicity and small size of annual social nests—typical of bumble bees and paper wasps—is testimony to the absence of parallel task performance.

None of this happens in *perennially* social species like the honey bee, where the colony exists in a perpetuating social state with uninterrupted division of labor and parallel task performance. It is no accident that these conditions are associated with the largest and most complex insect societies.

The age-based sequence of honey bee worker tasks generally holds to the pattern of nest tasks while young, then shifting to foraging tasks with age (Toth and Robinson, 2007). Young bees are physiologically optimized to produce brood food—proteinaceous products of their mandibular and hypopharyngeal glands metabolically derived from pollen and fed to larvae. Similarly, the expression and senescence of wax glands is age-dependent. There is variation within these task bounds, depending on needs of the colony and genetic proclivity of any particular worker. Some workers are hyperspecialists and perform only one or a few related tasks their entire lives, never graduating to different task categories.

Division of labor, this paramount feature of worker honey bees, will be covered in detail in chapter 8, but suffice it for now to say that worker honey bee task specialization is a recapitulation of that ancient tried and true recipe for success—cooperation and specialization. The same kind of thing happened when ancient eukaryotic cells massed together and specialized to form the first multicellular organisms.

Worker honey bees also occur in physiologically distinctive forms called "summer" versus "winter" bees. The production of these individuals is closely tied to season. Summer bees are short-lived (25–40 days) whereas winter bees are long-lived (> 250 days), with intermediate life spans in spring and fall [reviewed by Knoll et al. (2020)]. Broadly speaking, the task of the summer cohort is to restore the colony's workforce and food supply after the period of reproductive colony fission (swarming) in spring. The task of winter bees is to energetically sustain the superorganism over winter and rear new bees in late winter/early spring in preparation for the next springtime fission event ("Honey Bees as a Colony" section).

There are no visible differences between summer and winter bees, but winter bees have significantly lower levels of juvenile hormone (JH) and higher levels of blood proteins (Fluri et al., 1977), in particular vitellogenin. Juvenile hormone is one of the most biologically active hormones in insects. It is most basically associated with "retaining juvenile-ness," i.e., delaying metamorphosis between molts during development. In honey bees, JH is a behavioral pacemaker, regulating the speed of worker maturation and transition from nest to foraging duties (Robinson and Vargo, 1997), as well as a regulator of gene expression between queen-like or worker-like traits during larval development (Asencot and Lensky, 1976, 1988). Vitellogenin is an egg yolk precursor associated with egg formation in the queen (Engels et al., 1990) and nursing behavior in workers (Guidugli et al., 2005).

Unlike workers of some ant species who have permanently lost their ovaries (Bourke, 1988), worker honey bees retain vestigial yet minimally functional ovaries. The eggs of worker honey bees are the products of ordinary meiosis, gametogenesis, and oogenesis; however, because workers have lost the ability to mate and store sperm, their eggs are by extension unfertilized, haploid, and, as true to all Hymenoptera, can only develop into males.

Worker ovary development is normally restrained by the physiological effects of pheromones produced by immature bee brood and the colony's queen. However, if these constraints are removed, whether by the death of the queen or dense colony populations that restrict equitable flow of queen pheromone, then some workers' ovaries will activate and produce eggs. These so-called *laying workers* invariably abandon work on behalf of the colony, selfishly coopting the resources of the nest to produce their own sons. If a colony becomes permanently queenless, then laying workers rapidly increase in number and the colony begins an inevitable decline. The propensity to activate ovaries is higher in young workers (Delaplane and Harbo, 1987a) and in certain genetic lineages (Châline et al., 2002).

Laying workers do not have the behavioral repertoire that results in orderly egg deposition of the kind seen in queens. Instead, laying workers deposit eggs in a haphazard fashion with multiple eggs per cell (figure 3.5). Likewise, they apparently cannot discriminate worker-sized from drone-sized cells and so lay their eggs in each cell type. Because there are far more worker cells than drone cells, worker-laid drones are disproportionately reared in worker cells and are much smaller than queen-produced drones.

The workers' readiness to engage in direct reproduction the moment their constraints are removed is testimony to the fragility of the conflict resolutions worked out in the superorganism in which the soma is composed of organisms. This is in contrast to organisms who have achieved high conflict resolution through the use of a single-cell reproductive bottleneck, resulting in a soma of clonal cells.

THE QUEEN

Compared with workers, the queen (see figure 3.3) is behaviorally far less interesting. Most queens are produced during what beekeepers call swarm season—early spring when the colony is at peak population and preparing to divide by reproductive fission and form an offspring colony. The mother queen has a small degree of control over the production of her replacement. Throughout the year, there are cells on the comb with slightly enlarged bases, often oriented in a downward posture (figure 3.6). In preparation for

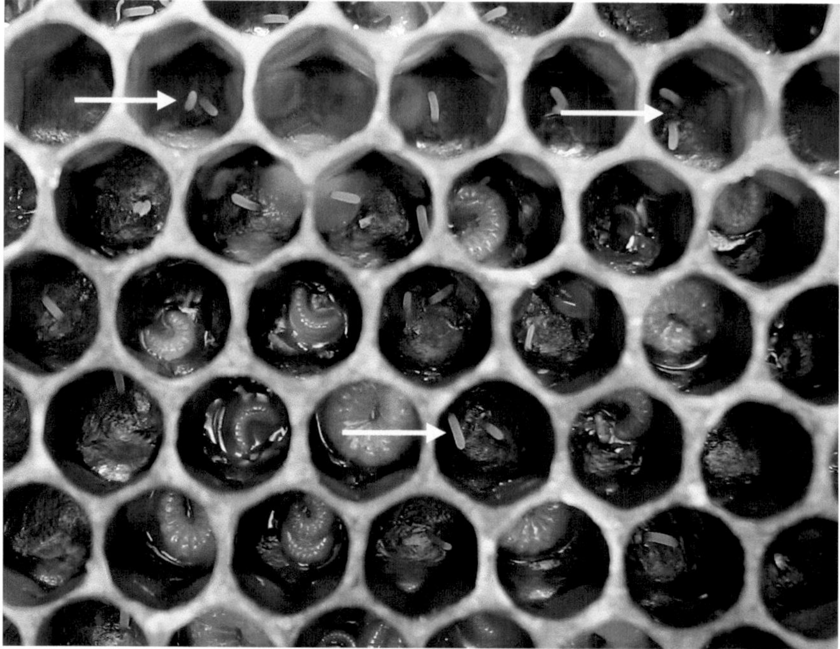

FIG. 3.5. Eggs deposited by laying workers are haphazard, with multiple eggs per cell and some laying on their sides. White arrows point to a few examples. PHOTO: LONNIE FUNDERBURG.

swarming, the mother queen deposits fertilized eggs in these *queen cups*, and eggs deposited in these cups have a disproportionately higher chance of being reared into queens. However, workers ultimately choose larval fate because workers are the ones who preferentially feed queen-destined larvae the royal jelly that switches on the genes responsible for "queenness." But larvae in queen cups get preferential treatment.

Colony fission is a chaotic time. It is typical for the mother queen to accompany the prime swarm[1] when it departs. This means that succession to the home throne is up in the air. Numerous sister queens are emerging from their cells at nearly the same time. The first to emerge engages in a spate of fratricide, systematically seeking out and destroying her rivals while they are still ensconced in their cells. The combatant opens the cell from the side, stinging

1. The mother queen accompanies the first, or "prime" swarm. This is the largest and most ecologically fit of fissions the colony will produce that season. An exceptionally large colony may produce one or more secondary swarms. Each of these is smaller and accompanied by one of the unmated daughter queens produced among the many queen cells reared during swarm season.

FIG. 3.6. A queen cup built into the normal brood rearing area of a comb. These special cells exist year-round, and larvae emerging from eggs deposited in them by the queen have a higher than average chance of being fed royal jelly and developing into a queen cell. PHOTO: GEORGIA P. ZUMWALT.

her teneral sister and personally engaging in removing her dead body from the cell (figure 3.7). Workers may intervene in this process, even to the extent of protecting some queen cells from a murderous sister, presumably in anticipation of a tertiary swarm. But whether there is one prime swarm, a secondary, or a tertiary, ultimately one new daughter alone inherits the home colony.

Within the second week of her adult life, this young queen engages in one or more mating flights. She encounters mates in drone congregation areas (DCAs) and copulates on the wing with an average of 12 males (Tarpy et al., 2004b), storing their sperm in an invagination of her median oviduct called the *spermatheca*. There, the sperm are oxygenated and nourished, remaining viable for the life of the queen, who retains muscular control over their release as an egg passes down her median oviduct.

This choice of whether to release or withhold sperm is mediated by the queen's systematic examination of every cell before she deposits an egg in it. In a ritualized behavior, the queen inserts her head into a cell, gauging its diameter with her antennae. If the cell is "worker dimensioned," the queen will about-face, insert her abdomen in the cell, and deposit an egg, releasing sperm as it passes the spermatheca (figure 7.2). If the cell is "drone dimensioned" [about 20% wider in European *A. mellifera* subspecies (Seeley and Morse, 1976)], she will perform the ritual minus the release of sperm.

This is what queens do. Over and over. With a normally functioning queen in peak season, i.e., mid- to late spring, this means about 2,000 fertilized (worker) eggs a day (Harbo, 1986). But there's a big range in the quality

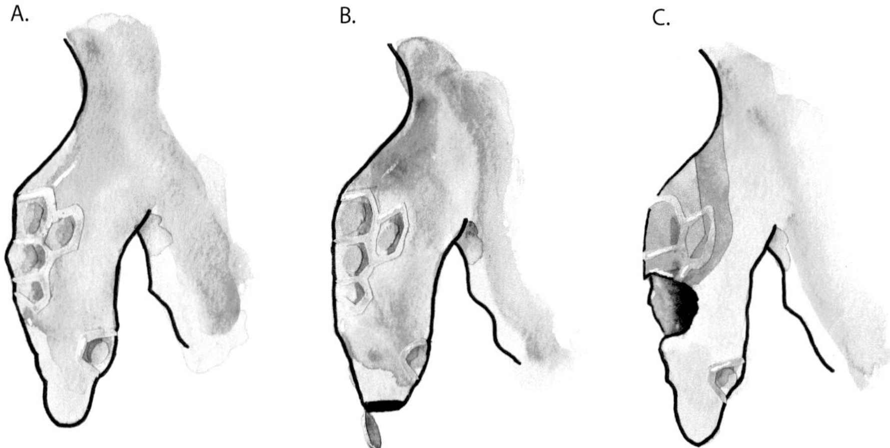

FIG. 3.7. Queen cells in three descriptive states, from left to right: (**A**) a normal cell under development, (**B**) a cell from which a live queen has emerged normally by chewing out at the tip, nearly completely severing the cap which in this case is dangling at the end, and (**C**) a cell that has been aborted; a rival queen opens the cell at the side and lethally removes her younger sister. In colonies that swarm more than once, workers may hold virgin queens protected as virtual prisoners inside their cells. It is believed that this gives advantage to the inmate either by allowing time for a rival to depart with a secondary swarm or by granting the prisoner time to mature and strengthen to do lethal battle with her rival.

by which this most basic of tasks is performed. An optimally performing queen fills adjacent cells in a solid pattern (figure 3.8), and as any beekeeper knows, these solid brood patterns translate into large populations of workers. Failing queens deviate from this pattern, skipping adjacent cells, resulting in a brood pattern that is spotty (figure 3.9). Sometimes this failure is from age and senescence, but other times, spotty brood expresses because the breeding population is inbred, i.e., has lost many of the alleles for the sex locus—and by extension probably a lot of other alleles too. Nurse bees recognize homozygous male eggs and eat them ("Some Things Common to All Honey Bees" section).

A close second in importance behind the queen's role as egg layer is her role as producer of pheromones that perform many functions regulating nest homeostasis and worker behavior and physiology. As chemicals that move throughout the nest from social contact (Naumann et al., 1992), queen pheromones stimulate worker foraging, increase worker longevity, optimize colony food stores, promote worker colony defense behavior (Delaplane and Harbo, 1987b), suppress queen cell construction and associated swarming

FIG. 3.8. A high-functioning queen will lay eggs in a solid pattern with brood of similar or nearly-similar age in contiguous cells.

(Winston et al., 1991), and, along with brood pheromones, inhibit worker ovary development (Hoover et al., 2003).

With social insects, when it comes to effecting reproductive submission, there seems to be an evolutionary vector from direct physical aggression in simple colonies—queen to worker and worker to worker—toward indirect suppression via pheromones in more complex species (Wilson, 1971). We see this when we compare the annually social bumble bees, whose queens and workers antagonize workers with developed ovaries (Van Doorn and Heringa, 1986), with the perennially social honey bees, who rely on brood and queen pheromones. As we will see in chapter 7, however, worker honey bees still play an important direct role in regulating who gets to reproduce by eating the eggs of other workers, a behavior called mutual *policing*.

DRONES

The male honey bee, also called drone, has a relatively straightforward biology (figure 3.4). As a hymenopteran, he is the product of an unfertilized egg, and thus his genome is wholly derived from his mother. However, because of

FIG. 3.9. A failing queen will deviate from the ideal form in figure 3.8—showing skips in cells or brood of dissimilar stage next to each other. Spottiness can be mirrored in an unrelated problem—high rates of inbreeding in a breeding population. Workers can detect inbred diploid male eggs and selectively eat them, removing diploid males from the population in the long term but creating a spotty brood pattern in the near term. PHOTO: GEORGIA P. ZUMWALT.

chromosomal recombination (see figure 2.5) during formation of his mother's eggs, he and his brothers are not clones. It is entirely reasonable to think of drones, in the memorable words of one of my professors at Louisiana State University, as "sperm with wings." In terms of biological organization, each is an organism complete with motility, but in terms of function, each is also a gamete of his mother, a haploid sampling of her total genome. One peculiarity associated with a drone's haploidy is the fact that visible mutations are more commonly seen in drones than in workers. Drones have no father's genome to mask recessive mutations his mother's genome may be carrying. In fact, the drone shown in figure 3.4 is carrying the visible eye mutation *ivory*.

Drones famously fly from their natal nests and join airborne aggregations of drones pooling together from many neighboring colonies to form a drone congregation area (DCA). Drones in these DCAs participate in "flyways" spanning a mile (1.6 kilometers) or more in circumference, 50–100 meters wide, and closely aligning with strong landmarks. The literature on DCAs is large and spans decades but has only recently been critiqued for what may be a longstanding error in execution. A large majority of DCA studies have used tethered queens or queen pheromone lures to detect aggregations of drones, raising the possibility that DCAs have been created by the very means used to detect them. Loper et al. (1992) made an important corrective to this pattern by using X-band radar to seek and detect DCAs near a commercial apiary without the use of lures. Within a 5 × 2-kilometer area, they documented 18 kilometers of shared flyways in which at least 26 DCAs existed. The flyways and DCAs were stable, being reestablished day after day and year after year for the four years of study. Flyways aligned with visible landmarks such tree lines in washes, but the flyways also branched, and most DCAs formed at these branches.

By 2021, technology had improved to the point that flight behavior of individual drones could be tracked. Woodgate et al. (2021) glued transponders onto the backs of individual drones and used harmonic radar to track DCAs and the flight patterns of individual drones in them. Monitoring the same agricultural landscape for two years, this team confirmed earlier reports of the constancy of flyways, with drones from numerous colonies converging on the same routes over successive years. Individual drones engage in periods of straight, direct flight interspersed with episodes of looping, convoluted flight. As numerous drones directly converge on the same spot and thereupon engage in looping flight, these concentrations likely constitute a local DCA. Drones from all monitored colonies visited all DCAs, and it was common for a drone to visit and perform a looping flight at more than one DCA per flight. DCAs were shown to have roughly symmetrical cores of 30–50 meters in diameter.

Regrettably, queens proved to be "less amenable to radar tracking than drones," and Woodgate et al. (2021) could not draw conclusions on the behaviors queens use to search for mates. In any case, one can hypothesize from evolutionary theory that because drones are cheap and young queens expensive, natural selection should favor high drone densities across a wide landscape so that young queens can be detected and inseminated as soon as possible.

The number of colonies contributing to DCAs can be quite large; a genetic analysis of one DCA in Germany estimated that its drones were the product of 240 colonies (Baudry et al., 1998). This study, however, was conducted in a landscape dense with managed colonies. A much more conservative estimate was derived from genetic analyses of 13 DCAs across the natural range of *A. mellifera* with the median number of contributing colonies ranging from 26 (Jonkershoek, South Africa) to 2 (Gotland, Sweden) (Jaffé et al., 2010). A companion study indicated a range of 12–72 colonies (queens) contributing to a single *A. m. scutellata* DCA in South Africa (Jaffé et al., 2009).

This is a good place to point out that there are three sources of genetic variation engineered into the reproductive biology of a honey bee colony: the high incidence of chromosomal recombination in the formation of honey bee gametes (Beye et al., 2006) (see figure 2.5), the queen's habit of mating with many males (*polyandry*), and the high genetic diversity among males represented by a typical DCA.

The morphology of drones reflects adaptions for optimizing mating efficiency under flight conditions. First of all, their body mass is significantly greater than that of workers, roughly matching that of queens, with robust musculature to ensure strong sustained flight. It is no accident that large body size is positively associated with mating success (Berg et al., 1997). By extension, this means that the smaller drones produced by laying workers are at a reproductive disadvantage; however, these small drones are genetically fully functional, and it is likely that a significant number of them still participate in local breeding populations.

Secondly, drones possess the largest compound eyes in the colony, with optical adaptations that favor spatial resolution and contrast sensitivity under conditions of bright ambient light (Menzel et al., 1991). Obviously, such optics are favorable for discriminating a queen's small silhouette against a bright sky.

Drones are well known for their infidelity to natal nest site. Drones enter alien colonies, encountering little or no resistance from guard bees, in a behavior called *drifting*. In managed apiaries in which colonies are crowded closely together (figure 3.10), the occurrence of non-natal drones in a colony can reach 42% (Pfeiffer and Crailsheim, 1998). This horizontal movement

FIG. 3.10. High colony densities typical in commercial-scale beekeeping.

of drones is considered an important expedient of parasite transmission between colonies and a driver of the bee health problems plaguing beekeeping, but it is also possible that drone-mediated disease spread is an artifact of the artificially high colony densities of modern beekeeping and thus is without context in evolutionary history.

Honey Bees as a Colony

Having described in the last section the actors on the stage, we now move to the main event, the pertinent level of biological organization for talking about *Apis mellifera*—the superorganism, the colony.

The importance of *colony* as the relevant level for *Apis mellifera* was brought home to me in an ecological context—blueberry pollination. It was this awakening, in fact, that led to my interest in the social foundations of organismal evolution and ultimately this book.

When I showed up at the University of Georgia as a young Assistant Professor in 1990, the beekeepers in south Georgia were agitated with late news out of the agricultural college (and perennial football rival) Auburn University in our neighboring state of Alabama. My Auburn University colleague[2] Dr. Jim Cane had recently published convincing evidence that honey bees are poor pollinators of rabbiteye blueberry (Cane and Payne, 1990), an important commercial crop and native species of the southeastern United States. Being a poor pollinator meant that a single bee visit to a blueberry flower is insufficient at delivering enough pollen to result in a marketable berry. It so happened that a solitary soil-nesting bee also native to the southeastern United States, *Habropoda laboriosa*, had a 35% chance of fertilizing a blueberry flower with one flower visit, whereas a single visit by a honey bee worker had only a 1% chance of producing a marketable berry. Dr. Cane's research was widely publicized at grower meetings. Blueberry growers were taking notice, and the upshot was that commercial honey beekeepers were losing contracts to pollinate blueberry orchards *en masse*.

It took me over 10 years to mobilize the resources to revisit this matter, and the answer took the form of my Ph.D. student Selim Dedej, former Professor at the Agricultural University of Tirana and Deputy Minister of Agriculture for the government of Albania. Selim arrived at my lab in 2000 and quickly settled down to a research project with the blueberry orchard that grew outside my lab's doors.

Selim recognized that Dr. Cane's analysis had not taken into consideration one important fact: social, colony-making honey bees do not visit flowers as singletons but rather in succession—as repeat visitors from one colony. In other words, the *total number of flower visits* that a species makes should be considered in the analysis.

Selim went about designing his dissertation research to analyze the impact of repeat flower visits at effecting *fruit-set*—the pollination of a flower that results in a marketable fruit. He tented blueberry bushes with honey bee colonies ranging in populations from 0 to 12,800 bees and compared fruit-set among them with that of untented bushes open to all pollinators. Predictably, the number of flower visits per minute increased as bee populations in tents increased; more importantly, fruit-set also increased as bee popula-

2. Dr. Cane subsequently transferred to the USDA Pollinating Insects Laboratory at Logan, Utah, where he continued his distinguished career.

tions and flower visitation increased (Dedej and Delaplane, 2003). Apparently, repeat flower visits by an inefficient pollinator can still get the job done if there are enough of them. Selim concluded that the social honey bee is in fact a very satisfactory pollinator of blueberry *when considered at the level of colony, not individual bee.*

This experience opened to my mind the realization that the honey bee, and presumably other superorganisms, interacts meaningfully with its ecosystem at the level of *colony*. Moreover, the life history of this colony looks nothing like the life history of its parts—the workers, the queen, and the drones. In fact, the honey bee colony recapitulates the kinds of qualitative changes that happen at every major evolutionary transition when new forms of individuality come into existence.

It was not just a professional revelation but a personal one: to discover in mid-life that my stock in trade—honey bees, a diversion that my parents introduced me to when I was 13 and caught me up into a lifelong interest, graduate school, and eventually a professorship—were a window into the cooperative ways my own body and other multicellular organisms came into being. Surely, this story deserves a wider telling, and so it is to the life history of the relevant level of honey bee life that we now turn.

TEMPERATE HONEY BEE LIFE HISTORY

The typical temperate-evolved honey bee colony in the wild is contained inside a tree hollow of roughly 30–60 liters in volume (Seeley and Morse, 1976) (figure 3.11). Inside this hollow are up to eight parallel beeswax combs suspended from the ceiling and side walls of the cavity. The cells in these combs are structurally integrated, with each cell roughly the volume of one worker bee. It is in these cells that all food is stored and the brood (immature bees of all stages) reared. Summertime worker populations in natural temperate colonies have been recorded at around 16,000 (Seeley and Morse, 1976), although population maxima in managed colonies are much higher than this, in the neighborhood of 29,000 (Delaplane and Hood, 1999) to 45,000 (Moeller, 1961). But for our purposes, the hollows in forest trees are the arenas in which the evolution of the temperate honey bee colony life cycle is played out.

Figure 3.12 shows the main evolutionary and environmental drivers that shaped this history. We begin with a state of our animal received from its deep

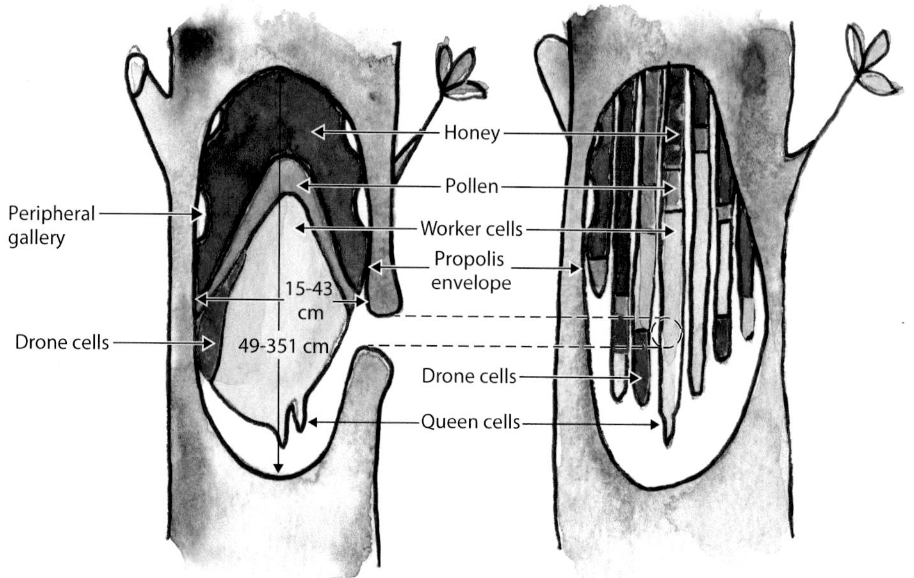

FIG. 3.11. Median and cross-sectional view of a stylized natural bee nest in a tree hollow, redrawn after Seeley and Morse (1976).

past, its evolutionary inertia, if you will: the fact that honey bees live together as mutually dependent castes, committing the species to uninterrupted year-round eusociality. In the modern honey bee, we have workers who have abandoned mating and producing fertilized eggs and instead evolved morphology and behaviors suited to work. We have the queen who has abandoned all colony work and evolved efficiency at mating and producing fertilized eggs. Neither can complete a life cycle without the other.

Committed to year-round social life, the ancient honey bees also had to solve the problem of feeding a permanent homeothermic population. This required storing food and preserving it. The food storage space that that we see in bee nests today is the outcome of a *syndrome* of related evolutionary pressures—effects so closely intertwined that it is nearly impossible to disentangle them. In this case, optimum nest cavity size and comb storage space are the products of interactions among ambient temperature, minimum

obligate perennial eusociality

Permanent homeothermic populations

food stores, reproduction by swarming (colony fission)

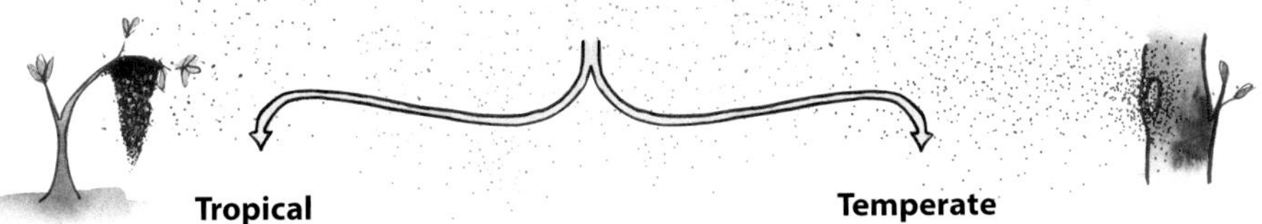

Tropical

frequent nectar flows

emphasis on many smaller swarms

Temperate

long dearths, cold winter

emphasis on fewer larger swarms

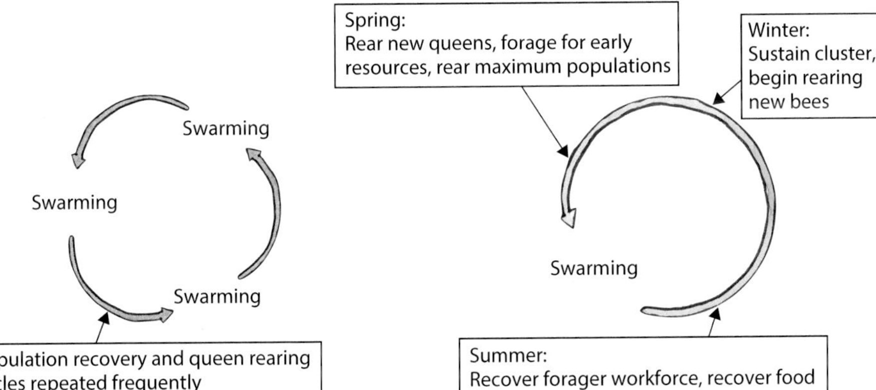

Tropical characteristics

- Small populations
- Small honey stores
- Small, exposed nests common
- Swarming rate high
- Worker lifespan short
- Absconding common
- Migration possible

Temperate characteristics

- Large populations
- Large honey stores
- Nests almost always in cavities
- Swarming rate low
- Worker lifespan long
- Absconding rare
- Migration unknown

FIG. 3.12. Life history options for tropical and temperate *Apis mellifera*. Commitment to mutually dependent castes is an ancient shared character between the two geographic groups. Each is therefore committed to sustaining a year-round homeothermic adult worker population and reproducing by social colony fission. In a reproductive swarm event, up to half of the worker population flies away from the home nest, accompanied by the mother queen (or daughter if a subsequent swarm) and lands on an object—often a tree limb—to form a temporary bivouac before resettling to a new permanent hollow. In tropical zones, a comparatively steady supply of nectar allows bees to invest more heavily in reproduction, issuing several small swarms a year. In temperate zones with extreme seasonality of food and long winter dearths, the colony invests more heavily in the soma, issuing fewer and larger swarms per year.

worker populations necessary to sustain homeothermy and nest function, and periodicity of food availability.

The nectar-feeders that occupied these nests also evolved at the same time the capacity for acidifying and dehydrating nectar to make *honey*—a concentrated high-calorie carbohydrate resistant to spoilage. Likewise, early honey bees acidified the plant pollens they collected to convert them to a spoilage-resistant protein called *bee bread* which can be stored alongside honey in their durable combs made of *beeswax*—a worker bee glandular exudate derived from the metabolism of nectar.

This is a good place to pause and think about the ecological and community implications of a vegetarian diet for a bee that is perennially eusocial. Consider the well-known observation that highly specialized and efficient plant pollinators are overwhelmingly represented by solitary bee species (or eusocial species in their solitary queen phase) (Winfree et al., 2008). It is plausible that it is critical for these species, which often have active seasons measured in days or weeks, to be good pollinators for whatever plant species are blooming at the same time. It is in their selfish interest to ensure the reproduction of those plants so that they will be available next year to sustain their offspring. Matters are less urgent for a perennially eusocial species like the honey bee, who has the leisure, so to speak, of visiting the whole season's panoply of flowers. Westerkamp (1991) went so far as to say that catholicity in flower visitation is in fact a precondition for perennial social life in the honey bee. For such a flower shopper, there is little or no selective pressure to become a pollinator specialist for any one.

But I think we need to be cautious on this point. What does matter is quantity of pollen and enough successive flower flushes (plant species) to provide it. Because the colony is actively rearing brood the entire growing season, it must extract between 37–75 pounds of pollen (16.7–34 kilograms) from its foraging range (Keller et al., 2005). In modern Europe, this is known to occur with at least 31 significant pollen-bearing plant species. However, a recurring theme is that when examined locally, the bulk of bee-collected pollen comes from relatively few species; across Europe, five plant species are responsible for 60% of annual pollen yield (Keller et al., 2005). This raises questions about the extent to which a generalist flower visitor receives positive selection for being a good pollinator. A narrow taxonomic range of pollen plants, as data for modern Europe seem to indicate, would suggest that selection exists. However, it is an open question whether this condition applied in flora of the pre-agricultural Europe during which *Apis mellifera* evolved.

I'd like to hypothesize a second idea about vegetarianism—whether it is conducive to, or even a precursor for, the evolution of perennial aboveground eusocial life in cold latitudes. Most eusocial hymenopterans living in cold climates defer to an annual life cycle: most famously, the bumble bees experience complete die-off in autumn except for newly-mated queens, who revert to solitary hibernation for overwintering. Alone among the perennially eusocial cold-climate aboveground hymenopterans stands *Apis mellifera*. Is there something unique about vegetarianism that enables *food storage*? Because this is really what we're talking about with a group of insects that live together, forego hibernation, and weather a long food dearth. The only case of food storage I can find for perennial wasps is in fact nectar (Kojima, 1996), but the equivalent need for protein for a cold-weather perennial wasp would be the storage of meat—which doesn't seem to happen. I admit that the problem of storing meat in cold climates doesn't sound very difficult, but we must remember that evolution never tells us "what could have happened," only what did. For whatever reason, acidifying pollen proved a successful strategy for storing protein over winter, and thus honey bees were able to preempt other challengers for the niche of "cold-climate, aboveground perennially eusocial hymenopteran."

Mixed up in the syndromes driving life history evolution in *Apis mellifera* was the problem of colony reproduction. Commitment to uninterrupted eusociality meant that reproduction had to be a social event—a form of fission: colony splitting. In honey bees, colony fission is called *swarming*. Fission in the superorganism means the separation of the soma (workers) between parent colony and daughter colony and provision of germ cells (queens) for each. Being a social event, it is qualitatively more complex than reproduction by individual pioneering like we see in the termites and some ants.[3]

3. The termites and some ants return briefly to near-solitary states at colony foundation, with solitary mated females or pairs of males and females dispersing to form new incipient colonies.

For the honey bee lineages that moved into temperate Europe, the problem of food storage took on the added urgency of seasonality in the availability of nectar and pollen. With winters long and nectar flows brief, the life history of temperate *Apis mellifera* became a high stakes game of balancing dynamics that are oppositional and fatal.

To begin, year-round sociality demands a certain minimum population biomass for ensuring group homeothermy. For this reason, a honey bee colony's population is normally in the range of tens of thousands, achieving peaks of 50,000–60,000 workers just before reproductive colony fission. In late summer and autumn, a colony rears a sizeable cohort of winter bees that are physiologically primed for long life. These are the bees that will sustain the colony and its queen through winter by clustering in the center-most combs and generating heat by shivering their flight muscles. In autumn, all brood rearing ceases, and bees allow the core nest temperature to drop as low as 18°C (64.4°F). This number is significant because it represents the minimum temperature at which adult bees can shiver and generate heat (Allen, 1959); it also illustrates the razor's edge of efficiency by which bees consume their precious reserves. Such standards of efficiency help temperate honey bee colonies routinely survive temperatures of –30°C (–22°F) or colder (Gates, 1914).

But as impressive as this efficiency is, what bees do next in the dead of winter once days begin lengthening borders on the unbelievable. This is when the wintering bees resume rearing brood. Energy demands skyrocket, not only from worker bees producing brood food to feed developing larvae, but from the extra heat generation required to incubate them. Core nest temperatures now jump to 34.5°C (94°F), rarely deviating thereafter by more than one degree [reviewed by Seeley (1985)]. All the while, this surge in energy consumption is underwritten by the same finite food supply held in reserve from the previous summer.

So why do bees engage in this profligate energy expenditure at the most vulnerable time of year? It can only be understood as a strategy for optimizing reproductive success of the future swarm. Although overwintered colonies, especially populous ones, may produce two or more swarms during spring, the first or *prime* swarm is the largest and the only one accompanied by the tried-and-true mother queen. There are strong positive correlations among earliness of swarming, population size of the swarm, and subsequent measures of its growth rate and survivorship [reviewed by Winston (1987)]. In other words, the bigger and earlier the swarm, the better its chances of surviving a temperate habitat. Earliness is critical in order to permit the new swarm colony and its parent colony as much summer foraging time as possible to recover a worker forager force and collect new honey stores in time for next winter.

Only reproductive benefits of an overarching magnitude could explain such a reckless adaptation as commencing brood rearing in the middle of winter. Brood rearing begins modestly but expands rapidly once new workers emerge to help rear more of their sisters. Bees are sensitive to food stores and adjust brood rearing accordingly—even to the point of cannibalizing brood if necessary to sustain adults during a food dearth (Weiss, 1984). Some temperate latitudes have early floral nectar sources—in Georgia, we have nectar from red maple (*Acer rubrum*) as early as January—and these flora provide critical supplements to dwindling food stores. But these resources tend to be consumed immediately and do not add meaningfully to the stores. It remains that late winter/early spring is the most dangerous time for starvation.

In a successful outcome, the colony will pull off the balancing act between rearing a large population while optimizing its food consumption rate. Gradually the declining food balance plateaus and is eventually reversed as natural springtime nectar flows come online. As colony populations peak, the mother queen begins to decelerate egg laying, during which time she loses body mass, an adaptation to permit her the ability to fly again.

A simultaneous precursor to swarming is the production of queen cells. In the lead-up to swarming, nurse workers select female larvae within their first 12 hours post-hatch and begin feeding them an uninterrupted diet of royal jelly, which activates queen genes and down-regulates worker genes. Individuals who complete their entire larval career on royal jelly will mature into the queen caste. Just as worker brood may be cannibalized, queen cells may be torn down and restarted in a fitful effort to synchronize availability of queen cells with a period of sustained nectar flow.

There is strong adaptive advantage for colony swarming to happen on a waxing nectar cycle. Such environmental conditions favor the consumption of nectar and activation of worker wax glands necessary for the production of beeswax. There is no higher priority for a new swarm than to rapidly construct its beeswax combs in which food is stored, brood reared, and all the social activities of the colony acted out. Bees dispersing with a swarm are engorged with nectar from their home colony—their foreguts or honey stomachs are elastic and adapted for carrying liquids—and this finite food reserve constitutes the sum of food available to the new swarm. A sudden cessation of nectar flow before a new colony has built and stocked its new combs spells certain death by starvation.

Once the optima of maximum population, presence of queen cells, and sustained nectar flow converge, the colony awaits only a warm sunny day to launch the swarming process. Half of the worker population leaves on the wing, accompanied by the mother queen. This winged cohort lands on a temporary bivouac site such as a tree limb (figure 3.13) and within a few hours chooses and relocates to a new permanent nest site. Back at the home site,

FIG. 3.13. A reproductive swarm in the temporary bivouac stage. Scout bees will perform recruitment dances on the surface of this cluster, setting the stage for the colony's ultimate decision on a new permanent nest site (chapter 15). PHOTO: HOLLI KIRCHER.

daughter queens eventually emerge, fight to the death to establish one new heir, and the victor mates and resumes the egg-laying functions of the colony. Swarm colonies settle on average greater than 300 meters from the parent nest, which is recognized as an adaption for avoiding landscape-scale resource competition (Seeley and Morse, 1978).

After the parent colony and swarm have sundered ways, each begins the business of rebuilding a full foraging force by which it will collect the next winter's food supply. To that end, the honey bee has evolved colony-level adaptations to expedite the extraction of available calories from its habitat. Chief among these is the honey bee worker's semi-stereotyped career of age-based task specializations (chapter 8), the recruitment dance language (chapter 14), and colony-level regulation of foraging effort (chapter 15).

The temperate honey bee colony's life cycle is like a path through a minefield—a pathway worked out by the evolutionary trial and error of generations. It's an exercise at balancing conflicting imperatives: producing the earliest and biggest swarm possible while building up and sustaining year-round the worker workforce (soma) necessary to build it, all against a backdrop of extreme seasonality in the availability of food. Thanks to the demographic work of Tom Seeley from the 1970s, some of it revisited as recently as 2017, we have a good idea of just how dangerous this game can be. In temperate, cold-weather New York state, for example, only 23%–24% of new incipient swarm colonies survive until the following spring. But if a colony survives its critical first year, then its chances of surviving to a subsequent year jump to 78%–84%. All told, once a colony survives its first 12 months, its expected life span is 5.6–6.2 years (Seeley, 1978).

TROPICAL HONEY BEE LIFE HISTORY

A prominent contender for the ancestral home of *Apis mellifera* is Africa (Tihelka et al., 2020, Whitfield et al., 2006). This scenario implies that the temperate adaptations I described in the preceding section are adaptations from an ancestral tropical state—a state I'll describe here.

First of all, African *Apis mellifera* share with their European *conspecifics* (belonging to the same species) the ancestral states of mutually dependent castes and obligate perennial sociality, year-round homeothermic worker population (soma), and colony reproduction by fission. This means that when we witness life history differences between the two, we are seeing the outcomes of two different sets of ecological selectors on a common evolutionary starting point (see figure 3.12). Or put another way, because the two groups are conspecifics, we can be confident that the differences we see between them are caused by recent ecological selection rather than differences in deep evolutionary history.

Compared with the temperature extremes of temperate latitudes, seasonality in the tropics revolves around cycles of rain and drought. In Africa, the dry period tends to be the season of nectar dearth. Nevertheless, many regions in Africa report more than one major nectar flow a year (Fletcher, 1978; Silberrad, 1976); moreover, African subspecies of honey bee can engage in true migration, the seasonal colony relocation in pursuit of nectar flows—a character wholly absent in European bees. The sum of this means that African subspecies have more frequent access to nectar flows and accordingly swarm more often per year. It is generally true that tropical African nectar flows are smaller and more frequent, whereas temperate European nectar flows are fewer and larger.[4]

It is generally believed that predators played a more active role in the evolution of *Apis mellifera* in Africa than in Europe (Fletcher, 1978), and of those predators, none was more dangerous than the bipedal ape *Homo sapiens*—a status it retains in parts of Africa to this day (Kajobe and Roubik, 2006). If frequent nectar flows provided the energy for frequent swarms, then high rates of colony mortality by predation provided the kind of population turnover that selects for high reproduction rate.

Figure 3.12 summarizes the colony life history differences between temperate and tropical honey bees. Tropical *Apis mellifera* exhibits the kinds of adaptations that favor rapid reproduction. Smaller populations and smaller honey stores reflect the reduced need for large homeothermic insulating populations and the large food stores to sustain them. With lower thresholds for viable fissionable populations, recovery time is faster between swarms, with

4. These generalizations must be treated cautiously. In South America, the wet seasons can be the dearth period, and nectar flows can be quite large by managed standards. But for our purposes, I am focusing on African conditions because these are most pertinent to our discussions on honey bee natural history.

queen replacement, forager force recovery, and food store recovery concentrated into shorter cycles. Tropical worker life spans are shorter because there is little need for a worker subcaste physiologically primed to carry the colony through a long dearth. The availability of diverse, unpredictable, small nectar flows has made it profitable for tropical *Apis mellifera* to practice resource-based absconding and migration, whereas in temperate latitudes, large predictable nectar flows with long intervening winters have selected for the opposite—strong commitment to one large heavily-invested nest.

Honey Bees and Life History Theory

If one were to summarize these differences, one could say that tropical *Apis mellifera* invests heavily in reproduction, whereas temperate *Apis mellifera* invests heavily in the soma. The outcome of the first strategy is many smaller offspring, and the outcome of the second is fewer larger offspring. Starting from the same social foundations, life histories of tropical and temperate honey bees have diverged significantly, and it is compelling to think that environment is the differentiator.

Understanding patterns like this is the domain of *life history theory*, the branch of evolutionary biology that studies how environment regulates life strategies adopted by an organism—basic demographics like age of first reproduction, birth rate, and death rate. The most robust studies are comparative: making predictions about birth rates and death rates for different populations of the same species across different environments (Kochin et al., 2010). On this count, the honey bee is an ideal model system owing to its ancient origin in tropical or northern Africa and its movement into temperate Europe as recently as 300,000 years ago (Tihelka et al., 2020; Wallberg et al., 2014).

In its formative years, life history theory was heavily influenced by the theory of *r*- and *K*-selection championed by MacArthur and Wilson in their 1967 book, *The Theory of Island Biogeography*. These authors envisioned life history being driven by habitat differences and how those differences translate into density-dependent population regulators, mainly resource limitation.

Imagine an unstable habitat where a population frequently experiences high mortality from density-independent factors—mortality agents that operate independently of population size like erratic weather, drought, flood, habitat destruction, or intense predator pressure. A residual population recovering after such a calamity would find their circumstances rather favorable: there would be abundant resources and little competition from conspecifics. Such a habitat would select for populations with rapid population growth. MacArthur and Wilson called this *r*-selection—invoking a parameter from a logistic equation for per capita population growth. Such *r*-selected species are predicted to excel at rapidly acquiring resources and converting them to offspring. At first read, African honey bees appear to fit this description.

At the opposite pole is a more stable, predictable habitat in which populations hover near their *carrying capacity*—a population's maximum size given the resource constraints of its habitat. Such populations are sensitive to density-dependent mortality—things related to population size such as resource scarcity, disease, and parasites. Such habitats select for populations with strong competitiveness and efficiency at acquiring, retaining, and using resources. MacArthur and Wilson called this *K*-selection, using the mathematical term for carrying capacity.

The *r*-/*K*-selection theory was enthusiastically embraced by ecologists during the 1970s and 1980s. There was much to like, chiefly the intuitive logic of categorizing habitats along a continuum of *r*-selecting to *K*-selecting and the irresistible urge to catalog species somewhere along that continuum. Temperate honey bees had their turn, and during the heyday of *r*-/*K*-selection's hegemony, they were described as showing signs of *K*-selection (Seeley, 1978).

But by the 1990s, the winds began blowing against *r*-/*K*-selection (Reznick et al., 2002). The tendency of *r*-/*K* to dichotomously lump habitats (and their denizens) along a single continuum was deemed simplistic, if not arbitrary. In the case of honey bees, it was never obvious that either temperate or tropical habitats could easily fit an *r*- or *K*- label, at least in a way that was consistent with the theory's predictions and observed bee life histories. For example, in supposedly *K*-selecting temperate Europe, *Apis mellifera* population densities are at least two times *lower* than in Africa (Jaffé et al., 2010; Wallberg et al., 2014). In addition, the notion that predation rates are lower in Europe than Africa has been disputed, given that each range historically included the world's most dangerous honey bee predator, *Homo sapiens* (Rinderer, 1988).

It is true that experiments with fruit flies (*Drosophila melanogaster*) tended to support MacArthur's and Wilson's premise that natural selection is sensitive to density-dependent mortality factors. Fruit flies reared in *r*-type habitats (low density) evolved the capacity for higher reproduction at low densities, but they could not repeat this performance at high densities. Conversely, flies reared in *K*-type habitats (high density) were more competitive, as expressed by higher feeding rates and by pupating higher above the surface of the feeding medium (pupae on the surface risk being buried by feeding larvae) (Mueller and Sweet, 1986). But showing that natural selection is responsive to density-dependent mortality is not the same thing as showing a link between habitat type and life history strategy, however one may rationalize how different habitats affect density-dependent mortality.

In the end, *r*-/*K*-selection was not so much rejected as assimilated into more sophisticated models that recognize more factors than density-dependent regulation in the evolution of life history (Reznick et al., 2002). The leading successor to *r*-/*K*-selection is *demographic theory*, which, as its name suggests, focuses on a population's age structure and how this regulates its survival rates, fertility, and investment in reproductive effort alone or in interaction with other factors. Density-dependent factors are now inputted into models alongside age-specific mortality and density-independent factors such as habitat change, food availability, and climate change to arrive at mechanistic links between environment and life history choices (Reznick et al., 2002).

Honey bees have been included in demographic studies, mostly at the scale of colony. But some colony-level demographics are suggesting themselves as important enough to drive regulation at population (inter-superorganismal) scales.

It has long been known that when colonies suffer large losses of forager bees, young workers will precociously jump ahead to foraging duties. Moreover, this is recognized as a natural phenomenon in colonies that have suffered predation, nest damage, or even the trauma of normal swarming (Winston and Fergusson, 1985).

This precocious shift to foraging is not without cost. Rueppell et al. (2007) showed that age of first foraging is a pivotal moment in the life of a worker bee and that mortality peaks rapidly thereafter. It has traditionally been assumed that this mortality comes from extrinsic risks associated with foraging, but when investigators artificially reduced foraging activity and its associated hazards, mortality in this cohort remained high. There is, apparently, an intrinsic physiological switch toward reduced longevity in workers who have made the shift to foraging duties.

A connection between this demographic shift and colony fate was shown by Khoury et al. (2011), who modeled how forager death rates influence colony mortality. Chronic loss of foragers creates a negative feedback loop that imperils colony survival. As young bees precociously advance to foraging duties, their life spans are truncated as a matter of course, while the cohort normally responsible for rearing brood is diminished. With lower adult replacement rates, the colony begins a fatal downward spiral. Workers normally forage about 6.5 days of their life, but the model of Khoury et al. (2011) showed that if this time is reduced by chronic stressors to 2.8 days, then the colony will likely fail.

It is intuitively impossible to ignore environment when one compares tropical- and temperate-evolved bees, and a simulation conducted in 2013 (Russell et al., 2013) included an environmentally sensitive parameter, seasonality of food collection rate (milligram/bee/day) while corroborating the aforementioned results of Khoury et al. on worker task shift. This simulation reinforced that colony survival is sensitive to seasonal swings in food availability and food demand from brood rearing.

To the extent that habitat affects within-colony demographics, we can expect differences across habitats in the ways natural selection has optimized these dynamics for colony survival. But to my knowledge, only one paper has looked at the question across biological levels from worker to colony to landscape population. Al-Khafaji et al. (2009) modeled how worker state (i.e., life span), queen state (her survival dependence on workers), and colony state (population size at swarming and inter-swarm time intervals) interact to affect landscape-scale population changes. The positive feedbacks, negative feedbacks, and auto-correlations are dizzying.

For example, landscape population growth is affected by average inter-swarm intervals and the probability of queens (and their workers) surviving those intervals. Queen survival is positively correlated with her supporting worker population size, including the size of the cohort that accompanies her when she swarms. A *longer* inter-swarm interval permits the colony time to produce a larger worker cohort with positive effects on the subsequent swarm. With a jump start on a large population, such a swarm is primed for a *shorter* inter-swarm interval and faster production of its subsequent swarm, leading to increasing landscape scale populations. However, if this fast turn-around is purchased at the expense of inviably small fractionated worker populations—for the swarm or original colony—then this could jeopardize survival of each. How do habitat differences sort all this out?

First, consider one enigmatic parameter mentioned above—a queen's survival dependence on workers. It's indeed reliable that larger worker populations correlate to queen survival, but this effect is stronger or weaker depending on habitat. A benign habitat in which fewer workers are needed to collect food or defend the nest selects for low queen dependence on workers, and queen survival is comparatively high across a range of starting swarm population sizes. The reverse is true in a harsh environment in which many workers are needed to thermoregulate a nest and collect food. The queen's survival dependence on workers in this case is high and, correspondingly, there is selection for longer worker life, larger starting swarm populations, and longer inter-swarm intervals. In general, I find this consistent with observed differences between European and African *Apis mellifera* (see figure 3.12).

There is growing awareness in life history theory that the project should focus on principles that are common across population ecology, evolutionary biology, and life history choice (Coulson et al., 2010). This reminds me of our discussion in chapter 2, namely, that we should hope for principles that are common across all stages of evolutionary transition, not treating any transition as a singular event. But in the case of a global species like *Apis mellifera*, one species whose natural range crosses nearly 100 degrees of latitude, it seems optimistic to expect such a holistic and parsimonious explanation. The present state of science indicates that honey bee life history theory will instead be informed by an amalgam of demography and habitat descriptors.

Summary

The honey bee, *Apis mellifera*, exists as a superorganism, arguably the paradigmatic superorganism, given its entrenchment in the human imagination as a model for research and agent for agriculture. Being a superorganism means that the honey bee is subject to natural selection on at least two levels: the individual worker bees that are themselves organisms, and the colony for which workers subordinate their selfish reproductive interests and labor for its common fitness. As such, honey bees exemplify the evolutionary transition to a new form of individuality.

Honey bees are holometabolous insects, passing through developmental stages of egg, larva, pupa, and adult. In contrast to the eusocial tropical *Meliponini*, *A. mellifera* practices progressive feeding of young, which may impart efficiencies that favor rapid colony growth after seasonal dearth periods. In common with all Hymenoptera, honey bees express haplodiploid sex determination by which each male is haploid—a product of an unfertilized egg—and thus inherits and transmits only its mother's alleles. Females, in contrast, are diploid, possessing, like mammals, a 2N complement in somatic cells and 1N in reproductive cells. This fate choice is dictated by the queen, who possesses muscular control of the release of sperm from her spermatheca, thus rendering an egg haploid or diploid. All diploid female eggs are genetically qualified for one of two developmental commitments—becoming a worker or becoming a queen—a choice mediated by workers in response to colony needs. Workers feed queen-destined female larvae a special diet called royal jelly, which triggers upregulation of queen-like genes.

A typical colony possesses 10,000–50,000 workers depending on time of year, one queen, and males or drones varying in number from a few thousand in spring to zero over winter. Among the colony's members, workers possess the richest behavioral repertoires, each worker advancing through a predictable sequence of in-nest to out-of-nest tasks as she ages.

Temperate *Apis mellifera* evolved under conditions of harsh winters and strong seasonality in food availability, thus selecting for large colonies and heavy investment in a single reproductive swarm. In contrast, tropical *Apis mellifera* evolved under conditions of high predation and smaller, more frequent nectar flows, thus selecting for smaller colonies and higher rates of reproductive swarming. Life history differences such as these are a product of environment interacting with colony demographics such as age structure of members, worker life span, a queen's dependence on her workers, colony population size at swarming, and inter-swarm time intervals.

CHAPTER 4

First Came the Nest

IT IS ALMOST AXIOMATIC to say that society presumes shared space. This presumption may need to be held loosely if Strassman and Queller (2010) are right that it's possible to extend organismality to non-contiguous cells like pathogens in the body of their host (chapter 2, footnote 2). Likewise, shared space may no longer apply to modern human societies, given our obsession with technology-assisted global connectedness. But these cases excepted, we still tend to think of societies in terms of assemblies of individuals attached by culture or genetics to a place—*Cuban* society, *British* society; even among professional societies we have the Entomological Society of *America*, and so on. The colonial insects seem to feel this way too—or at least the entomologists who study them, because insect societies invariably fit the description of individuals sharing space: a nest.

The Soil-Nesting Precursors

To begin, let's envision our ancient honey bee ancestor as a solitary reproductive female—no need to call her a queen yet—with the cognitive capacity for searching for and finding a suitable nest, recognizing landmarks, navigating about the landscape, and returning home. What I have just described is a herculean feat for a brain of any size, and the vast majority of insects can't do it. Most insects are free-living in their local habitats with at most only a brief and transitory association with a place of any kind, which for most is a sheltered spot to deposit their eggs.

But in the insect order Hymenoptera—the ants, wasps, and bees—nesting is well-represented. It may not look like much—a hollow reed or twig (figure 4.1), but among the wasps, it is common, and in the bees, their descendants, it is universal except in the *kleptoparasitic* (cuckoo) species, who have abandoned nesting in preference to occupying nests of their hosts. Nesting is an ancient shared character of bees and if not exclusive to social species (even solitary bees do it), it is nevertheless a prerequisite to sociality. Nesting allows for parental care of young, defense, food storage, and shelter—all of which are stabilizing forces friendly to the evolution of colonial life. And although some modern species of nomadic ants (most famously the army ants, *Eciton* spp.) have abandoned nests in favor of transient aboveground group bivouacs (Wilson, 1971), it's hard to imagine sociality taking off without the initial benefits of a permanent nest. Under the most advanced insect societies, nesting has evolved beyond tunnels and hollow reeds into elaborate triumphs of animal architecture like the nest of the honey bee (figure 3.11).

It is this nest, of course, that interests us most. But it is a product of historic precursors of much simpler designs and functions. Most certainly, the ancestral state for all bees was tunneling in soil given that this character is

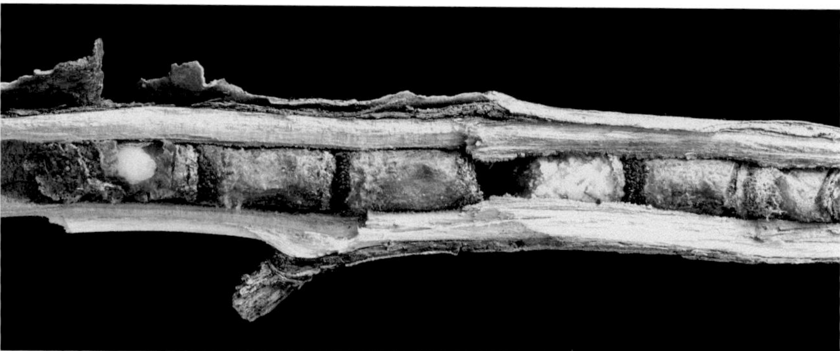

FIG. 4.1. An exposed nest of the megachilid *Hoplitis* spp. in a hollow twig of *Sumac* spp. showing the repeating linear pattern of mass-provisioned brood cells. PHOTO: JAMES CANE.

common in the sphecoid wasps from which the bees diverged (Michener, 2000), and fossil soil nests, even possessing features suggesting eusociality, are known from as early as the Late Cretaceous (66–145 million years ago) (Genise et al., 2002), coincident with the beginning of bees (see figure 1.3). From these fossil nests, we can infer that our honey bee ancestors' nests were simple tunnels elaborated over time with increasingly compartmentalized cells for food storage and developing young.

For soil nests, the stereotyped pattern is a central shaft with lateral tunnels leading to individualized brood cells (figure 4.2), individual cells being an ancient character suggesting that protecting the developing young against trampling by nestmates was strongly selected for. The majority of solitary and primitively social species practices mass provisioning. Within each cell is deposited a ball of pollen and an egg; the emerging immature is thus housed with all the food it needs and receives little or no subsequent parental care (figure 4.3). The alternative feeding mode is progressive provisioning, the kind practiced by honey bees, in which cells are kept more or less open and the larvae fed many small meals continuously.

Depending on species, the interior walls of excavated cells are lined with either fine compacted clay, tediously smoothed by the occupants, or with a glistening film derived from the Dufour's gland near the sting apparatus, a lining which imparts resistance to water, pathogenic fungi, and invertebrate soil predators. It is tempting to think that the abdominal wax glands and beeswax in the social bumble bees, stingless bees, and honey bees derive from the ancient Dufour's gland and its secretions. But beeswax glands share no morphological homology with the Dufour's gland (Landim, 1963) and must be viewed as independent innovations.

It is possible to detect signal of sociality in the structure of fossil (and modern) soil bee nests. First of all, there is consensus that nests with very large numbers of cells, and by extension individuals, cannot be the work of one female and must therefore reflect some degree of social cooperation by their inmates (Breed, 1976; Elliott and Nations, 1998; Packer, 1992; Thackray, 1994). Others have argued that nest designs with short or non-existent lateral tunnels encourage high rates of contact between individuals, promoting sociality (Kukuk and Eickwort, 1987). By these measures, one may,

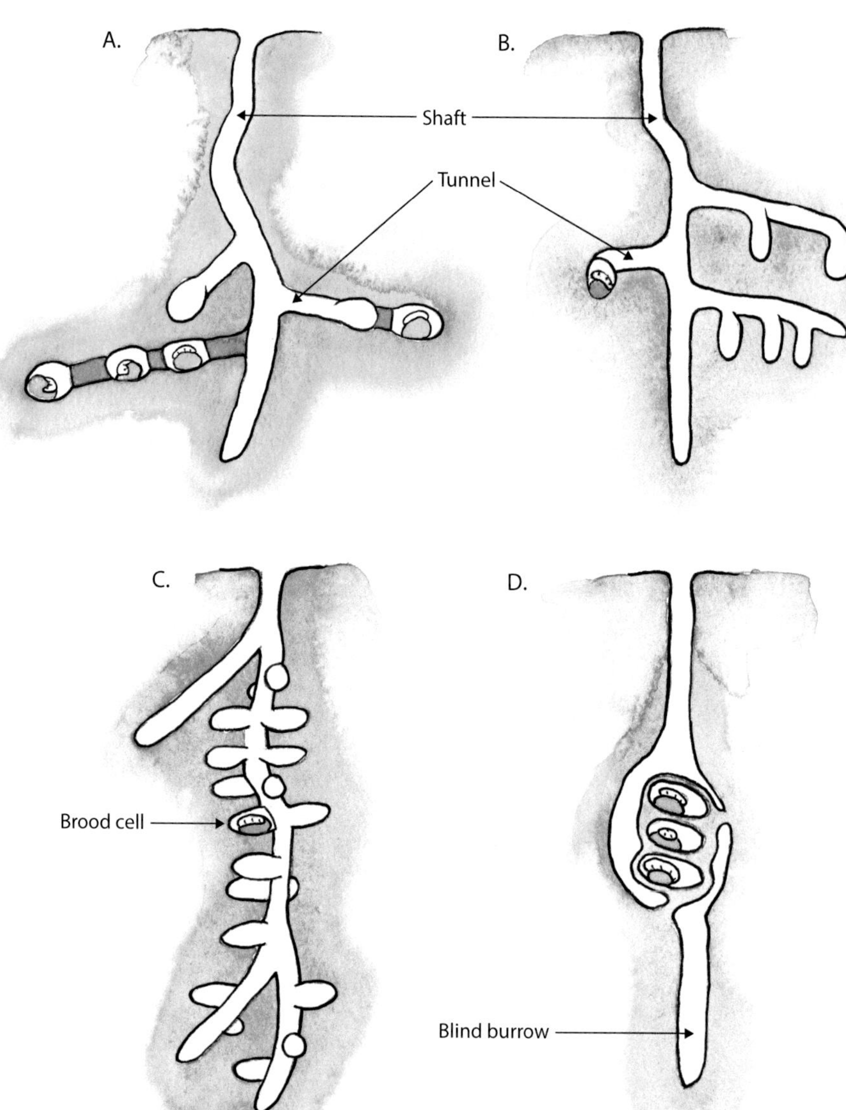
A.
B.
Shaft
Tunnel
C.
D.
Brood cell
Blind burrow

FIG. 4.3. A mature larva of the halictid *Nomia melanderi* in its brood cell with pollen provision. PHOTO: JAMES CANE.

for example, hypothesize that sociality has reached a higher pitch in the design of figure 4.2(C) than in figures 4.2(A) or 4.2(B). But, in contrast to these opinions, Stephen et al. (1969) concluded that for the soil-nesting bee family Halictidae, there is no correlation between nest design and degree of sociality.

The gulf between ancestral soil-nesting and the nests of modern *Apis* is wide, and no formally constructed phylogeny has attempted to bridge it nor

FIG. 4.2. (FACING PAGE) Diagrammatic representation of some bee soil nesting designs, showing in each nest at least one complete brood cell with pollen ball and larva. Nest categories and features after Hasiotis (2003). The *main burrow* leads from the entrance to brood cells either directly or by means of one or more lateral burrows. A main burrow that is vertical is called a *shaft*, and nearly horizontal burrows are called *tunnels*. Main burrows lead to one or more *brood cells*. A *blind burrow* is often present at a level below the brood cells and acts as a sump for water and waste or a source of excavated soil for use elsewhere in the nest. **A.** A *linear* nest of the communal halictid, *Pseudagapostemon divaricatus*, redrawn and modified from figure 6.1 in Michener (1974). **B.** A *branched* nest of the communal andrenid *Oxaea austera* from Argentina, redrawn and modified from figure 19 in Sarzetti et al. (2014). The lateral tunnels of branched nests usually each terminate in one brood cell, but this specimen is distinctive for having multiple brood cells per tunnel. Relatively lengthy laterals, as in this example, are considered by some authors evidence for lower grades of sociality. **C.** A *branched* nest of a colonial halictine, *Halictus ligatus* from Trinidad, redrawn and modified from figure 7-3 in Michener (2000). The extreme truncation or absence of laterals, as in this example, is considered by some authors conducive to evolution of sociality. **D.** A *chambered* nest of *Augochlorella striata*, redrawn and modified from figure 7-5 in Michener (2000). In this design, the bees construct cells of earthen linings, joining them into a common clump in the center of an excavated cavity.

identify the *polarity*[1] of character state changes involved. In fact, the only hypothesis-driven phylogenies using nest architecture to my knowledge are restricted to the Neotropical Meliponini (Camargo and Pedro, 2003; Rasmussen and Camargo, 2008; Wille and Michener, 1973). It may prove difficult to reconstruct a deeper history, given the experience of those working with meliponines who have found many independent convergences and reversals of nest characters within this one large group (Wille and Michener, 1973). Adding to the problem is the fact that a single species may exhibit nest building *plasticity*—variable expression of a single character.[2] Such confusion on identity and polarity of character states in the Meliponini hints that the situation will be no neater in their fellow corbiculate tribe, the Apini, the "true" honey bees (see figure 1.3).

But this doesn't mean that we can't use current biology to speculate on plausible transition points in nesting biology from primitive soil nesting to the arboreal, cavity-nesting, comb-building habits of modern *Apis mellifera*. I offer below a thought experiment on these transitions.

But first, three things: Here and always we must resist the temptation to see either a qualitative superiority to sociality or a necessary vector in that direction. The vast majority of modern bee species (1) remain soil nesting, and (2) live at a sub-eusocial grade, putting the lie to any notion that evolution is progressive. Second, in talking about nest evolution, I cannot avoid talking about social evolution, as the two circularly co-reinforce. So although I will below flit past some evolutionary mileposts of colossal importance, the reader can rest assured that they will get their due in course. And third, because of these caveats and to restrain us from wandering too far across the sprawl of bee biology, I formally introduce the reader to the *corbiculate* bees, those four tribes comprising the subfamily Apinae: the orchid bees

1. *Polarity* speaks to the direction of evolutionary change for a heritable character. If two variants of character x exist, x_1 and x_2, it is the phylogenist's job to determine if x_2 derived from x_1, or vice versa.
2. Beekeepers of European *Apis mellifera* are familiar with one example of character plasticity—those times when bees build a full set of combs and live on a tree limb or otherwise open to the weather. A phylogenist of the distant future, encountering such a singular fossil could be excused for considering "open nesting" a discrete character state when in fact it is a deviation from normal "cavity nesting." Driven by cavity scarcity, open nesting is almost always maladaptive in temperate latitudes.

Euglossini, the bumble bees Bombini, the stingless bees Meliponini, and the true honey bees Apini (figure 1.3) (see also chapter 21). These four tribes share a common morphological feature—a *corbiculum* or complex pollen basket on their hind legs (figure 4.4), and within these four tribes, we see the full range of social and nesting possibilities known for bees. One way or

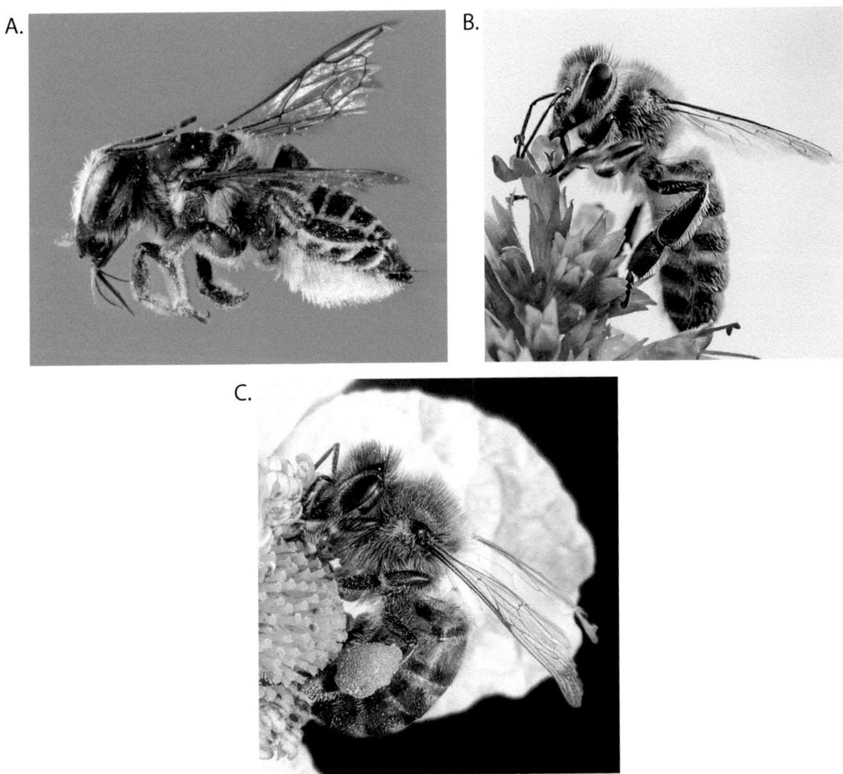

FIG. 4.4. The term *scopa* refers to pollen-carrying hairs that occur on female bees; the term is reserved for hairs used to carry pollen back to the nest, not hairs used to rake pollen off a flower. The scopa can occur as rows of stiff parallel hairs on the underside of the abdomen as in this specimen of solitary bee *Megachile* sp. (**A**) or as bristles on the hind legs. If a fringe of scopal hairs surrounds a nearly hairless flat or concave area on the hind leg in which pollen is packed for transport, the structure is called a *corbiculum*. The corbiculae of the namesake corbiculate tribes Apini, Bombini, Euglossini, and Meliponini is on the outer side of the hind leg tibia with associated structures on the widened basitarsus (see Michener 2000, p. 48 and Winston 1987, p. 24). This figure shows an empty corbiculum on a foraging honey bee worker (**B**) and the same structure on another bee loaded with pollen (**C**). PHOTO A COPYRIGHT: ANTHONY VAUDO, USED BY PERMISSION. PHOTOS B AND C: GEORGIA P. ZUMWALT.

another, corbiculate bees will occupy our attention for the rest of this book, permitting us at this juncture to bid adieu to the non-corbiculate bees and their own eminently fascinating life histories, solitary, social, subterranean, and above ground. With such caveats understood, let us proceed.

A Plausible Evolutionary Scenario from Soil Nesting to Arboreal Nesting

We know from the fossil record and extant species that soil nesting is compatible with social evolution in bees. In fact, the mere existence of a nest is, as Charles Michener (1974, p. 61) put it, "an invitation to other bees to enter and live colonially." This gets us to an important feature of a nest beyond its obvious role as shelter. The nest as a focal point in space triggers a kind of emergent property called *stigmergy*—the positive feedback caused when an actor leaves a trace in the environment upon which the same or subsequent actor can act. It can be as simple as one nestmate beginning to excavate a brood cell but then abandoning it, and another discovering it and finishing the job. The first bee is the first actor; the unfinished cell is its trace, and the second bee the second actor. The nest provides countless opportunities for indirect social interactions like this. As a result, task efficiency improves at the group level: no one individual need perform all jobs, and multiple jobs can be done at once. Stigmergy has been recognized as a force in social insect evolution since at least the 1950s (Grassé, 1959).

We will call the first transition point in our thought experiment the stigmergic synergies that led to such innovations as short lateral tunnels and high-density brood cells that in turn encouraged more social interactions and complexity (figure 4.2(C)). In what may be the second transition point, the plasticity of the soil medium permitted an ancient precursor to comb—the chambered nest (figure 4.2(D)) in which cells are clustered together into a contiguous mass. There are efficiencies of space, motion, and labor to be gained by consolidating the fundamental units of a burrow—the brood cells; and these efficiencies translate into larger populations and increasing social complexity. To the extent that the clustered cells share the same walls, they may be properly called *combs* (Hasiotis, 2003). The architectural regularity of comb, in turn, provides a stimulating stigmergic template inviting more bees to participate in its construction.

From the innovation of consolidating cells, it was an easy progression to the innovation of segregating cells by function—brood cells in one area of the nest and food storage cells in another, a modification driven by efficiencies in incubating brood (Seeley, 1982). Segregation of cells by function persists into modern taxa, even though nest and cell designs of Meliponini + Bombini differ strikingly and visibly from those of the Apini (see figure 1.3).

Let's next address the transition to arboreal life. Bees as a group are most abundant and diverse in warm temperate and xeric habitats around the world (Michener, 1979). This environmental niche bias was probably influenced by ancient hazards associated with soil nesting. Dry climates afford more relief from nest inundation and high humidity that would otherwise spoil nest provisions and encourage pathogenic fungi. Balanced against these risks are the benefits of soil's ubiquity as an available substrate, its malleability to design innovation, its insulating properties, and its spatial infinity. In fact, these advantages are formidable, and I have long thought that soil nesting is the reason that the ants and termites, not the bees, are the champions of social complexity.

Nevertheless, there were interactions between local habitat and life history that made it adaptive for some corbiculate lineages to abandon soil for aboveground nesting. Compared with the omnivorous diet of ants and cellulose diet of termites, stored pollen was probably more vulnerable to spoilage microbes in soil. Some bee lineages may have gained adaptive benefit from aboveground brood incubation temperature optima, allowing more than one generation of brood per year. In addition, the concentration of a centralized brood-rearing area in the form of chambered nests/proto-combs may have increased risk from soil predators, selecting for bias toward protected aboveground cavities.

I think it's significant that cavity nesting is near-ubiquitous in the corbiculate tribes, and those cavities occupy the full range of available sites from soil to arboreal to manufactured substrates such as termite mounds (Michener, 2000, p. 650). The social bumble bees, tribe Bombini, occupy abandoned rodent burrows at turf depths (figure 4.5); the social honey bees Apini occupy hollow trees and mini-caves, and the uber-diverse social Meliponines occupy virtually everything else (Rasmussen and Camargo,

FIG. 4.5. An exposed nest of a bumble bee *Bombus*. Among corbiculate bees, Bombini, Euglossini, and Meliponini retain single pot-like cells of beeswax, whereas Apini make combs of contiguous cells. Meliponini makes both combs and pots, compartmentalizing each into different parts of the nest. PHOTO: SYDNEY CAMERON.

2008). It is self-evident that cavity architecture of the kind seen in chambered nests/proto-combs is applicable across a range of substrates, and we have already noted that it is an architectural design that accommodates social life, although the cavity-nesting non-eusocial Euglossini remind us that there is no necessary connection between the two.

But because we are interested in this book in a particular lineage of the tribe Apini, that one leading to the western honey bee *Apis mellifera*, we are now ready to become more particular in our thought experiment. I suggest that the third major transition from soil to arboreal nesting for this lineage involved an evolutionary *syndrome* of characters—a suite of related, inseparable, and interacting adaptations.

Imagine, if you will, a soil-nesting bee population comfortably on its way to eusociality or already there. Its inherited comb-making habit has synergized its social evolution, but moisture-related disease pressure and temperature-related reproductive advantage are rewarding those members of the population who move their nests into cavities at or above ground level. The abandonment of soil demands a shift in architectural strategy: brood cells are less and less excavated and more and more constructed. The behavioral antecedent for this is the ancient cell linings derived from the Dufour's gland, but morphologically something new happens instead—abdominal wax glands (Landim, 1963) that excrete a malleable waterproof construction medium, beeswax. The architectural regularity of the comb template encourages stigmergic amplification of comb size and, by extension, population size and social complexity. With a consolidation of biomass into combs, predation risk pushes the population into more secure cavities, and tree hollows and caves afford cavities that are resistant to water and temperature extremes, as well as predators who are becoming increasingly vertebrate, intelligent, and bipedal.

The Cavity-Nesting Modern Forms

With modern Apini, the true honey bees, we are on firmer ground in our analysis of nest evolution because in the living members of *Apis* (the only living genus in the tribe) we have a range of designs for aboveground beeswax comb nests. Restricted to a more narrow taxon and armed with a range of expressions for nesting, phylogenists have a better chance at determining history and relationships of the taxon's members. But as the reader will see, there are conflicting opinions on such fundamental questions as character state polarity and even geography of origin for the genus.

To begin, bees in the genus *Apis* can be divided into three subgenera[3] according to their body size and nesting habits (see figure 1.3): (1) *Apis* (*Micrapis*), small bees with a single exposed comb (figure 4.6); (2) *Apis* (*Megapis*), large bees with a single exposed comb (figure 4.7); and (3) *Apis* (*Apis*), medium-sized bees with multiple combs in cavities (figure 3.11).

It has long been held that the polarity of nesting state in *Apis* is from ancestral single comb open nests to derived multiple combs in cavities (Garnery et al., 1991; Raffiudin and Crozier, 2007). This premise has gone hand in hand with belief that the genus originated in tropical southeast Asia (Ruttner, 1988), where today its largest number of species is to be found. By this view, it has been held that cavity nesting was an adaptation that, among other things, permitted *Apis mellifera* and its sister species *Apis cerana* to expand from their tropical homelands into cold latitudes (Hepburn et al., 2001). There is indeed much to like in this interpretation, not least of which

3. Subgeneric names are rarely used except when the context makes it useful. When they are, convention calls for the name to be italicized, capitalized, and in parentheses between the generic and specific names; hence, for example, *Apis* (*Micrapis*) *florea*.

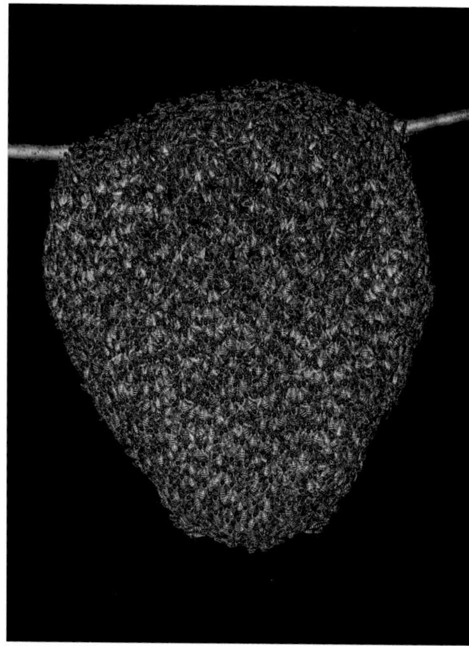

FIG. 4.6. A colony of the small, single-comb open nesting "dwarf" honey bee of Southeast Asia, *Apis* (*Micrapis*) *florea*. Notice how the comb (unseen under the bees) wraps over the top of the limb; this provides a "dance floor" upon which foragers perform their recruitment dances. PHOTO: JEFFREY C. MILLER.

is the way it has been used to set the narrative for other areas of honey bee biology such as evolution of colony size (chapter 13), symbolic dance recruitment language (chapter 14), thermoregulation (chapter 11), swarming (chapter 16), and modern species and subspecies distributions (chapter 22).

Arguments have been made for alternative views. Kotthoff et al. (2013) have argued for a central European origin for *Apis* on the basis of fossil evidence, and Koeniger et al. (2011) made a case for cavity nesting as a primitive state with single-comb open nesting as derived. A European origin for the genus remains a vigorous argument (Carr, 2023), but work subsequent to Koeniger et al. (2011) has sustained the traditional view that single-comb open nesting is more ancient (Bossert et al., 2019; Fouks et al., 2021).

Evolutionary ghosts of its tropical origins still linger in modern *Apis mellifera* to be noticed and puzzled at by modern observers. I offer as one exam-

FIG. 4.7. A colony of the large, single-comb open nesting "giant" honey bee of Southeast Asia, *Apis* (*Megapis*) *dorsata*. Aside from being much larger, the comb of *A. dorsata* is conspicuous from *A. florea* for not wrapping around the top of the limb to which it is attached. Uncommitted to a narrow limb, *A. dorsata* is free to occupy diverse nesting sites such as cliff overhangs, large limbs, even eaves of buildings. PHOTO: MICHAEL BURGETT.

ple the rare phenomenon of open-air nesting (see figure 11.3). Every now and then, a swarm settles on a limb or bush and stays there, building comb and going through all the activities with every indication of permanence. But depending on one's latitude, this is typically suicidal behavior because the colony cannot withstand the next winter without the shelter of a cavity. Colonies that do this are presumably reacting to an extreme shortage of nesting sites. The second example is the curious behavior known as "festooning," in which bees form chains or curtains with their bodies over the surface of comb (see figure 3.2). This seems to me an evolutionary throwback to exposed combs in the Asian tropics. Modern *Apis dorsata* and *A. florea* do this, forming curtains of living bees over their single combs as a kind of proxy shelter against predators and temperature extremes. Michener (1974) called such curtains "an inefficient use of bees," but also noted that this behavior may have been a preadaptation that later permitted *A. mellifera* and *A. cerana* to enter dark cavities as nesting sites. If this is accurate, then cavity nesting had the added benefit of freeing up large fractions of the population to do other tasks.

At this point in our discussion, the connection between nest and social evolution becomes so inextricable that we must defer more modern innovations in nest design to later chapters in which nest interacts to create new socially evolved characters. For now, we are where we need to be: the corbiculate bees are a foment of innovation in terms of nest design and social evolution; among them is the tribe Apini and its single living genus, *Apis*; within *Apis* we have three subgenera distinctive for their body size and nesting habits. It is within the milieu of these facts that our discussion on social evolution will unfold.

The modern honey bee nest is far more than a shelter but rather an integral and dynamic participant in the evolution, behavior, health, and social interactions of its inmates. At its most primitive, the containment afforded by a nest provokes more social interactions than would otherwise occur in free living organisms. At its most derived, the nests of highly social species like the honey bee are as information-rich as a human city, generating stigmergic stimuli, feedback loops, emergent properties, and opportunities for individual decision-making, the collective weight of such decisions adding up to optimized actions at the level of group—actions which may or may not be good enough to ensure the survivability of the group. And so it is that the colony—the individuals that make it up in context with each other and the nest substrate—becomes Darwin's unit of natural selection (Moritz and Southwick, 1992; Seeley, 1997), with all of its novelties and innovations judged together under the impartial test of What Works.

Summary

Nesting is an ancient shared character of bees and a prerequisite to sociality. Nesting allows for parental care of young, defense, food storage, and shelter—all of which are stabilizing forces for the evolution of colonial life. The modern honey bee nest with its cavity-enclosed repeating combs is a marvel of animal architecture, but it is derived from much simpler precursors. The ancestral state for all bees is tunneling in soil, and fossils show a pattern of simple tunnels elaborating over time with increasingly compartmentalized cells for storing food and housing young. The stereotyped pattern is a central shaft with lateral tunnels leading to individualized brood cells, individual cells being an ancient character for protecting the developing young against trampling by nestmates. The interior walls of cells are lined with fine compacted clay or a Dufour's gland secretion that resists water, pathogenic fungi, and invertebrate soil predators. Modern beeswax glands share no morphological homology to the Dufour's gland and must therefore be viewed as independent innovations.

There are features of fossil and modern soil nests that suggest varying degrees of social cooperation by their inmates—features such as large numbers of cells, suggesting large populations, and short or non-existent lateral tunnels which encourage high rates of social contact between individuals. The gulf is wide between ancestral soil nesting and the arboreal cavity nests and repeating combs of the modern honey bee, but in this chapter I offer a plausible evolutionary sequence for this transition. Modern *Apis* are divided into three subgenera, *Apis* (*Micrapis*), *Apis* (*Megapis*), and *Apis* (*Apis*), according to body size and nesting habits such as single versus multiple combs or open versus cavity nesting. Because nest and social evolution are so tightly interacting, discussion of these later nesting habits will be covered in future chapters in context to particular social developments. Throughout the evolution of the honey bee lineage, nesting has provided stigmergic stimuli, feedback loops, emergent properties, and opportunities for individual decision-making—all of which have been powerful synergists for social group formation and maintenance.

CHAPTER 5

Grades of Sociality

IT'S TIME THAT WE BEGIN zeroing in on the social part of social evolution. Along the way, we'll see that "social" and "sociality" have become mutable words, accumulating qualifiers and descriptors, just like "organism" did in chapter 2, as our understanding of the principles governing group dynamics has matured.

Degrees and Qualities of Sociality

We should start by acknowledging the anthropocentrisms that have infused the writings on social insects, the ghosts of which linger to this day. From the earliest musings of Aristotle to that 1609 bestseller, Charles Butler's *The Feminine Monarchie*, human writers have seen in the teeming populations of insect colonies and body enlargements of the reproductive castes direct corollaries to human feudal society, monarchs, and top-down organization. To casual biology readers and recreational beekeepers, such terms as "queen" and "society" may still serve as stumbling blocks to some of the ideas in this book—ideas such as organismality, individuality, group decision-making, heterarchical organization, emergent properties, and superorganism.

But the label "social" is intuitive, powerful, and persistent, and even non-entomologists recognize that the colonial insects demand some kind of a social descriptor of an extraordinary order. Here is something that even if not *social* is nonetheless very much like it—and of a higher state than transient social exchanges like mating, fighting, territorialism, aggregations of young dispersing from egg clusters, or congregants at a shared waterhole.

Definitions of sociality and its degrees of expression have been in revision ever since entomologists in the 1960s and 1970s began laying down terms that are still in currency today. I'm reminded of the old joke among taxonomists about whether one is by temperament a "lumper" or a "splitter"—prone to pool categories or carve up new ones. Evidence of either approach is apparent in the summary of terms in table 5.1.

It was E. O. Wilson (1971) who synthesized earlier authors' ideas on the matter and presented a set of three criteria or "qualities of sociality" building up to *true* sociality or *eusociality*—criteria that have had an enduring influence on the literature ever since. During at least some point in its annual life cycle, a eusocial species must simultaneously express (1) cooperative brood care, such that individuals of the same generation help tend the common brood, (2) reproductive division of labor, such that some individuals abandon reproduction in order to help others reproduce, and (3) overlapping generations, such that some or all offspring remain at the nest to help their parent(s) rear more of their siblings.

These criteria systematized by Wilson (1971) and sources therein, chiefly Michener (1969), allowed for a kind of hierarchy or "degrees of sociality," given that a species may express fewer than the full number of criteria, or none at all beyond mating behavior. Table 5.1 gives these categorical possibilities plus two definitions of eusociality developed outside the constraints of Wilson's criteria.

Evolutionary Routes toward Eusociality

Wilson's criteria, once formulated, also gave a vocabulary for understanding the two plausible evolutionary routes toward eusociality. Wilson called these the *parasocial* route and *subsocial* route (table 5.2). The key difference is this: the subsocial route involves the union of parents and offspring and

TABLE 5.1. Definitions and categories of social behavior

DEGREES OF SOCIALITY	VARIANTS	QUALITIES OF SOCIALITY			NOTES
		COOPERATIVE BROOD CARE	REPRODUCTIVE DIVISION OF LABOR	OVERLAPPING GENERATIONS	
Solitary*		–	–	–	
Subsocial*		–	–	–	Parental care of young
Communal*†		–	–	–	Individuals of same generation share nest
Quasisocial*†		+	–	–	
Semisocial*†		+	+	–	
Eusocial	Primitive	+	+	+	Reproductive and worker castes morphologically indistinguishable (Michener, 1969); workers capable of mating (Wilson, 1971)
	Simple (Bourke, 2011)	+	+	+	Workers reversible to reproductive form
	Annual	+	+	+	Colony retreats to a solitary state in the form of overwintering, mated females who single-handedly found new colony next spring
	Advanced	+	+	+	Morphological variation between castes (Michener, 1969); workers incapable of mating (Wilson, 1971)
	Complex (Bourke, 2011)	+	+	+	Workers adapted for work, irreversible to reproductive form
	Perennial	+	+	+	Colony sustains a eusocial state year-round

Facultatively eusocial (Crespi and Yanega, 1995)	+	+	+	Reproductive caste totipotent or potentially autonomous
Obligately eusocial (Crespi and Yanega, 1995)	+	+	+	Worker and reproductive castes mutually dependent
Alternative, Sherman et al. (1995)	+	NA	NA	Care of non-descendent young under influence of kin selection
Alternative, Bourke (2011)	NA	+	NA	Any society with reproductive division of labor regardless of reversibility

*Presocial describes any state short of eusociality. †Parasocial describes those presocial states in which individuals of the same generation occupy one nest and includes communal, quasisocial, and semisocial groups.

Note: From Kocher and Paxton (2014), Michener (1969), Wilson (1971), and other references as cited. The last two entries offer alternative definitions of eusociality that fall outside Wilson's classical three qualities of sociality.

TABLE 5.2. Evolutionary pathways to eusociality

	QUALITIES OF SOCIALITY		
DEGREES OF SOCIALITY	COOPERATIVE BROOD CARE	REPRODUCTIVE DIVISION OF LABOR	OVERLAPPING GENERATIONS
Parasocial route (via same generation), either egalitarian or fraternal			
Solitary (communal)	—	—	—
Quasisocial	1	—	—
Semisocial	1	2	—
Eusocial	1	2	3
Subsocial route (via offspring), fraternal			
Solitary	—	—	—
Intermediate subsocial I	—	—	1
Intermediate subsocial II	2	—	1
Eusocial	2	3	1

Note: Numbers indicate presumed order in which transition is achieved and retained. Degrees and qualities of sociality by Wilson (1971).

hence is kin-based, whereas the parasocial route involves the union of same-generation individuals and is therefore open to the union of non-kin. The parasocial pathway is what Michener (1974, p. 61) had in mind when he called the nest entrance "an invitation to other bees to enter and live colonially." In the interest of harmonizing this usage with the evolutionary transitions in chapter 2, one may think of the parasocial route as egalitarian and the subsocial as fraternal.

But as interesting as all this may be, the evidence seems to weigh against the parasocial (egalitarian) route as a viable pathway to eusociality. This is not to say that communal living is not successful; on the contrary, communal living is widely represented across taxa. There are many ecological benefits to be gained from communal living (figure 5.1), some of which we touched upon in chapter 4 in the context of nests. To those we can add some others.

Communal nesting, and by extension cooperative nest construction, may aid cooperators if the blooming season of their obligate food plants is brief. Speedy construction of a shared nest allows more time for foraging and building brood cells. Second, communal living may grant a cooperator "advantage of assured fitness returns" (Gadagkar, 1990), which is a kind of life insurance: if a communal female dies before the end of the active season,

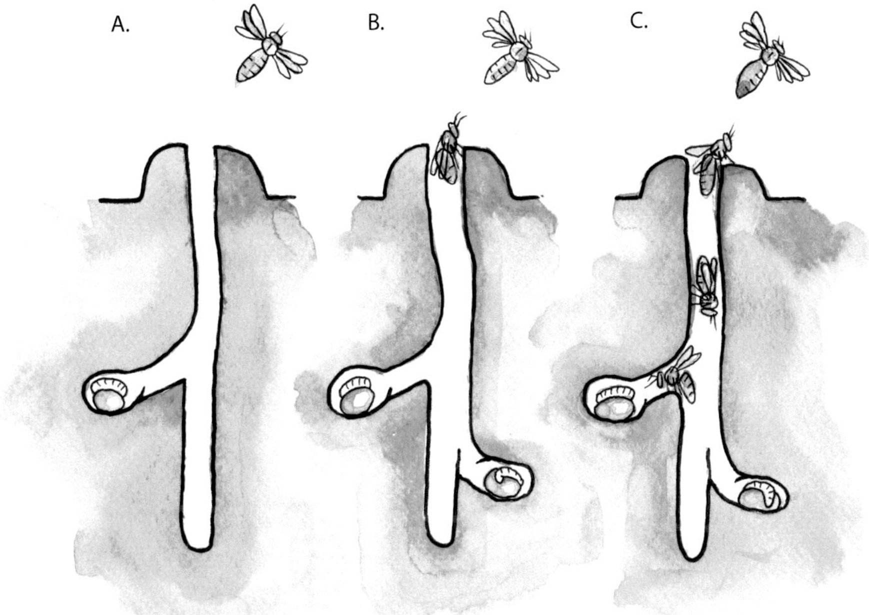

FIG. 5.1. Communal living—multiple females, related or not, living in one nest—is a simple and successful social life strategy represented across taxa. Among other benefits, communal living advantages participants with economy of labor in the construction and defense of a nest. There is no evidence, however, that communal living by non-kin is a pathway to eusociality. In (**A**), we see the state of a solitary female, single-handedly making a nest and provisioning a brood cell. In (**B**), we see evidence of communal and possibly quasisocial sociality (table 5.1)—two females and two brood cells, admitting the possibility for cooperative brood care. In (**C**), we see evidence of semisocial sociality—more females than brood cells, suggesting that some females have abandoned reproduction in preference to work on behalf of the group. This figure is discussed further in a different context in chapter 17.

her brood cells may still survive to adulthood thanks to the nest defense activities of their mother's nestmates.

But in spite of these ecological advantages, there is little evidence that egalitarian communal living has ever led to eusociality (Bourke, 2011; Danforth et al., 1996; Paxton et al., 1996; Tierney et al., 2008). When communal living does lead to eusociality, it invariably follows the fraternal route—communal living of near-kin (Bourke, 2011, p. 96). Moreover, whether it's multicellularity (Bourke, 2011, p. 97) or eusociality (Bourke, 2011, p. 101 and table 4.2) complex social groups overwhelmingly follow the subsocial fraternal route. Communal living is a successful, stable, and widespread mode of living, but it's an evolutionary dead-end when it comes to eusociality

(Paxton et al., 1996; Tierney et al., 2008). In this book, I will trace an evolutionary history of the honey bee that follows a fraternal subsocial route.

Simple and Complex Eusociality

At this point, we need to make explicit an important feature in table 5.1: two major grades of eusociality are to be discerned from what is likely a continuum in nature. These two grades can be called *simple* and *complex* eusociality. All of the variants of eusociality in table 5.1 can be lumped into one of these two, and these labels have the advantage of avoiding the value-encumbered usage of "primitive" and "advanced" found elsewhere.

What makes a social group, whether a multicellular organism or eusocial colony, simple or complex is a mixture of characteristics, some quantitative and more or less obvious—such as cell number and type—and others more qualitative and foundational. Moreover, these characteristics seem to track each other in parallel between multicellular organisms and eusocial colonies, itself a clue that common principles of sociobiology undergird the evolution of life on this planet. Table 5.3 presents some of these parallels.

Table 5.3 indicates that complex multicellular and eusocial groups share in common a strong qualitative descriptor: a robust segregation between germlines (reproductives) and the soma (workers). In simple groups, there is reproductive potentiality in cells (or members) that must be restrained if higher orders of biological organization are to happen. This strikes at the heart of the emergence of individuality as we discussed in chapter 2. It may be helpful to reiterate the relevant point from that chapter:

> An *individual* is an entity composed of formerly independent parts who now share a high degree of integration and cooperation; whose genetic interests are aligned or at least subsumed to the interest of the group; whose genome is transmitted in common; and who express near complete eradication of within-group conflict.

In multicellular organisms, this alignment of genetic interests is achieved through the evolution of sex and the genetic bottleneck represented by the fertilized zygote that renders all somatic cells clones of one another. But in complex eusocial groups, we are dealing with members who are already organisms, so clonality is not an option. Instead, genetic interests are aligned

TABLE 5.3. Comparisons of simple versus complex social groups: multicellular organisms and eusocial colonies

MULTICELLULAR ORGANISMS			EUSOCIAL COLONIES		
	Simple	Complex		Simple	Complex
Morphological difference between germ line and somatic cells	Low	High	Morphological difference between queens and workers	Low	High
Reproductive potential of somatic cells	High	Low	Reproductive potential of workers	High	Low
Single lineage of germline cells confined to one part of body	Absent	Present	Segregation of brood by reproductive potential	Absent	Present
Commitment to multicellularity	Can be facultative	Usually obligate	Reversibility to solitary state	Possible	Impossible
Timing of divergence of germline and somatic cell lineages during development	Late	Early	Timing of divergence of immatures into reproductive or non-reproductive forms	Late	Early
Number of somatic cell types	Low	High	Number of morphological castes among workers	Low	High
Size (number of component cells)	Low	High	Size (number of component members)	Low	High

Note: Adapted from Bourke (2011) and Fisher et al. (2013).

by a compromise between self and group. Although this will be covered in more detail in chapter 7, for now, it can be understood as follows:

Complex eusocial groups evolve extreme behavioral and morphological divergence between the reproductive caste and workers—an evolutionary milepost called by Wilson and Hölldobler (2005) "the point of no return." At this point, the reproductives become so specialized for mating and laying eggs that they can no longer perform colony work, and the workers become so specialized for colony work that they can no longer mate or lay eggs. Their states are irreversible. Neither can revert to primitive solitary life. Each has become mutually dependent on the other, rendering the social group a Darwinian unit of selection.

The only (and temporary) exception to mutual caste dependence in complex eusocial groups occurs in those species that practice *independent* colony founding in which a young queen (or royal pair in termites) disperses from her home colony and founds a new nest in solitude. She performs all colony functions until her first clutch of worker daughters emerges and frees her to focus solely on egg laying (this is further discussed in the chapter 16 section "Independent versus Dependent Colony Founding").

With the colony now an adaptive unit, group selection promotes one of two pathways for generating high intra-colony genetic variation, where genetic variation is rewarded at the group level for improving task specialization and group competitiveness (Brown and Schmid-Hempel, 2003). Those two pathways are *polygyny*—the presence of multiple queens in one colony, with each queen mated with one or more males; and *polyandry*—the presence of one queen in one colony, who is mated with multiple males. Ants and termites practice polygyny, whereas social honey bees exclusively practice polyandry. A female practicing polyandry possesses an organ called the *spermatheca* that stores and nourishes the sperm of her numerous mates for her entire life, rendering her colony a mosaic of *patrilines*—numerous families of workers, each family sired by one male. In the Hymenoptera, workers sharing the same father are called *supersisters*, whereas workers with only their mother in common are called *subsisters*.

One consequence of polygynous or polyandrous colonies is a reduction in average relatedness among members of the colony. In the case of honey bees, a given worker is laboring to produce siblings with whom she shares on average 0.27 genes in common, which is significantly less than the 0.5 she could propagate by directly producing offspring as a solitary individual (chapter 7). But "reduced average relatedness" is simply the flip side of "increased genetic diversity," and colonies expressing high diversity are well-known for exhibiting greater size, parasite resistance, and ecologic robustness (see table 9.2). Large complex colonies are not a monopoly of species practicing polygyny or polyandry—exceptions exist in the highly social stingless bees of the tribe Meliponini. But it is also true that species expressing high within-colony genetic diversity have achieved efficiencies of size, scale, and specialization that mark them as masters of terrestrial ecosystems. Ants and termites alone make up only 2% of the estimated 900,000 insect species known, yet together account for more than half of all insect biomass (Wilson and Kinne, 1990).

Thus, a complex eusocial group represents an uneasy compromise of interests, the resolution of which we will cover in chapter 7. The queen wins the genetic lottery prize—the maximum compliment of genetic transmittal, 0.5, available to a sexually-reproducing diploid animal, a specialized morphology to do little else but lay eggs, and a factory/fortress of nest plus workers (Oster and Wilson, 1979) to help her rear them. The workers, on the other hand, accept an average reduction in gene transmittal, decreasing from 0.5 to 0.27, in return for the ecologic benefits of colonial living. There's no getting around the fact that this sacrifice to direct reproduction is a high bar for any taxon to breach. If eusociality is a rare event in evolution, then complex eusociality with its irreversible sterile castes is even rarer. To the seven independent origins of complex social groups in insects reported by Bourke (2011) including ants (three), wasps (more than one), and termites (three), we may add the two (Apini and Meliponini) (Cardinal and Danforth, 2011) and one (Allodapini) (Rehan et al., 2012) reported for bees, to make up a total of at least 10.

Against this backdrop, we can only pause to appreciate the innovation of sex in the history of multicellular life. Through the single-celled bottleneck of the fertilized zygote, complex multicellularity rendered the somatic cells clones, sweeping away the conflicts and compromise that worry eusocial superorganisms (chapter 2). Note that I said "complex" multicellularity: just as polyandry is associated with the largest and most complex superorganisms, so is sex associated with the largest and most complex multicellular organisms. In any case, the lopsided representation of complex organisms over superorganisms in the world's history is testimony to the successful stabilizing power of sex. But the parallels between the two levels shout for attention and testify to the common sociobiological principles that bear them up.

What Is a Society?

Having spent this chapter talking about descriptors of sociality, perhaps we should end with an attempt at defining a *society* and the limits of the word's appropriate use. I think "society" is one of those members of the entomological lexicon overdue for an overhaul.

To begin, consider a traditional definition put forward by E. O. Wilson in 1971 (p. 5), who called a society "a group of individuals that belong to the same species and are organized in a cooperative manner." A little later, he adds "[r]eciprocal communication of a cooperative nature" as another "essential intuitive criterion of a society" (Wilson, 1971, p. 6).

By 2011, Bourke (pp. 15, 20) is ready to call a society (*sensu stricto* a "social group") "a stable group of any entities that cooperate in ways that make the group a potential candidate for consideration as an individual." With this definition, we catch a whiff of 40 years' worth of progress on understanding major evolutionary transitions and the sociobiological principles that undergird them. With this definition, we also intuit the distinction between a unit of individuality and the component members that make it up. What is that distinction between the two?

I propose that reproductive autonomy is the critical distinction. In Darwin's economy, if transmittal of one's genes is coin of the realm, then participation is limited to those that can reproduce as a contiguous genome—unfettered and uncompromised. With our modern understanding of evolutionary transitions as the subsummation of lower levels of biological organization into higher, then this is synonymous with the subsummation of lower levels of reproductive autonomy into higher.

A society is a group of autonomous reproductives held together by place and cooperation. To the extent that each of a group's members retains reproductive autonomy, the group is a society. Conversely, to the extent that a group's members acquiesce reproductive autonomy to the group, the group becomes something else. If the group's members are themselves organisms, it is a superorganism; this is why "superorganism" is such a useful word.

Summary

A species is considered truly social, or *eusocial*, if at some point in its life cycle it simultaneously expresses three qualities: cooperative brood care, reproductive division of labor, and overlapping generations. Because a species may meet some, all, or none, of these criteria, this admits the possibility for degrees of sociality, and these categories have been summarized in table 5.1. The three qualities of sociality are useful in framing two evolutionary routes toward sociality—the *subsocial* route, which imagines groups forming fraternally from offspring lingering at the nest to help their parents rear more siblings, and the *parasocial* route, in which members of the same generation (related or not) join in egalitarian fashion to form a social group (table 5.2).

The classical eusocial categories of table 5.1 can be collapsed into two of overarching evolutionary and ecologic significance—simple social groups in which members retain some degree of reproductive autonomy versus complex groups in which specialization has proceeded to the point that reproduction and work on behalf of the group are segregated by morphological castes. With its members irreversibly and mutually dependent, an entity at this stage is subject to strong group selection and can be thought of as having breached the evolutionary transition from organism to superorganism. Complex multicellularity and eusociality have emerged from subsocial evolutionary pathways characterized by high kin-based relatedness. On a numeric basis, complex multicellularity is the dominant life form on Earth thanks to the stabilizing effects of sex and the single-celled zygote that renders all somatic cells clones with harmonized genetic interests; however, complex eusociality has achieved different but adequate means of reducing somatic conflict, achieving along the way large colony populations and sophisticated division of labor, giving eusocial species, especially the ants and termites, an ecologic significance disproportionate to their species numbers.

Given the intellectual foment surrounding social behavior and its degrees of expression, the word "society" struggles to retain currency common across human usage. I propose that "society" be reserved for groups of reproductively autonomous individuals bound together by place and cooperation. A nest of communal bees is a society. A nest of honey bees is a superorganism. Each occupies a different place on the hierarchy of evolutionary transitions.

CHAPTER 6

Altruism and the Emergence of Simple Eusociality

NO GREATER IMPEDIMENT is to be found for the recruitment of new beekeepers than the fact that bees sting. In fact, it's a joke in beekeeping circles that it's a good thing bees sting or else honey producers would be overrun with competitors.

Whether this is *actually* a good thing or bad is fodder for another conversation, but for now, this gets us to an important point about the honey bee worker—that she is the consummate altruist. For not only does she sting, but she mortally injures herself in the process. Her barbed stinger implants in the skin of her adversary like a harpoon, tears away as she leaves, and the poison sac attached to it throbs away under involuntary muscles pumping out every last dreg of venom. Meanwhile, the injured worker renews her assault with more determination than ever. She flies recklessly at the intruder, burrowing into fur or hair with a buzzing designed to distract and deter. She flies straight at the eyes. It's as if she knows she's a goner and has nothing to lose. If she survives the encounter and returns to the nest, she has only a few hours to live before her internal injuries catch up with her and she dies, only to be unceremoniously disposed of by an undertaker bee.

Now I must confess that in my early days of beekeeping, I took vengeful comfort in this knowledge, especially at the end of a hot day with my hands sore and swollen and my sweaty shirt pinned to my back. But if we can dissociate ourselves from the personal insult, we may appreciate the necessity and efficiency of such a repertoire of behaviors. From the perspective of a large predator, the honey bee nest is a lottery prize of protein and calories, and this stationary target requires extraordinary protection. The detachable stinger/poison sac essentially doubles the defensive output of every single worker, freeing its possessor to return to the attack with additional intimidations. Multiply this one terrifying performance by tens, scores, or even hundreds of assailants and the result can be effective—the successful deterrence of a predator. And lest we humans get too complacent thinking ourselves somehow detached from this story, it's good to remember the co-occurrence of *Apis mellifera* with ancient humans and our immediate ancestors on the African continent. It's not far-fetched to believe the bees' defensive behaviors are in fact evolutionary adaptations *to us*, arguably the most dangerous predator they've encountered in their natural history.

The Problem of Altruism

But this self-sacrifice, impressive enough, is amplified when we remember that this same worker has abandoned direct reproduction in order to labor on behalf of another—her reproductive queen. In human terms, the worker bee is saying, "Not only will I give up reproduction in order to help my mother reproduce, but I will die if necessary to do so." This is *altruism* practiced to a pitch rarely seen anywhere. At the beginning of things, it was an affront to Charles Darwin's theory of natural selection. How can one pass on any genes whatsoever, favorable or not, if one isn't even capable of reproducing? In Darwin's economy, the only currency is reproduction, that is—getting one's genes passed on to the next generation.

The awkward problem of the sterile worker was not lost on Darwin (1859) himself, and in chapter 7 of *The Origin of Species*, he confronts the problem, and I quote:

> I will . . . confine myself to one special difficulty, which at first appeared to me insuperable, and actually fatal to my whole theory. I allude to the neuters or sterile females in insect-communities: for these neuters often differ widely in instinct and in structure from both the males and fertile females, and yet, from being sterile, they cannot propagate their kind. (p. 352)

Hamilton's Rule and Inclusive Fitness Theory

At this point, it is interesting to pause and consider one of history's great *what ifs*: During the 1850s, while Darwin was putting the final touches on his great theory in the south of England, a contemporary of his, the Augustinian monk Gregor Mendel, was untangling the basic laws of hereditary genetics 1,000 miles (around 1,600 kilometers) away in Brno in what is today the Czech Republic. The two never met, and it is unclear to what extent either was aware of the other's work. As a result, Darwin went to his grave ignorant of the mechanisms of inheritance, the consequences of which he nevertheless intuited. More practically, this historical near-miss meant that the great neo-Darwinian synthesis between genetics and evolution had to wait until the 1930s and 1940s. If Darwin had known even a rudimentary level of Mendelian genetics, he might have propelled evolutionary biology forward by decades and solved his insuperable problem as well. As it stands, he still hit pretty close to the mark, and again I quote:

> This difficulty, though appearing insuperable, is lessened, or, as I believe, disappears, when it is remembered that selection may be applied to the family, as well as to the individual, and may thus gain the desired end. (Darwin, 1859, p. 354)

With this sentence, Darwin anticipated the neo-Darwinian explanation for eusocial altruism that would emerge over 100 years later in the form of a two-part paper published by 26-year-old W. D. Hamilton at University College, London (Hamilton, 1964). It was Hamilton, advantaged with Mendelian genetics, who mathematically showed that the self-sacrificing behavior of an individual can be adaptive if it promotes the survival and reproduction of near-kin who possess the same genes, or to put it in his words:

> [A] gene may receive positive selection even though disadvantageous to its bearers if it causes them to confer sufficiently large advantages on relatives. (Hamilton, 1964)

Mathematically, Hamilton's insight looks like this:

$$c < br \qquad \text{(equation 1)}$$

Genes for altruism are predicted to spread in a population if the cost (c) of an actor's altruism is less than the benefit (b) to the recipient times the relatedness (r) of the actor and recipient. *Cost* is the negative change in offspring number borne by the actor, and *benefit* is the positive change in recipient's offspring number enabled by the actor's altruism. *Relatedness* (r) is the probability that the two individuals share the same genes at any given loci relative to the average probability for those same genes in any two members randomly drawn from the population, familial relationship being the most common reason for high r. Even though r ranges from zero to one, a value of $r = 0$ does not mean that two individuals have no genes in common, only that their chance of sharing the same gene at any focal locus is no different from the gene's average population frequency (Bourke, 2011). This is a convenient place to point out that an increase in an allele's frequency in a population is evidence for positive selection: if members of one generation contribute progeny whose allele frequencies are unchanged from the population average, by definition no selection has occurred.

Let's dive into an example with a normal diploid human family. The relatedness (r) of a pair of human siblings is 0.5 and can be derived by

$$\begin{aligned} r_{\text{sib}} &= (G_{\text{mother}}\, P_{\text{mother}}) + (G_{\text{father}}\, P_{\text{father}}) \\ &= (0.5 \times 0.5) + (0.5 \times 0.5) = 0.5 \qquad \text{(equation 2)} \end{aligned}$$

where G_{mother} = fraction of one's genome shared with their common mother; P_{mother} = proportion of mother's genome shared; G_{father} = fraction of one's genome shared with their common father; and P_{father} = proportion of father's genome shared. For diploids, P_{mother} and P_{father} average to 0.5 because of the chromosomal reshuffling that happens in the formation of eggs and sperm (see figure 2.5).

By definition, the relatedness of a clonal offspring and its parent is one. The relatedness of a sexually produced diploid individual and its diploid parent is 0.5. To determine one's relatedness to a nephew/niece or cousin, equation 2 is modified by recognizing the halving that occurs at each level of relatedness removed:

$$r_{\text{niece}} = (G_{\text{sib}}\ P_{\text{sib}}) = (0.5 \times 0.5) = 0.25 \qquad \text{(equation 3)}$$

$$r_{\text{cousin}} = (G_{\text{aunt}}\ P_{\text{aunt}}) = ((0.5 \times 0.5) \times 0.5) = 0.125 \qquad \text{(equation 4)}$$

Although cost can be levied against lifetime production of offspring, cost can be conservatively set at simple self-replacement or $c = 1$ in order to show minimum benefit necessary to balance out self-sacrifice. Accordingly, by solving for b, one can use Hamilton's rule to calculate the number of kin progeny an actor's altruism must achieve to offset the actor's unrealized self-replacement (table 6.1).

In other words, if the actor's altruism sacrifices its ability to produce one offspring ($c = 1$), then its altruism must enable its sibling to produce $b = 2$ offspring, its niece to produce four offspring, or its cousin eight offspring before the cost to intergenerational gene transmittal is recovered. Obviously, the efficiency of the altruism is greater the higher the r of the beneficiary toward whom it is directed. But because Hamilton's rule is an inequality, theory predicts that genes for altruism will only spread in a population if the benefit *exceeds* parity.

It is this outcome of Hamilton's rule that John Haldane humorously anticipated, however imprecisely,[1] when he famously said, "I would lay down my life for two brothers or eight cousins" (Haldane, 1932).

With Hamilton's insight, deriving logically from Mendelian rules of inheritance, he formalized the concept of *inclusive fitness*—the sum fitness an actor realizes by its own reproduction plus the reproduction of near kin who are aided by the actor's altruism; this, in distinction to individual or *direct* fitness. Hamilton opened up the resources of Darwin's theory of natural selection to biologists working with social insects and their hordes of sterile workers. In one stroke, Hamilton unloosed a new science onto the world—

1. Haldane was thinking literally of self-replacement, i.e., $r_{\text{self}} = 1$. Benefits would have to exceed those in table 6.1 for altruism to spread.

TABLE 6.1. Solving for *b* in $c < br$ to determine parity

RELATIONSHIP	c	r	b
Cousin, great-niece/nephew	1.0	0.125	8
Niece/nephew	1.0	0.25	4
Sibling	1.0	0.5	2

inclusive fitness theory—the idea that individuals in social units can be predicted to behave in ways that promote transmittal of their genes even if those genes are in the bodies of near-kin.

The Haplodiploid Hypothesis for the Emergence of Eusociality

Hamilton's math is persuasive, but when it comes to the social Hymenoptera, of which bees are a part, Hamilton's math appears positively compelling. Readers will recall that males of the order Hymenoptera are haploid—possessing only one half (1N) of the species's genome (chapter 3's "Some Things Common to All Honey Bees" section). This means that there is no recombination of chromosome pairs in gamete formation (figure 2.5); for a male bee, a chromosome has no homolog mate with which it can exchange sections of DNA. Every sperm that a drone produces transmits his entire genome. By rendering $P_{\text{father}} = 1$, male haploidy has huge implications for relatedness among hymenopteran sisters, revising equation (2) as follows:

$$r_{\text{sib}} = (G_{\text{mother}}\, P_{\text{mother}}) + (G_{\text{father}}\, P_{\text{father}})$$
$$= (0.5 \times 0.5) + (0.5 \times 1.0) = 0.75 \qquad \text{(equation 5)}$$

Two such worker siblings, sharing a common father and 0.75 of their genome in common are the *supersisters* we talked about in chapter 5. Consider just one consequence of this: A worker bee, although functionally sterile, is nevertheless prone to develop her ovaries and produce eggs parthenogenically if she is removed from the ovary-inhibiting influence of her queen's pheromones. Being herself diploid, her eggs undergo chromosomal recombination in the normal manner so that each is genetically unique to another ("Workers" section in chapter 3). If this worker succeeds in producing a male egg and if her colony rears it to adulthood, that worker has essentially succeeded in the game of diploid reproduction—passing on 0.5 of her genome to the next generation.

But if that same worker altruistically forsakes direct reproduction and instead helps her mother produce supersisters, she stands to pass on 0.75 of her genome with each supersister she helps rear. This is as near to a Darwinian no-brainer as it gets. Hands-down, 0.75 will carry the day.

Except it's not quite that simple.

Such *relatedness asymmetries* are indeed a hallmark of the social Hymenoptera and the few other groups, mostly insects and mites, in which haplodiploidy predominates. Hand in hand with Hamilton's inclusive fitness theory, relatedness asymmetries were the impetus behind an early and persuasive meme in the scientific literature called the "haplodiploid hypothesis"—the idea that eusociality is overrepresented in haplodiploid taxa. But subsequent years have uncovered enough independent emergences of eusociality in diploid lineages to show that the principles of social group formation are general and not restricted to haplodiploids (Choe and Crespi, 1997). No better proof exists than the fully diploid termites, whose modern species today rival the ants in social complexity and ecologic dominance. The haplodiploid hypothesis is still considered an open question by some (Bourke et al., 1995; Crozier, 2008), but there's no question it has lost traction (Alpedrinha et al., 2014, 2013; Gardner et al., 2012; Queller and Strassmann, 1998).

Part of the reason involves the other progeny class we haven't talked about yet—the males. The same haplodiploid condition that allows a helping worker to generate supersisters with whom she shares 0.75 genes in common also generates brothers with whom she shares only 0.25 genes in common (figure 6.1), which averages out to 0.5 gene transmittal, the same as for a diploid.

So, thinking we had arrived at the threshold of altruistic sociality, we instead find an apparent setback, or at least a formidable question: Why should any individual, haplodiploid or diploid, forsake direct reproduction (0.5 genome transmittal) in favor of altruistically rearing kin when even for haplodiploids, self-sacrificial altruism appears to offer no improvement over 0.5 genome transmittal?

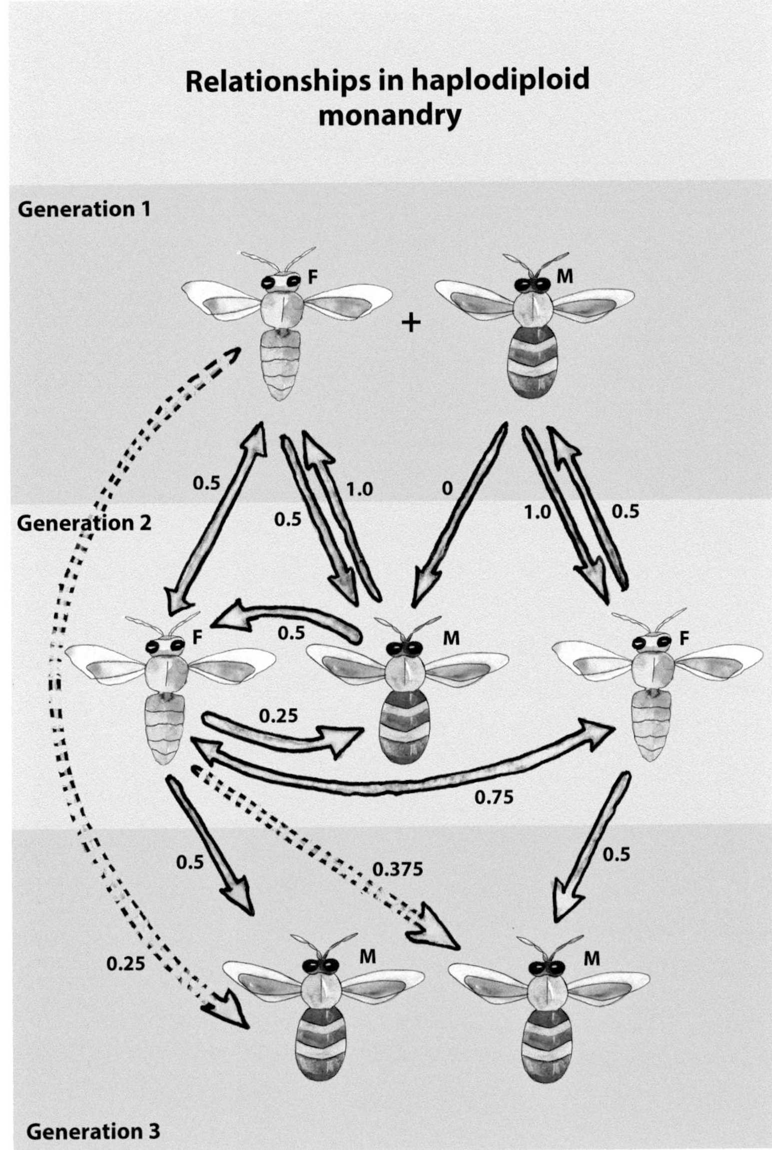

FIG. 6.1. Genetic relatedness—the probability that two individuals share the same alleles at any given loci—in a haplodiploid monandrous family in which the mother is mated to one male. F = female; M = male. Hatched arrows show relatedness by indirect descent.

Sex Allocation, Split Sex Ratios, and Reproductive Value

The haplodiploid hypothesis must confront the relatedness asymmetries that abound in a hymenopteran species experimenting with eusociality. These asymmetries mean that mothers and daughters are in potential conflict over the desired ratio of females to males in the colony. Classical sex allocation theory (Fisher, 1930) predicts that in freely mating populations and absence of inbreeding, selection will favor population sex ratios of 1:1 females to males. In the social Hymenoptera, sex ratio means reproductive queens and males (but see the "Restraining Sex Allocation Conflict in the Honey Bee Superorganism" section). By 1976, Trivers and Hare noted that if a 1:1 sex ratio held for the helping daughter of a singly mated hymenopteran mother, then the outcome of helping rear a supersister (0.75) and a brother (0.25) averages to a net result of 0.5 gene transmittal, the same as if the daughter rejected altruism, stayed solitary, and reared her own offspring (Trivers and Hare, 1976).

A haplodiploid proclivity for altruism may be rescued, Trivers and Hare (1976) proposed, if workers bias their efforts toward helping rear supersisters. The queen, passing on 0.5 of her genes with every egg she lays, is expected to favor a 1:1 sex ratio in her progeny, whereas workers, facing a 0.75/0.25 relatedness asymmetry with their siblings, are expected to prefer rearing sisters over brothers by a ratio of 3:1. This scenario presumes the ability of daughters in early monogamous groups to distinguish male from female eggs and to privilege females (Nonacs and Carlin, 1990); such aptitude for discrimination is known in modern honey bees (Haydak, 1958), but it is not clear whether this is an ancestral ability or a later derived state. In any case, this discrimination seems important for sustaining the hypothesis because, as pointed out by Craig (1979), helping daughters would otherwise be rearing equal numbers of $r = 0.75$ supersisters and $r = 0.25$ brothers, simply reflecting their mother's 1:1 egg output.

This rescue hypothesis is supported by field data and is therefore not wholly discredited (discussed further in chapter 7's "Sex Ratios in Complex Eusocial Colonies" section). But it also fails to take into account differences within sex-skewed populations in *reproductive value* of individuals—an individual's probability of contributing genes to the population. In the case of a 3:1 population female bias, reproductive value is inflated for males (Craig, 1979), and here's why:

Imagine a random hymenopteran gene in the future and trace it back to the present generation. There is a one in three chance it comes from a male and two in three chance it comes from a female owing to the fact that females carry two sets of chromosomes and males only one (if it were a diploid population, it would be a 50:50 chance a random gene comes from a male or female). These ploidy-based probabilities hold regardless of a population's sex ratio.

Now imagine a population in which workers have inflated the female population by preferentially rearing supersisters. The same one-in-three reproductive value is now spread across proportionately fewer males, increasing the value of any one. With fewer competitors, each male's mating success improves: his reproductive value is higher.

Not to be outdone, Trivers and Hare (1976) suggested the possibility of split sex ratios, i.e., the idea that sex ratios may approach 1:1 if they are averaged across population and if colonies within that population vary in sex allocation. Workers at one colony may benefit from rearing a female-biased brood of supersisters as long as other colonies in the population are rearing male-biased brood (Gardner et al., 2012; Godfray and Grafen, 1988; Taylor, 1981). In this manner, genes for kin-based altruism could spread in a population. Situations in which colonies rear male-biased brood are familiar to any beekeeper and generally constitute dissolutive states of a society—nonmated queens, egg-laying workers, or queenlessness. Therefore, such a scenario must appeal to *balancing selection*—when evolution arrives at compromises between opposing dynamics, in this case, the spread of altruism genes at the expense of some colonies' fitness.

In general, the idea of split sex ratios has been invoked to reconcile Hamilton's appealing haplodiploid hypothesis (Hamilton, 1972) with plausible early routes toward eusociality. The attempt is not unmerited because split sex ratios do occur in nature (West, 2009), and theoretic modeling identifies conditions under which they can be expected to promote early eusociality—namely, periods of queen virginity and replacement during which there is no worker-produced brood (Gardner et al., 2012). But these restrictions are very narrow and perhaps not realistic: workers without a normally functioning queen commonly produce male brood. When these restrictive assumptions are removed, models tend to marginalize the importance of split sex ratios as drivers of eusociality (Alpedrinha et al., 2014, 2013) especially compared with more powerful drivers such as lifetime monogamy (Hughes et al., 2008b; West and Gardner, 2010).

The Monogamy Hypothesis for the Emergence of Eusociality

In studying any model for the emergence of eusociality, it's important to acknowledge that the bar is high for any taxon to achieve it. The number of times eusociality has independently evolved in the history of Earth depends on the liberality of one's definition of eusociality (table 5.1), but by any measure the number is not large. Wilson and Hölldobler (2005), using Wilson's (1971) narrow criteria, named 12 independent origins of eusociality among the arthropods, whereas Bourke's broader criteria admitted 22–26 origins in the same phylum [Bourke (2011)'s table 1.4]. Eusociality is largely the domain of insects, the only other known examples being limited to shrimps (*Synalpheus* spp.) and rodents, the naked mole-rats of Africa (Heterocephalidae), and provisionally even to humans (Foster and Ratnieks, 2005; Wilson, 2012). The fact is, the vast majority of life on Earth has rejected eusociality.

Attempts at answering what makes a species opt for sociality have historically orbited around one of two camps of evidence—one stressing kinship criteria and the other stressing ecologic benefits. Or, to put it in terms of Hamilton's rule: the kinship camp looks for explanations that elevate $r > 0.5$ relative to siblings, whereas the ecologic camp looks for explanations that elevate the value of helping (high b/c ratio).

In the kinship camp, we have already seen that haplodiploid relatedness asymmetry ultimately devolves to average relatedness of only $r = 0.5$ between altruists and their sibling beneficiaries. A haplodiploid advantage, however, might be restored if it can be shown that workers bias their efforts toward female siblings, a situation called sex-ratio biasing. Even though there is evidence in favor of sex-ratio biasing, late theoretic modeling has tended to devalue it as a relevant driver of early social evolution (discussed more in the "Sex Ratios, Split Sex Ratios, and Reproductive Value" section).

In the ecologic camp, it's clear that factors other than high relatedness are friendly to group formation. We've already talked about some of them—the task efficiencies afforded by division of labor, the economy and security of a common nest, and life insurance, amounting to the vicarious protection delivered to a mother's brood by her roommates even if she dies. To these may be added environmental variation in food or nest site availability that selects for non-dispersal from a favorable site—"non-dispersal" being tantamount to "staying with a group." Group living is also a buffer against environmental stressors, predators, and parasites [reviewed in Bourke (2011)'s table 4.3]. And finally, zeroing in on Hymenoptera in particular, we can add genetic preconditions for sophisticated parental brood care and a sturdy sting that is easily co-opted for group defense (Strassmann and Queller, 2007). Indeed, such non kin-based selectors are so compelling that Nowak et al. (2010) have argued controversially in support of these alternative drivers, even to the point of challenging the project of inclusive fitness theory altogether (discussed more in the "Greenbeards, Relatedness, and Genes for Altruism" section).

So, returning to our honey bee's incipient social ancestor, let's admit that 0.5 genome transmittal is not a bad deal, and it wouldn't take much to tip the balance toward altruistic helping. The alternative for our honey bee's ancestor is to go it alone—to forego altruism, make her own nest, and rear her own brood. But the ecologic benefits of cohabiting and group living are not to be lightly dismissed. As Ratnieks and Helanterä (2009) put it, "When relatedness is high, even small asymmetries in costs versus benefits that favour rearing siblings instead of offspring can select for helping." Neither does r need be exceptionally high, only greater than zero (Bourke, 2011, p. 66).[2]

This gets us to the heart of a persuasive alternative explanation for incipient sociality called the *monogamy hypothesis* (West and Gardner, 2010). It is persuasive because it makes the obvious connection: fusing the kinship and ecologic camps into one synthetic model.

Whether in a diploid or haplodiploid species, a monogamous (singly mated) mother creates the potential for daughters to be related equally to

2. It is helpful to remember here that in inclusive fitness theory a value of $r = 0$ does not mean that two individuals have no genes in common, only that their chance of sharing the same genes at any focal loci is no different from that for any two members randomly drawn from the population.

one's siblings, the product of one's helping ($r_{help} = 0.5$), as to one's own offspring ($r_{offspring} = 0.5$). When $r_{help} = r_{offspring}$, any small efficiencies that favor helping over rearing one's own offspring will favor altruism. These efficiencies can fall on either side of Hamilton's inequality: high relatedness ($r > 0$) or favorable b/c ratios, bearing in mind that there is no theoretically stable basis for altruism outside of kin-based relatedness (discussed more in the "Greenbeards, Relatedness, and Genes for Altruism" section).

Imagine a monogamous bee mother and her supersister daughters (figure 6.2) who are preferentially helping their mother rear more supersisters ($r_{help} = 0.75$): if enough bee families (it is still too early to call them colonies) in the population are producing male-biased broods, then there's a chance that these supernumerary supersisters will successfully mate and found nests; because they themselves are products of altruism, their genes for kin-based altruism will spread in the population (discussed more in the "Sex Ratios, Split Sex Ratios, and Reproductive Value" section). On this count, haplodiploidy may grant Hymenoptera and other haplodiploid lineages a small incipient proclivity toward sociality.

But the more powerful driver is the simple fact of the mother's monogamy. Once $r_{help} = r_{offspring}$, we may add any of the numerous ecologic benefits of group living, that is, favorable b/c ratios, some of which are distinctives of Hymenoptera, such as parental care and a robust sting, and we have a sufficient reason for a daughter to stay at her natal nest instead of strike out on her own: the bar for sociality has been breached.

So for our honey bee's incipient social ancestor, the portrait that begins to emerge is this: an erstwhile solitary corbiculate female, probably soil-nesting, definitely monandrous, with her supersister daughters experimenting with a subsocial domestic arrangement (see table 5.2), choosing to remain at the nest and help their mother produce more supersisters (figure 6.2). Two hypotheses can explain this choice. The haplodiploid hypothesis, supported with split sex ratios, sees natural selection rewarding these early altruists with an immediate boost to fitness—an inflated albeit indirect transmittal of genes to supersisters. Here, this choice is far from an evolutionary dead-end because all females at this stage, including the altruists' daughters, are still reproductively competent, free to leave the nest and strike out on their own.

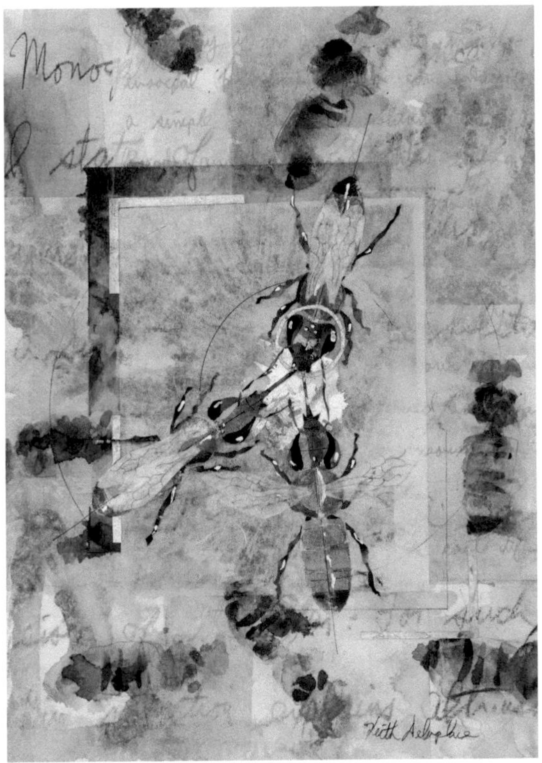

FIG. 6.2. A solitary female corbiculate bee, singly mated, and her cohabiting daughters exemplify the cusp of eusociality. If such a daughter stays at the nest to help her mother rear siblings, her average relatedness to those siblings will be $r_{help} = 0.5$, the same as if she were to strike out on her own and rear her own offspring $r_{offspring} = 0.5$. Even very small ecologic efficiencies afforded by group living may be enough to tip daughters into remaining at the nest. Such a situation constitutes the subsocial route to eusociality in table 5.2.

This strategy, however, depends on other colonies in the population producing male-biased broods to offset the reduced female reproductive value in female-biased populations, with male-biased sex ratios being associated with colonies in states of dissolution maladaptive to the colony. Nevertheless, in what may be a case of balancing selection, the genes for altruism spread in the population because the altruism is based on kinship.

In contrast, the monogamy hypothesis, elegantly parsimonious, highlights that the mother's monogamy means that for her daughters, $r_{help} = r_{offspring}$, and any small ecologic benefit of group living is sufficient to reward genes for altruism and tip daughters toward group living. For either hypothesis, when we say "genes for altruism," we're talking about things like non-dispersal, self-sacrificial nest defense, and care for related brood—things that benefit all of the actor's genome carried in the bodies of her beneficiaries. In other words, the interests of the altruistic genes and the actor's whole genome are aligned (discussed more in the "Greenbeards, Relatedness, and Genes for Altruism" section).

The increase in nestmate number that follows a decision to not disperse has immediate social benefits in the form of rudimentary division of labor (discussed more in chapter 7's "Division of Labor and Evolution of Caste" section), the emergence of which signals the first opportunity for natural selection to shift its attention from the bees to the group. The extent to which any group's task efficiencies deviate from the average for other groups is an opportunity for evolution to reward genes that benefit survival of the group. The criteria for these task efficiencies are associated with ecologic rewards for sociality: how well the group exploits available nest sites, constructs nest sites, extracts calories from the environment, thermoregulates the nest, and wards off parasites and pathogens. On this count, the social insects have been wildly successful, reaching no less than ecologic dominance on Earth, and the credit for this falls squarely on division of labor (Fittkau and Klinge, 1973; Wilson, 1987).

Our corbiculate honey bee ancestor is taking its first steps from subsociality toward simple eusociality through Intermediate Steps I and II in table 5.2. There is overlapping generations and cooperative brood care, marking the first expressions of altruism. Simple division of labor, offering new efficiencies to the group, offers evolution the first chance to select at the level of group. Something new is beginning to emerge, a new level of individuality (chapter 2). We are at 78–95 million years ago, the dawn of eusociality in the corbiculate bees (see figure 1.3).

Greenbeards, Relatedness, and Genes for Altruism

Wilson and Hölldobler (2005) and Nowak et al. (2010) have mounted a controversial objection to inclusive fitness theory with an alternative model that puts primacy on ecologic selectors for sociality along with putative genes for altruistic behavior. By these authors' understanding, ecologic benefits of colonial life, genes for altruism, and simple group selection are sufficient drivers of eusociality, with high relatedness a result of, not a cause of, social living.

It is beyond my purposes in this book to offer a full defense of inclusive fitness, a job convincingly accomplished by others (Abbot et al., 2011; Bourke, 2015; Fisher et al., 2013; Kay et al., 2020; Queller, 2016). However, I want to emphasize here the importance of altruism based on relatedness as a general basis for group formation.

Hamilton's inequality presumes that genes for altruistic behavior in fact exist and are propagated by near kin. But just because one can do the math and show high relatedness between a worker and the objects of her helping doesn't compel the worker to perform altruistic acts. Included in that mix of shared genes must be genes that express altruistic behavior at the individual level, to resurrect the word's technical meaning from chapter 2. I stress individual level because the fundamental selfishness of genes is not to be denied. As long as the self-sacrificing altruism is *expressed* in only certain individuals (i.e., an altruistic actor), then in compliance with Hamilton's rule, those same altruistic genes will be inherited by others in the population (the beneficiaries) and not go extinct with the actor's self-sacrifice. The individual's soma is the vehicle not only for its genes for altruism, but its genes for everything else. Hence, the only stable way for self-sacrifice to evolve and spread in a population is for it to be based on high relatedness: the interests of the altruism genes being aligned with the interests of the carrier's whole genome.

Altruistic *behaviors* are self-evident, and we have already talked about some of them—things like non-dispersal, self-sacrificial nest defense, and care for related brood. But a fairly late addition to the inclusive fitness literature has been explicit searches for *genes* encoding for altruistic physiologic pathways. Several have been found, a significant number of them from the honey bee, including genes regulating worker sterility (Oxley et al., 2008), selfless colony defense (Arechavaleta-Velasco et al., 2003; Hunt et al., 1998), and queen-like versus worker-like caste development in workers of the parasitic Cape bee of South Africa (Aumer et al., 2019). Thus, a mechanistic thread is established from high relatedness to altruistic behavior to heritable altruistic genes.

But debate lingers over the extent to which high genomic relatedness and altruism are necessarily connected. Imagine the possibility of a population with a radically explicit gene for altruism. Possessors of this gene can recognize one another and direct altruistic help toward one another with no regard to any other genes in common (Dawkins, 1989; West and Gardner, 2010).

Such a possibility was recognized by Hamilton (1964), but it was Richard Dawkins who memorably named it the "greenbeard" phenomenon (Dawkins 1989)—as if the gene's possessors sport green beards by which they spot one another in a crowd. Greenbeard genes cause altruism toward *non*-kin, or to put it more narrowly, altruism based not on kinship but rather genetic relatedness at the greenbeard locus. The existence of such independent genes for altruism is important because it suggests an alternative starting point for colonial life—the pairing up of non-relatives whose only gene in common is a greenbeard, with high relatedness across the rest of the genome following afterward if at all. It is the kind of altruism inferred in the critiques of inclusive fitness by Wilson and Hölldobler (2005) and Nowak et al. (2010).

Although it is logically feasible to draw such a nuanced distinction between kinship and genetic relatedness, this approach marginalizes the obvious fact that kinship is the most common reason for genetic relatedness. Although greenbeards are now confirmed in nature [reviewed by West and Gardner (2010)], their inheritance is complicated by the fact that their interests are uncoupled from the rest of the genome. Consider this: if a greenbeard altruist helps a greenbeard recipient, all the other genes of the altruist's are hazarded in the transaction. In this case, the greenbeard gene is an outlaw, and other genes can be expected to select against it and its altruistic behavior. In contrast, no such resistance is expected among genes of the greenbeard recipient because they enjoy the free benefits of altruism.

Thus, we see two sources of altruism, i.e., "alien" altruism in the form of greenbeards that may (in the case of the recipient) or may not (in the case of the altruist) serve the interests of the rest of the individual's genome, and kin-based altruism, which arises in response to high relatedness across the genome. Hands down, kin-based altruism is the more stable—and common—of the two because it proceeds without any selection conflicts within the genome.

The Generality and Preeminence of Kin-Based Group Formation

The monogamy hypothesis is attractive because it simplifies our understanding of incipient group formation, unifies what were formerly competing hypotheses, and eliminates distractions. For example, it is no longer necessary to demand $r \geq 0.5$ among players experimenting with group formation, only average $r > 0$. The monogamy hypothesis also has received strong empirical support from a phylogenetic analysis of eight lineages of colonial bees, wasps, and ants that showed that the ancestral state for each was monogamy (Hughes et al., 2008b)—a singly mated mother and her brood, a situation constituting the highest state of relatedness across nestmates.

A high view of kin-based relatedness in the formation of groups is intuitive because it reduces genetic conflicts among members of the group and aligns their genomic interests. It has long been assumed that high group relatedness is not some idiosyncrasy of the social insects but rather a descriptor that applies at every major evolutionary transition—from genes to genomes, unicellular organisms to multicellular organisms, and organisms to superorganisms (see table 2.1). If so, then the twin pillars of the monogamy hypothesis—high kinship and ecologic benefits—may constitute a principle of general evolutionary importance.

An important breakthrough came in 2013 with a comparative phylogenetic analysis of 168 species representing all but two of Earth's multicellular lineages (Fisher et al., 2013). The authors compared the types of multicellularity that emerges from groups that are clonal (somatic cells $r = 1$) versus non-clonal (somatic cells $r < 1$) and showed that species that form groups clonally are more likely to have transitioned to *obligate* multicellularity,[3] have more cell types including sterile "altruistic" cells, and tend to have more cells in a group. On those occasions when multicellularity emerged from non-clonal lineages, it was exclusively *facultative* multicellularity with much narrower ranges of organismal size and complexity.

3. With *obligate* multicellularity, the unicellular stage is finite and cannot complete a life cycle without returning to a multicellular stage; the unicellular stage usually takes the form of the single-celled zygote that divides to form a soma of clonal cells. Obligate multicellularity is a feature of animals, plants, fungi, red algae, and brown algae. With *facultative* multicellularity, the unicellular stage is indefinite and only periodically reverts to a multicellular form. The poster child for facultative multicellularity is the slime molds, genus *Dictyostelium*. Species of this genus can persist several generations as unicellular amoebae in soil. Under certain environmental conditions, the amoebae come together to form mobile multicellular slugs which in turn form a multicellular fruiting body. The stalk of the fruiting body is made up of sterile "altruistic" cells. Spores are released, and the species returns to its unicellular form. The terms *obligate* and *facultative* correlate closely to the *complex* and *simple* multicellularity of table 5.3 but are not synonymous. It is possible to have simple obligate multicellular species as well as relatively complex facultative species with division of labor and sterile cells.

Key to appreciating this finding is remembering that clonal reproduction proceeds by means of cells joining parental cells in a subsocial fashion (resulting somatic cells $r = 1$), whereas non-clonal reproduction proceeds by means of cells aggregating together in a parasocial fashion (somatic cells $r < 1$). To put it plainly, high relatedness from kinship is the rule, not the exception, for incipient formation of the kind of multicellularity that led to honey bees and human beings. Genetic conflict arising from the lower levels of relatedness featured in parasocially forming groups is an insurmountable obstacle to the transition to obligate multicellularity (Fisher et al., 2013).

In chapter 5, we applauded the innovation of sex and saw how the single-celled bottleneck of the zygote renders the somatic cells of complex multicellular organisms clones with zero genetic conflict, a fact that probably explains the stability of the organismal level of organization and its numeric dominance on Earth. Moving up one level to the superorganism, however, we are now dealing with the assimilation of group members who are themselves organisms. Genetic non-conflict to the extent afforded organisms is no longer available, but the principle of kin-based relatedness still applies to group formation at this level, only now the mechanism is lifetime monogamy of the common mother (Hughes et al., 2008b).

And so what about ecologic benefits at the inception of multicellularity? When the relatedness of an incipient eusocial group is described by $r_{help} = r_{offspring}$, then the monogamy hypothesis predicts that even small ecologic efficiencies (favorable b/c ratios) can tip the members toward altruism and eusociality. The same kinds of ecologic efficiencies happen when we're talking about evolution of cellular sociality. These efficiencies can take the form of cells clumping to evade predators (Boraas et al., 1998) or to share public goods such as nutrients released by neighboring cells (Koschwanez et al., 2013). Consistent with evidence that specialization is an inevitable outcome of groups (discussed more in chapter 7's "Division of Labor and Evolution of Caste" section), division of labor rapidly follows the clumping of cells (Ratcliff et al., 2012) and quickly leads to such benefits as the separation of incompatible functions—photosynthesis from nitrogen fixation in cyanobacteria (Kaiser, 2001), for example, or locomotion from cell division in flagellated cells (Grosberg and Strathmann, 2007).

In conclusion, we see that the same dynamics—high kin-based relatedness and ecologic benefits—described by the monogamy hypothesis for the transition from organism to superorganism apply to the more basal transition from cells to multicellularity. Perhaps we need a more inclusive label to capture the conditions for stable group formation common across levels of biological organization. The term *sociogenesis* has been put forward to capture this (as discussed in chapter 2, in the "Evolutionary Transitions" section), but the lexical semantics of this word seems to limit its use to explicitly *social* evolution. I propose *sociobiogenesis*. Sociobiogenesis exists in the scholarly literature, but only as a narrow proposition by humanist philosopher Frantz Fanon that "human beings materialize themselves biologically through a genesis of symbolic forms that in turn become autopoietically functioning systems—ones that actually serve as biological adaptations to the world around us" (Cornell, 2017). We cede to students of Fanon this apt descriptor of human agency, but I think sociobiogenesis can be coopted for more magisterial purposes.

Socio*bio*genesis makes explicit and publicizes a fundamental truth about the way life is organized on Earth: new levels of biological organization emerge from social processes. In future chapters, with practical examples from the honey bee, we will see that life is also transformed, complexified, and stabilized by social processes.

In the beginning was solitaries. Those solitaries who joined with near kin to better exploit limited ecologic resources saw their genomes multiply. The evolutionary dynamics that pushed the honey bee's corbiculate ancestor toward eusociality are the same dynamics that pushed eukaryotic cells toward cellular sociality—complex, multicellular organismality—creatures such as us. The evolution of life is the evolution of sociality. The social insects—or for beekeepers, the wards they tend in their hives—have become windows into the architecture of life.

Summary

Altruism and abdication of reproduction by part of a group's members is a defining feature of eusociality and complex cellular sociality. W. D. Hamilton's inequality $c<br$ forms the basis for inclusive fitness theory and predicts that genes for altruism will spread in a population if the cost (c) of an actor's

altruism is less that the product of the recipient's benefit (*b*) times their relatedness (*r*). Cost and benefit are measured in terms of total lifetime sacrifice or increase, respectively, in number of offspring. A group member's altruistic helping can increase the member's fitness if the member's helping increases reproduction of near kin who share its genes. In spite of relatedness asymmetries caused by haplodiploidy in the Hymenoptera, an altruist's relatedness to the siblings it helps rear (average $r_{help} = 0.5$) is no higher than its relatedness to its own potential offspring ($r_{offspring} = 0.5$) if it were to reject group living and live solitarily. The central question of eusociality, therefore, becomes "What makes members of a group tip in favor of altruistic helping over direct reproduction?"

This question has traditionally been answered by competing hypotheses—one that looks for explanations that increase kin-based relatedness (high *r*) and another that looks for ecologic benefits of group living (favorable b/c ratios). The monogamy hypothesis harmonizes these hypotheses by predicting that eusociality can emerge if $r_{help} = r_{offspring} = 0.5$ when even small ecologic benefits are present that favor group living ($b/c > 1$). The monogamy hypothesis has been reinforced by phylogenetic evidence that the ancestral state of colonial Hymenoptera was monogamy—one female, singly mated, generating the highest possible kinship across group members. The interacting dynamics of high kin-based relatedness and ecologic benefits of group living are not particular to the social insects but apply also to eukaryotic cells in the process of evolving cellular sociality—complex multicellular lineages like the animals and plants. I propose the term *sociobiogenesis* to describe this general principle. The evolution of life is the evolution of sociality.

CHAPTER 7

Complex Eusociality: Restraining Conflict in the Honey Bee Superorganism

EUSOCIALITY, ONCE ATTAINED, is a fragile thing. The proposition of subordinating one's reproductive autonomy to a group is a tough sell, and the overwhelming majority of Earth's organismal lineages have rejected it.

I say *organismal* lineages because the process of subordination was entirely common prior to that point. Among the ally-seeking nucleotides, the bundling chromosomes, the compartmentalizing prokaryotes, the assimilating eukaryotes, and the clumping multicellular organisms, there was a stampede of reproductive subordination to a higher order (see table 2.1). But by the time multicellular organisms came around, things were slowing down. Sexual reproduction offered every generation a reshuffling of genes into stabilizing new combinations; the segregation of reproductive cells allowed for explosive specialization of somatic cells into new tissues and organs; and the single-celled bottleneck of the fertilized zygote rendered all somatic cells clones, eliminating internal genetic conflicts. These are powerful assets and help explain the stability and ubiquity of organismality.

But in the eusocial superorganism, we see the subsummation of lower entities that are themselves organisms—those stable inventions of the previous transition. Lacking a single-celled bottleneck, the eusocial superorganism is inherently more vulnerable than the organism to internal genetic conflict, although we will see in this chapter and elsewhere that the superorganism has evolved adequate means of restraining social dissolution. That is the gist of this chapter—how the eusocial group maintains itself, resists threats from within and without, and reinforces group cohesion. Along the way, there is an almost inevitable complexification that emerges as large somatic populations of workers, each with considerable behavioral autonomy, create the conditions for emergent properties—the spontaneous order that emerges out of large groups that itself provides scaffolding for evolution to act upon. In this book, the honey bee is our recurring model for superorganismality.

In the last chapter, we left our corbiculate ancestral honey bee at the earliest and simplest stage of hymenopteran eusociality—one mother mated to one male and sharing a nest with her daughters who forego direct reproduction to stay at the nest and help their mother produce siblings (see figure 6.2). When one averages the relatedness of a helping daughter to the supersisters ($r = 0.75$) and brothers ($r = 0.25$) she helps rear, the gene transmittal gained by helping comes to $r_{help} = 0.5$, the same as if the daughter had directly reared her own offspring ($r_{offspring} = 0.5$). The monogamy hypothesis resolves this conundrum by predicting that eusociality can emerge under surprisingly minimal thresholds of $r > 0$ if even small ecologic benefits are present that favor group living (i.e., $b/c > 1$ in Hamilton's inequality). The presence of a nest and the stabilizing benefits of cooperation can be powerful rewards for the social option.

Restraining Genetic Conflict in the Honey Bee Superorganism

But just because one can do the math and show that $r > 0$ and $b/c > 1$, this does not compel workers to be altruistic and socially cooperative. For one thing, genes for altruistic behavior must really exist. The chances for this are good, but in the earliest stages, these altruistic phenotypes would have been for general altruistic states like non-dispersal from the nest and cooperative brood care, each of which would deliver collateral benefits to the actor's whole genome, reinforcing the kin-based foundations for altruism. Genes encoding for specific altruistic states have been confirmed in the honey bee for worker sterility (Oxley et al., 2008), behavioral cascades for colony defense (Arechavaleta-Velasco et al., 2003; Hunt et al., 1998), and female caste differentiation (Aumer et al., 2019), but these characteristics show a heavy stamp of social selection and cannot be considered relevant at the start of eusociality.

Granting the existence of primitive states of altruism, we must next imagine among nestmates a kind of *voluntary* altruism operating off the positive selection vectors set up by non-zero states of r and $b/c > 1$. To begin understanding this, let's consider the opposite of voluntary altruism, namely, selfish reproductive cheating. In the Hymenoptera, even an unmated daughter can activate her ovaries and produce sons—a perfectly viable option for passing on 0.5 of her genes. What makes such an egg layer selfish is not just the fact that she is using the resources of the group to rear her own sons, but that such daughters stop working on behalf of the group (Hillesheim et al., 1989). They become social outlaws, focused on laying eggs while relying on their sisters to pick up the slack (see figure 3.5). If every daughter pursued this course, the incipient colony would collapse with zero transmittal of genes for everyone, including the cheats.

Considering the early stages of eusociality we're talking about, a modern biology reader can hardly fault these cheats, given the hundreds of millions of years of evolutionary rewards preceding them for direct reproduction. Such a heritage cannot be lightly shaken off, so it should be no surprise to learn that theory describing benefits of direct reproduction predicts that even in a bee family whose mother is mated to only one male[1] that 14% of her daughters will reject voluntary altruism and selfishly produce eggs, even in the presence of altruistic genes (Wenseleers et al., 2004).

This prediction is noteworthy considering that monogamy is the ancestral condition of social Hymenoptera at the inception of eusociality, constituting the highest state of genetic relatedness possible for the early superorganism

1. Average relatedness among workers in a colony decreases if the mother mates with multiple males (discussed in the "Polyandry and Its Consequences" section). Accordingly, cheating among her daughters is expected to be lowest when the mother is mated to only one male and to increase as her mating number increases.

(Hughes et al., 2008b). This underscores the fact that no degree of familial relatedness can theoretically remove all possibility of selfish within-group conflict. It would take a family of clones with 100% of their genes in common to do that—a condition we have already admired among cells in the sexually reproducing multicellular organism and which is unavailable to the superorganism.

The question then becomes whether a conflict rate of 14% is an obstacle to the formation of a eusocial group. But this problem vanishes when we turn from theory to field data, where we find the curious fact that cheating in the honey bee colony is quite rare: only one in 1,000 workers attempts to selfishly lay eggs (Ratnieks, 1993), even in modern polyandrous honey bee colonies in which the predicted rate of cheating should be as high as 54% (Wenseleers et al., 2004). Clearly, the honey bee has solved the problem of within-colony genetic conflict, but not through voluntary altruism alone.

But what about *enforced* altruism?

The very act of coming together as a group sets the stage for social exchanges that could bias some individuals for work and others for reproduction (discussed in the "Division of Labor and Evolution of Caste" section). Reproductively coercive behavior is diverse in the social Hymenoptera and can be directed from mother to daughters, from daughters to daughters, and from daughters to mother. In the allodapine bee *Exoneura robusta*, the mother blocks the nest entrance to daughters who have mated (Bull et al., 1998). In ponerine ants, workers may punish an outlaw worker attempting to reproduce by immobilizing her, seizing her appendages, and holding her spread-eagled for hours or even days. If the outlaw survives the ordeal, she loses rank and abandons egg laying (Monnin and Ratnieks, 2001). Workers may mutiny against their mother's coercion to the point of committing matricide, as shown in vespine wasps and the European bumble bee *Bombus terrestris* (Bourke, 1994). And of course, the honey bee queen mandibular pheromone famously suppresses ovary activation by her daughters (Hoover et al., 2003; Strauss et al., 2008).

Such a range of coercive behaviors, some bearing a mark of social selection, suggests that reproductive coercion is a fixed feature of eusociality, complexifying in tandem with increasing social complexity of the species.

But for now, we should narrow our attention to coercive states that could be behaviorally or physiologically present at the onset of eusociality.

One such candidate is manipulation of the diets of larvae by their mother or sisters. Underfed female larvae may emerge as reproductively handicapped individuals for whom indirect reproduction is their best bid for fitness (Alexander, 1974; Eberhard, 1975). However, this hypothesis has received mixed empirical support, and its importance in the evolution of eusociality is considered doubtful (reviewed by Ratnieks and Wenseleers, 2008).

A more credible candidate is *policing* behavior—altruism enforced when a mother eats her daughters' eggs. Compared with parental brood manipulation, egg eating seems at least as ancient and available at the transition to eusociality given that it is known to occur in modern solitary wasps (Field, 1992). Moreover, policing has garnered empirical and theoretic support for its importance in initiating and maintaining altruism, especially in the honey bee.

Policing behavior can be directed from mother to daughters and from daughters to daughters, and it is to be assumed that the very first policing was directed by a mother against her daughters. However, inclusive fitness theory illustrates the conditions that make policing adaptive for all parties.

Under conditions of monandry, the initial state in the social Hymenoptera (Hughes et al., 2008b), a mother is doubly related to her daughters and sons ($r = 0.5$) over her daughters' sons ($r = 0.25$) (figure 6.1); under such conditions, theory predicts strong selection for mothers to police their daughters. Similarly, daughters in a monandrous colony are predicted to prefer rearing their brothers and supersisters (average $r = 0.5$) over nephews ($r = 0.375$), thus creating a positive selection vector for daughter-daughter policing.

However, a distraction emerges at this point. Inclusive fitness theory seems to predict equally that daughters will preferentially rear nephews ($r = 0.375$) over brothers ($r = 0.25$). Such a strategy may increase the actor's gene transmittal if she divides her altruism between nephews and supersisters (average $r = 0.56$), but it also rewards non-working cheating sisters and bloats the number of non-working males, reducing ecologic fitness of the group. In short, b/c is too low for altruism to spread in the population. The actor would be better off staying solitary and producing offspring directly ($r = 0.5$).

So we see that the ecologic benefits of altruism described by the monogamy hypothesis (chapter 6), the positive selection vectors for voluntary altruism, and the restraints against genetic conflict selected for by policing seem to provide a redundant and sufficient set of reinforcers for tipping a species into altruism at the earliest moment of eusociality. Policing resolves genetic conflict in the incipient social colony at a scale approaching the single-celled bottleneck in organismal development.

Let's now consider what happens once a mother begins practicing multiple mating. The mother's benefits for producing offspring remain unchanged ($r = 0.5$) from a monandrous state (figure 7.1), but the situation changes dramatically for her daughters, for whom average relatedness to each other ($r_{sisters}$) now decreases. Worker-worker relatedness as governed by their mother's mating number is described by

$$r_{sisters} = 0.25 + \left(\frac{0.5}{k}\right) \qquad \text{(equation 6)}$$

where k = number of males represented in the mother's spermatheca, assuming that each contributes equally to paternity. This formula flattens out to 0.26 after the 50th mating, a number that rarely occurs in nature, although published maximum mating numbers in *Apis mellifera* have been inching upward in recent years [see, for example, $k = 77$ in Withrow and Tarpy (2018)]. More typically, average mating number for *A. mellifera* hovers around 12, a number by which equation 6 yields $r_{sisters} = 0.29$. This value can be converted to r_{sibs} by averaging it with the fixed value for $r_{brothers} = 0.25$ for haplodiploid female self in Table 7.1 to yield $r_{sibs} = 0.27$.

Average relatedness of a daughter to sons of her sisters[2] ($r_{nephews}$) is a modification of equation 6 to

$$r_{nephews} = \left(0.25 + \left(\frac{0.5}{k}\right)\right) \div 2 \qquad \text{(equation 7)}$$

If we again assume an average mating number of $k = 12$, equation 7 yields an average relatedness of a worker to her nephews of $r = 0.145$.

2. In colonies with large worker populations, the contribution of a focal worker's sons can be functionally ignored, rendering $r_{nephews} \cong r_{males}$.

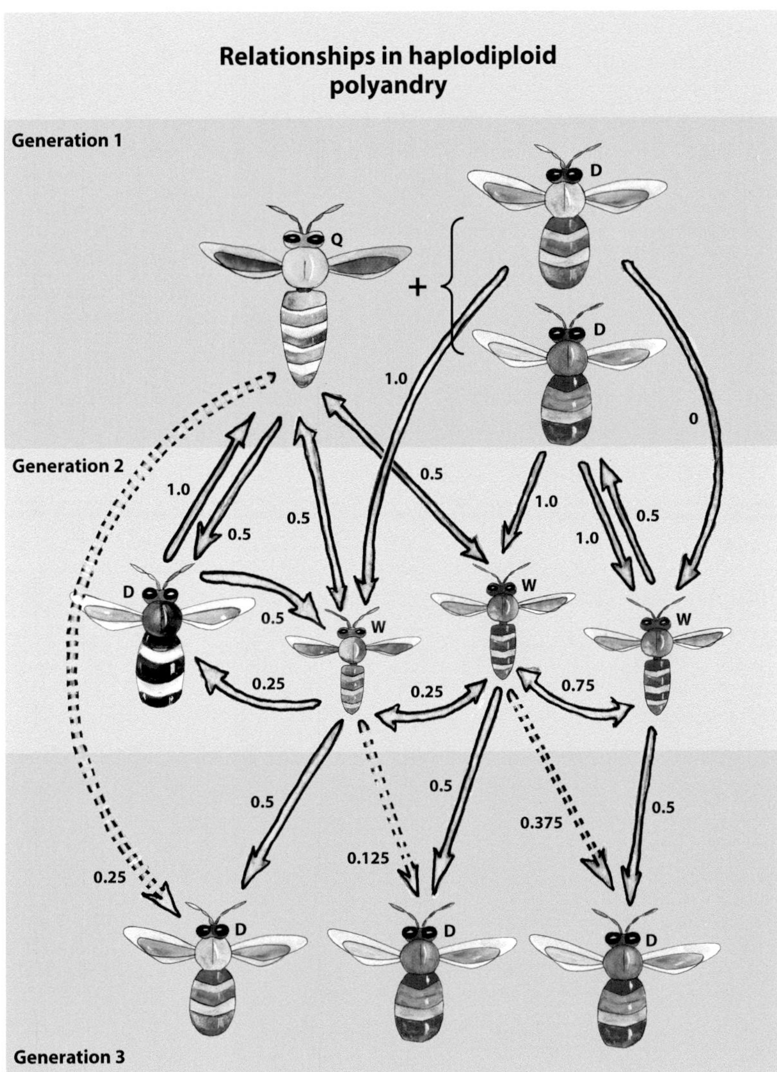

FIG. 7.1. Genetic relatedness—the probability that two individuals share the same alleles at any given locus—in a haplodiploid polyandrous colony in which the mother is mated to two males. We are assuming that divergent selection has produced two functional female castes; hence Q = queen, W = worker, and the male sex, D = drone. Relatedness is symmetrical (bidirectional arrows) between females, but asymmetrical (unidirectional arrows) between females and males owing to male haploidy (table 7.1). Hatched arrows show relatedness by indirect descent. Drones colored blue or red are represented in the queen's spermatheca and indicate fathers of their respective blue or red patrilines and descendants Generation 2 drone (black and white) derives entirely from his mother's genome and has no necessary kinship with spermathecal drones. Average relatedness among workers ($r_{workers}$) in a polyandrous colony is described by equation 6, and average relatedness of a worker to sons of other workers (r_{males}) is described by equation 7. In this example with $k = 2$ males, $r_{brothers} = r_{males} = 0.25$, but after k exceeds 2, worker-on-worker egg policing is predicted because $r_{brothers} > r_{males}$. A relationship of zero does not mean there are no genes in common, only that the relationship is no closer than that of the population average.

Armed with descriptions of the genetic relationships under the two kinds of mating conditions, we can now consider the relatedness differentials (Δ), or sacrificed gene transmittal, borne by daughters for tolerating cheats which can be expressed generally as $\Delta = \left(r_{\text{nephews}} - r_{\text{sibs}}\right)$. For monandry, the values are

$$\Delta = \left(r_{\text{nephews}} - r_{\text{sibs}}\right) = \left(0.375 - \left(\frac{0.75 + 0.25}{2}\right)\right) = -0.125 \quad \text{(equation 8)}$$

And for polyandry, assuming in this example $k = 12$, then

$$\Delta = \left(r_{\text{nephews}} - r_{\text{sibs}}\right) = \left(0.145 - \left(\frac{0.29 + 0.25}{2}\right)\right) = -0.125$$

Equations 8 and 9 show that the genetic stakes are unchanged for a policing worker whether one's mother is monandrous ($r_{\text{sibs}} = 0.5$) or polyandrous ($r_{\text{sibs}} = 0.27$). The value of one's nephews declines in equal proportion, so it's still adaptive—and at the same magnitude—for a worker to invest in siblings and altruistically work for her mother. Policing sustains its power as a stabilizing force for restraining genetic conflict. In fact, worker-worker policing occurs even more frequently in species in which queens practice polyandry or in which there are multiple queens,[3] as was shown in a comparative analysis of 109 species of eusocial Hymenoptera (Wenseleers and Ratnieks, 2006). This can be understood if we focus on the change in r_{nephews}. In a monandrous state, a worker will prefer nephews ($r_{\text{nephews}} = 0.375$) over brothers ($r_{\text{brothers}} = 0.25$), but this bias flips under polyandry, in which r_{nephews} plunges below r_{brothers} to 0.145, strengthening selection for worker-worker policing.

Thus, in both monandry and polyandry, we see a stable benefit to worker-worker policing, but this operates independently of the fact that the aggregate genetic fortunes of the workers have deteriorated. In working for $r_{\text{sibs}} = 0.27$ instead of $r_{\text{sibs}} = 0.5$, workers take a genetic hit once their mother

3. The situation being described here is anything that reduces average relatedness of a colony's workers. In bees, this means polyandry—one queen mated to multiple males. The functional equivalent in ants is polygyny—one colony with multiple reproductive females.

adopts polyandry. Discussing the costs and benefits of this situation will be the subject of chapter 9.

But for now, let's reflect on the fact that when workers are policed, whether by the queen or other workers, they are not directly punished for their criminal act, but neither do they get away with it. By eliminating rewards for outlaw egg laying, policing individuals select for group cohesion, preserve the fitness of the colony, and reduce the costs of altruism for all. Compared with colony collapse and zero gene transmittal, getting the chance to propagate $r \simeq 0.27$ of one's genes isn't such a bad deal.

To conclude, we see that altruistic behavior is sustained by high relatedness (voluntary altruism) in concert with social coercion (enforced altruism) in concert with sufficient ecologic benefits of group living (monogamy hypothesis). The phenomenon of a collapsing queenless colony familiar to any beekeeper is clear demonstration of what happens when the forces of social coercion are removed, in this case, the queen's inhibitory pheromones, a later derivation bearing a heavy mark of social selection.[4] The only forces for altruism remaining at that point are ecologic benefits of group living, the family's high overall relatedness, and any voluntary altruism that that relatedness elicits. Clearly, these are not enough to prevent social collapse. When it comes to constraining genetic conflict, coercion is to the superorganism what the single-celled zygote bottleneck is to the organism.

Restraining Sex Allocation Conflict in the Honey Bee Superorganism

In chapter 6, I discussed how hymenopteran mothers and daughters are in potential conflict over the desired ratio of females to males in the colony. Whereas Fisher's (1930) classic model on sex allocation predicts that fully diploid species will gravitate toward a 1:1 female to male population sex ratio, relatedness asymmetries in the social Hymenoptera mean that a helping daughter of a singly mated mother will prefer rearing her supersisters

4. There is debate over the extent to which queen pheromones are honest signals that indicate the presence and fertility of a dominant, reproductive female, or manipulative cues mothers use to coerce daughters into involuntary helping. More discussion has been given by Leonhardt et al. (2016).

TABLE 7.1. Life for life relatedness

FOCAL	INTERMEDIARIES	RECIPIENT	EQUATION	WITH VALUES	r
Diploids					
Self	NA	Clone	$G_{self} \times P_{self}$	1×1	1
Self	Mate	Offspring	$(G_{self} \times P_{self}) + (G_{mate} \times P_{mate})$	$(0.5 \times 1) + (0.5 \times 0)$	0.5
Self	Parents	Sibling	$(G_{mother} \times P_{mother}) + (G_{father} \times P_{father})$	$(0.5 \times 0.5) + (0.5 \times 0.5)$	0.5
Haplodiploids					
Female self	Mate	Offspring	$(G_{self} \times P_{self}) + (G_{mate} \times P_{mate})$	$(0.5 \times 1) + (0.5 \times 0)$	0.5
Female self	Parents	Supersister	$(G_{mother} \times P_{mother}) + (G_{father} \times P_{father})$	$(0.5 \times 0.5) + (0.5 \times 1)$	0.75
Female self	Parents	Subsister	$(G_{mother} \times P_{mother}) + (G_{father} \times P_{father})$	$(0.5 \times 0.5) + (0.5 \times 0)$	0.25
Female self	Mother	Brother	$r_{self, mother} \times r_{mother, offspring}$	0.5×0.5	0.25
Female self	Supersister	Son of supersister	$r_{self, supersister} \times r_{supersister, offspring}$	0.75×0.5	0.375
Female self	Subsister	Son of subsister	$r_{self, subsister} \times r_{subsister, offspring}$	0.25×0.5	0.125
Female self	Daughter	Grandson	$r_{self, offspring} \times r_{daughter, offspring}$	0.5×0.5	0.25
Female self	Daughter	Granddaughter	$r_{self, offspring} \times r_{daughter, offspring}$	0.5×0.5	0.25
Male self	Mate	Daughter	$(G_{self} \times P_{self}) + (G_{mate} \times P_{mate})$	$(1 \times 1) + (0.5 \times 0)$	1
Male self	Mother	Brother	$r_{self, mother} \times r_{mother, offspring}$	1×0.5	0.5
Male self	Mother	Sister	$r_{self, mother} \times r_{mother, offspring}$	1×0.5	0.5

Note: r = probability of sharing same alleles at a given locus) between focal individual (self) and recipient; G_x = fraction of indicated agent's genome contributing to recipient's genome; P_x = proportion of indicated agent's genome shared with focal individual. Relatedness mediated by out-mating is indicated with sums; relationships mediated by a single intermediary are the product of $r_{self, intermediary} \times r_{intermediary, recipient}$. Because haplodiploid males are entirely a product of their mother's genome, $r_{son, mother} = 1$, but due to diploid gametic recombination in the mother, the relationship is not symmetrical, i.e., $r_{mother, son} = 0.5$. Because haploidy renders gametic recombination impossible in males, his sperm are not only clones of each other, but r between a male and each of his sperm = 1 which means $r_{male, offspring} = 1$. Due to balanced diploidy, female-female relationships are symmetrical.

over brothers by a ratio of 3:1, as pointed out by Trivers and Hare (1976). There have been accommodating hypotheses such as split-sex ratios (chapter 6) and polyandry (chapter 9) put forward to explain this conflict and its resolution, at least at the crucial juncture of early group formation. But with the arrival of polyandry and complex eusociality, the selection imperatives surrounding sex allocation must be reexamined. We can ask whether conflict persists into the superorganismal rank with workers of numerous patrilines cohabiting a single nest.

To begin, we should consider that polyandry itself may bear on the evolution of colony sex allocation. As polyandry, genetic heterogeneity, and social complexity increased the environmental stability of colonies, selection would have rewarded sex allocation strategies that assured copious quantities of mates for females.

Second, we can ask, "What exactly should an investigator count if he or she is interested in honey bee sex ratio?" Does one count just the reproductives—queens and drones, excluding the non-reproducing workers? If so, then one gets terribly lopsided sex ratios in favor of males. Modern honey bees have among the most extreme male-biased sex ratios in the order (Boomsma et al., 2005) with roughly 20,000 males reared for every queen (Page and Metcalf, 1984).

However, beginning with Michener (1974, p. 76), then Hamilton (1976), and later Page and Metcalf (1984), it was pointed out that for a species like the honey bee that reproduces by colony fission, it is not legitimate to exclude the non-reproducing workers from sex ratio calculations. This is because divergent caste evolution has rendered queens incapable of founding nests independently of workers. Another reason for including workers is that workers *do* reproduce, as periodic laying workers, albeit at very low rates, but their sheer numbers contribute significantly to overall drone populations. As Michener (1974) put it, "For most social bees, if each worker were considered as having a realistic fraction of the reproductivity of a gyne [fertile female], it is likely that a sex ratio of 1:1 would be approached." Later experimental work backed this up: Page and Erickson (1988) showed that egg-laying workers were present in at least 38% of *queenright* colonies, and in queenless colonies, the laying workers were capable of producing up to

6,000 drones—certainly a non-trivial addition to the local breeding population. And Ratnieks (1993) pointed out that if even 1% of workers in a colony of 50,000 bees had developed ovaries and each laid 20 eggs per day (Perepelova, 1928), their male output could exceed that of the queen by orders of magnitude.

But a more general reason for including workers in sex allocation questions is that the worker population is inescapably constitutive of the colony's sex ratio outcome. As the sole providers of the energy, protection, and direct care needed to produce brood, workers are at least as important as the queen in determining the final numbers of males and females.

The problem then becomes how to differentiate the workers that produce males from the workers that produce females, or at least the equivalent resources allocated to producing either sex. Estimates for these values have, in fact, been made; Page and Metcalf (1984) analyzed published data for bee colonies in Davis, California, and Aberdeen, Scotland. These published data sets contained regular records for drone brood area, worker brood area, body mass (wet and dry) for workers and drones, comb area of both brood types, and weight of comb for both brood types—all this for an entire season. Page and Metcalf also made some simplifying assumptions: first, that total energy requirement during the lifetime of a drone is equal to that for a worker on a per-mass basis, including the energy required to rear it as an immature; and second, they ignored the possible population-scale inputs of drones from laying worker colonies.

Caveats aside, Page and Metcalf showed that queenright colonies in California and Scotland produced strikingly similar numbers and ratios of drones and workers. Next, they followed up by constructing a model using the energetic requirements to rear males, the energetic requirements to rear workers, the drone brood biomass, and the worker brood biomass to derive a sex investment ratio built upon these observed energetic commitments. They came up with an average investment ratio of females to males as 2.25:1 (wet weight basis) or 1.71:1 (dry weight basis).

If I may round their numbers for simplicity's sake, Page and Metcalf's ratios are nearly 2:1 in favor of females over males. These numbers are certainly closer to Fisher's 1:1 ratio than one would get from insisting on limiting sex ratio counts to virgin queens and drones. But still, they are not 1:1 but rather intermediate to 1:1 and 3:1. Apparently, a female sex bias prevails in spite of evidence that worker nurses cannot discriminate near-kin brood (Châline et al., 2005; Keller, 1997). What kind of driver is still pushing sexual investment toward females?

An obvious mechanism for this outcome is the numeric dominance of workers over their mother (Queller and Strassmann, 2002)—itself an outcome of demographic selection for large colonies and ecologic stability. Workers not only rear the brood but build the combs and determine the relative apportionment of worker-sized or drone-sized cells in them. Using data from Seeley and Morse (1976), one can show that workers construct worker-sized to drone-sized cells in natural colonies at a ratio of roughly 5.3:1. The queen is sensitive to cell diameter differences. She inspects every cell before depositing an egg in it (figure 7.2) and voluntarily releases sperm onto eggs she deposits in worker cells and withholds sperm from eggs in drone cells (Koeniger, 1970). Worker cell construction is thus antecedent to a queen's choice to produce a male or female, and sheer overrepresentation of worker cells will bias brood toward females.

But the structural reason for the persistent female sex bias is rooted in the relatedness asymmetries of haplodiploid inheritance. Consider the kin relationships in a polyandrous colony shown in figure 7.1. We can expect a

FIG. 7.2. A queen inspects a cell's diameter before she deposits an egg in it. Based on this inspection, she either fertilizes the egg if it's a worker cell (as in this photo) or withholds sperm if it is a drone cell.

worker to prefer to rear, in this order: a son ($r = 0.5$, male), a nephew ($r = 0.375$, male), any worker in the colony ($r = 0.29$,[5] female), and, lastly, a brother ($r = 0.25$, male). It's really those top two classes, both males, that conflict with the prediction of Trivers and Hare (1976) and field results of Page and Metcalf (1984). The resolution is to be found in the phenomenon of worker policing (discussed in the "Restraining Genetic Conflict in the Honey Bee Superorganism" section). Workers typically eat each other's eggs, thus mutually ensuring that no workers (or at least very few) become outlaws and selfishly lay their own eggs. With rewards removed for producing sons or nephews, the majority of workers content themselves with rearing female nestmates with average $r = 0.29$. Apparently this differential against brothers $r = 0.25$ is sufficient to tip the workforce of a polyandrous colony toward rearing females. It's noteworthy that whereas sex allocation is generally female-biased in single-queen hymenopteran societies, this does not hold in multiple-queen colonies where relatedness asymmetries cancel one another or in solitary bees or wasps where mothers have control over sex ratio (Queller and Strassmann, 1998).

Summary

Stability in the organism is sustained by the single-celled zygote that renders all somatic cells clones to one another, thus removing genetic conflicts. This stability is not available to the superorganism, whose constituent units are themselves organisms. When one averages the relatedness of a helping daughter to the supersisters ($r = 0.75$) and brothers ($r = 0.25$) she helps rear, the gene transmittal gained by helping comes to $r_{help} = 0.5$, the same as if the daughter abandons group living and directly rears her own offspring ($r_{offspring} = 0.5$). The monogamy hypothesis rescues group formation by predicting that eusociality can still emerge under minimal thresholds of $r > 0$ if even small ecologic benefits are present that favor group living (i.e., $b/c > 1$ in Hamilton's inequality).

However, theory also shows that 14% of helping daughters are expected to selfishly produce eggs ($r = 0.5$), even in the presence of altruistic genes that encourage them to prefer rearing supersisters ($r = 0.75$). Thus, voluntary

5. Equation (6), assuming k = 12.

altruism seems inadequate for explaining group formation and maintenance. The best alternative is a type of enforced altruism called policing behavior—the altruism enforced when a mother eats the eggs of her daughters or daughters eat the eggs of their sisters. By eliminating rewards for outlaw egg laying, policing individuals select for group cohesion, preserve the fitness of the colony, and reduce the costs of altruism for all.

A second threat to stability in the honey bee superorganism is conflict between workers and their mother over preferred sex ratios. Owing to hymenopteran relatedness asymmetries, a queen is expected to prefer a 1:1 allocation of females to males, whereas the daughters of a singly mated mother are expected to prefer rearing supersisters over brothers by a ratio of 3:1. A mother's polyandry reduces this conflict by lowering within-nest relatedness of her daughters. But there is evidence that conflict persists even in a colony with daughters of multiple patrilines. With colonies that reproduce by fission, it is appropriate to include female workers in sex allocation determinations because queens are not reproductively autonomous of their workers, and worker-laid males contribute significantly to local breeding populations. Field studies showed that colonies commit energetic resources in favor of rearing females to males at a ratio approaching 2:1. The mechanism for this sustained female bias is the numeric dominance of workers over their mother, itself an outcome of demographic selection for large populations and ecologic stability, whereas the structural explanation is policing behavior that grants workers a small bias in favor of rearing sisters ($r = 0.29$) over brothers ($r = 0.25$).

CHAPTER 8

Complex Eusociality

Division of Labor and Evolution of Caste

I HAD THE EXPERIENCE of working in Nepal in the early 2000s as a volunteer consultant to the beekeeping industry. During those weeks, I was able to witness first-hand the strange-to-me institution of a human caste system. Coming from a culture where we value self-actualization and economic mobility (or at least those ideals), I found it eye-opening to experience a culture where one's opportunities are defined by birth. Human caste resists easy definition, but it usually means a system of inherited hierarchical ranks with wide-ranging social, economic, and political implications. Members are expected to marry within caste, and certain vocations have caste-specific associations. It can be difficult or impossible to change one's caste. In the history of the West, perhaps the nearest we get to it is the feudal period in Europe, where sharp economic and social lines were drawn between nobility and the commoners. One's situation in life was inherited and fixed. For this bumbling American in Nepal, it meant the occasion for a steady stream of social *faux pas*. Try as I might, I was always saying or doing the wrong thing, including pointing with my feet or saying "thank you" too much. But my hosts were gracious and quick to laugh it off.

It was cases of human cultures such as these that led early writers about social insects to use the word "caste" to describe the radically different forms and functions of individuals occupying the same nest. In entomology, the term "caste" refers to variants within sex. It is sometimes stated that honey bees have three castes—queen, worker, and drone. This is not correct; honey bees have two sexes, the females of which occur as two castes. One of the measures of evolutionary complexity in insect societies is the number and variety of castes in a species. The ants and termites take the prize in this category as their colonies express large numbers of variants of both sexes.

Like a human caste system, the honey bee colony is anything but egalitarian. Of the 60,000 or so female honey bees in a colony at peak season, only one of them, the queen, gets to reproduce in any meaningful way. She is the mother of most of the drones and all of the workers, all of whom are reproductively, behaviorally, and morphologically distinct from the queen. How did things get so lopsided?

Division of Labor Defined

To begin, let's stress that division of labor is so not so much integral to eusociality as it is an inescapable property. Division of labor spontaneously emerges from groups, setting in motion synergies by which social complexity, group size, and caste number and diversity increase in tandem. It was by watching these hordes of teeming individuals, each busying herself on some important errand, that so many of us were captured into a life long fascination with social insects.

Let's refresh our memory on some terms. Michener (1974, p. 119) calls *division of labor* "any behavioral pattern that results in some individuals in a colony performing different functions from others, even if only temporarily." This is distinct from *reproductive* division of labor, which foregrounds those individuals privileged with egg laying and production of offspring. *Task specialization*, or being a task *specialist*, applies to any individual who performs one task at a rate higher than the population average for that task. And finally, let's restate the definition for *caste* given in chapter 3—a nest member who is a categorical variant, based on morphology or behavior, within sex. *Division of labor* is a property of groups, whereas *caste* and *specialist* apply to individuals engaged in specific labors. By definition, caste and specialist cannot apply to honey bee males for whom there are no known behavioral or morphological variants. Honey bees have two castes, both female—the queen and workers.

Just as increasing the number and types of somatic cells is a feature of complex multicellular organisms, the number and types of castes is a feature of complex eusocial superorganisms (see table 5.3). There is a trend for caste numbers and variety to increase in tandem with a eusocial species's average colony size (number of members). Bees are conspicuous for lacking morphological subcastes of the kind seen in ants and termites, but in the simple eusocial bees, we see a gradient of size difference with smaller workers restricted to nest duties, larger workers focused on foraging, and the largest females the queen. As colony size increases in evolutionary time, there is a trend for size differences between queens and workers to become exaggerated and bimodal (Bourke, 1999). In simple eusocial bee species, queens exercise reproductive dominance through overt aggressive acts toward their daughters, whereas queens of complex eusocial species exert dominance over their daughters by way of derived inhibitory pheromones (van Zweden, 2010; Wilson, 1971), which in the honey bee appear to operate by triggering programmed cell death and egg abortion in worker ovaries (Ronai et al., 2016).

In lieu of worker subcastes based on morphology as in the ants and termites, worker subcastes in the honey bee are based on age, with any given worker predictably succeeding through a series of tasks as she ages. This phenomenon is called *age polyethism,* and members of any task class at a given time are called a *behavioral* or *temporal* caste. On average, newly emerged bees concentrate on center-of-brood-nest activities like cell cleaning and feeding brood. After about one week of age, they transition to edge-of-brood-nest activities like comb building. Around the third week of life, they transition to outside duties like foraging (Seeley, 1982). To the extent we find metaphors useful, any of these task categories and the workers performing them can be considered analogous to specialized somatic tissues and organs of an organism.

Division of labor is not only a defining feature of eusociality but appears to be an immediate outcome of two individuals coming together. Rob Page

(1997) wrote that "at least the rudiments of division of labour are inescapable properties of groups. There never was a time in the history of the evolution of social insects where individuals shared nests and did not have a division of labor." Experimental examples seem to support this. When females of the normally solitary carpenter bee *Ceratina flavipes* are experimentally forced to share one nest, a reproductive division of labor ensues, with one female laying eggs and guarding the entrance while the other does the foraging (Sakagami and Maeta, 1987). Similar results have been found with experimental pairings of ant queens, in which one of the pair assumes egg laying while the other engages in nest excavating (Jeanson and Fewell, 2008).

It is important to stress here that these effects are immediate, not a product of ecologic time or selection, which suggests that pre-existing variation in solitary individuals is sufficient to explain division of labor at the earliest stages of sociality. From a human viewpoint, this makes sense: put together a committee, a parent-teacher association, or a beekeeping club, and it doesn't take long to figure out who's the leaders, the followers, the workers, and the hangers-on.

Note also that if it's true that division of labor is a spontaneous property of groups, then a division of labor *per se* cannot be a product of group-level selection. However, morphology and the specific features of task execution can be altered by natural selection to refine them in ways that improve group fitness (Calderone and Page, 1991). In other words, the emergent property "division of labor" provides a scaffolding upon which natural selection on the group can act. It follows, then, that group-level selection was operative at the inception of group formation. The varieties of morphological castes and socially integrated task syndromes we see today in the modern honey bee appear to be an example of group-mediated selection, with the colony being the level selected and workers the level adapted.

Let's now try to reconstruct a plausible path for the evolution of morphological and behavioral caste in the honey bee as it was modified by colony-level selection. We'll begin with the divergence of mothers and daughters into morphologically distinct queens and workers, a so-called *dimorphism*—literally "two forms," and then we'll follow this with a discussion of the evolution of the various task syndromes exhibited by modern honey bee workers.

Evolution of Queen/Worker Dimorphisms

The evolving dimorphisms between *mothers* and *daughters* led to the modern distinctions between *queens* and *workers*—the latter terms applying properly only after a species has evolved castes. In the case of female honey bees, we see a more specific kind of dimorphism called *diphenism*, which means the expression of two or more phenotypes from one genotype. Compared with queens, workers are short-lived and have two small ovaries containing 2–12 ovarioles (Winston, 1987), a vestigial spermatheca, a long tongue, complex pollen-carrying corbicula on the hind legs, and a barbed sting for nest defense. The queen is long-lived and has two large ovaries with around 175–200 ovarioles (Dedej et al., 1998), a shorter tongue, fully functional spermatheca, a non-barbed sting, and no corbicula. Workers win, hands-down, when it comes to complexity of behavioral repertoires and sophistication of cognitive functions (Chittka and Niven, 2009), a fact that can be a jolt to our anthropomorphic expectations.

The evolution of these dimorphisms is driven by socially controlled regulation of nutrition, growth, and development of the immatures. In the honey bee, this control is enhanced by the species's habit of progressive provisioning—feeding larvae in their individual cells small amounts of food continuously throughout their feeding period.[1] Although larvae can influence nurse bee behavior, especially under conditions of social parasitism (chapter 23), ultimately, the nurses control the quantity of food given and the ratio of its constituents, which together dictate a larva's metabolic fate toward queen-like or worker-like development. Chief among these food constituents are sugar—a feeding stimulant that regulates growth rate and differentiation (Asencot and Lensky, 1976, 1985, 1988) and royal jelly—a complex multicomponent (Kodai et al., 2007; Rembold and Dietz, 1966; Schmitzova et al., 1998) glandular excretion of nurse bees. Recently, one

1. The alternative is mass provisioning—feeding larvae, either singly or as a group, a single mass of food, usually a ball of nectar-dampened pollen, sufficient for their entire larval feeding career. Mass provisioning is common in bees across the continuum of social states, whereas progressive provisioning is the norm in all honey bees (Apini) and some bumble bee species (discussed in chapter 4's "The Soil-Nesting Precursors" section).

particular protein, major royal jelly protein 1 or "royalactin," was asserted to be the primary inducer of queen development (Kamakura, 2011), but this position has met resistance (Buttstedt et al., 2016), and current opinion holds that royalactin is one of many dietary components in royal jelly acting in concert to trigger the queen's developmental pathways (Leimar et al., 2012; Maleszka, 2018) or even a glue to hold the queen larva in its vertical cell (Buttstedt et al., 2018).

Here is the sequence of events that occur in the critical fate window of a female honey bee larva, starting with workers. Beginning at egg deposition, embryonic development inside the female egg is fixed and programmatic, but three days later, upon hatching (larval day = 1), the newly-hatched larva is bipotent for either the worker or queen pathway. During days 1–3, all larvae are fed brood food produced by nurse bees; however, overall sugar content is lower for worker-destined larvae. The food fed to all larvae at this stage has a clear fraction produced from the hypopharyngeal glands and a milky fraction from the mandibular glands. During days 1–2, worker-destined larvae are fed proportionately less of the mandibular component. Beginning day = 3, the mandibular component for worker larvae drops even more and the diet becomes admixed with honey and pollen, with pollen feeding peaking at day = 5, the day before the cell is capped and the larval feeding regime completed (Winston, 1987) (discussed more in chapter 3's "Some Things Common to All Honey Bees" section). Throughout their feeding career, developing worker larvae are in a general state of starvation stress from sugar deprivation, and this is accelerated by the further drop in diet quantity and quality that begins on day = 3 with the effect of decreasing body mass, reducing levels of juvenile hormone (JH) (Asencot and Lensky, 1976, 1985, 1988), and inducing programmed cell death in ovarioles (Schmidt Capella and Hartfelder, 1998, 2002). The result is a smaller individual with severely restricted powers of reproduction but curiously enhanced cognitive powers and morphological specializations for work.

In queen-destined larvae, the scenario is quite different. From the onset, their diet is richer in the milky mandibular gland fraction of brood food as well as sugars, which stimulate feeding and enhance growth. Moreover, queen-destined larvae are exempted from the rationing that begins on day = 3;

instead, they continuing enjoying an undiluted diet of mandibular and hypopharyngeal gland excretions now called, owing to its copiousness and qualitative differences, by the beekeeping term *royal jelly*. Indeed, it is not so much an "exemption from rationing" as it is a surfeit of bounty, as every beekeeper knows who has witnessed the seeming extravagance by which queen larvae are fed (figure 8.1). What looks to be wasteful excess to human observers is, however, apparently essential, because larvae fed lower quantities of royal jelly do not fully develop all queen-like features (De Souza et al., 2015; Kamakura, 2016).

In the end, the bounteous quantity of sugars and unadulterated royal jelly diet increase body size and increase levels of JH, major royal jelly protein 1, and other royal jelly proteins that deflect programmed cell death in ovarioles and retain other mother-like traits (Maleszka, 2018). An added bonus is that queen development time is faster than it is for workers, with natural selection favoring queens that develop as fast as possible. The first one to emerge kills her potential rivals while they are still in their cells.

An interesting feature of the queen-worker divergence is its diet-based plasticity during the 4–5-day feeding window. A larva is not irreversibly committed to either fate until days = 3.5–4 of larval life (Dedej et al., 1998; Shuel and Dixon, 1960), which means that any number of intermediate adult morphotypes will emerge from queen diets that are artificially delayed or truncated. The expression of queen-like characters increases as the length of time increases that the larva receives an uncompromised queen diet (Dedej et al., 1998), forming the basis for an old beekeeping recommendation for queen

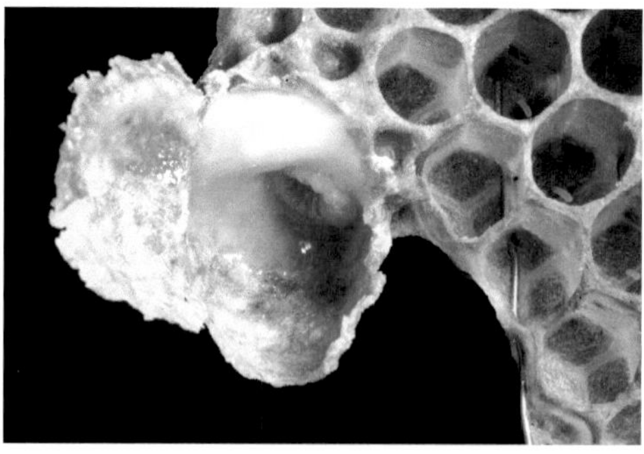

FIG. 8.1. A queen cell opened to show larva on its copious bed of royal jelly.

producers to begin rearing queens from larvae no older than a few hours post-hatch (Delaplane, 2007).

This plasticity is evidence that worker-queen divergence is driven by numerous developmental thresholds sensitive to nutrient triggers, not just a single on/off switch (Maleszka, 2018). Dedej et al. (1998), for example, showed that the age at which a larva begins a queen diet has strong effects on subsequent characters. The deadline for queen-like ovaries is early, by the first or second instars, whereas commitment to worker-like hind leg structures can be delayed until the fourth instar. A conspicuous spike in JH is apparent in early fifth-instar queen larvae but not worker larvae, a divergence that seems to be crucial for finalizing the two fates (Nijhout and Wheeler, 1982; Rachinsky et al., 1990).

This nutritionally driven divergence in gene expression is a hallmark of *Apis mellifera*; indeed over one-fifth of the species's genes is expressed differently in the brains of queens and workers (Weinstock et al., 2006; Grozinger et al., 2007). It is the most famous example of honey bee *epigenetics*—changes to phenotype that don't involve changes to the underlying genetic code but rather changes in gene expression, with many such changes being induced by environment (Berger et al., 2009). And although it is possible to demonstrate a range of intercastes with artificial feeding regimes, this serves more to demonstrate the different time points at which morphological divergences are triggered. In nature, the outcome is more fixed and divergent: two distinct castes, workers and queen. Intercastes are rare. The system is a tight integration between individual development and a colony-level social behavior, namely, the feeding regimes of nurse bees (Linksvayer et al., 2012, 2011).

SOCIALLY MEDIATED QUEEN/WORKER DIPHENISMS

How did such socially mediated queen/worker diphenisms evolve? The following explanatory scenario draws heavily from the review of Wheeler (1986) and the modeling predictions of Leimar et al. (2012).

We begin with states existing in early eusocial forms and ask how these states were coopted to initiate the transition of fully reproductive females into subordinate helpers. The most basic and pertinent ancestral state is the range in female body size that happens naturally from variation in seasonal food availability. This dietary variation was a matter of food quantity, not quality, because the ancestral diet was simply pollen and nectar, with glandular food from nurse bees a later derivation. In simple eusocial bees, female body size increases over the course of a season, a dynamic driven by the steadily increasing workforce and food reserves of the colony. In more ancestral states, these are differences in size, not form, but a large body size is positively correlated with dominance behavior (Breed et al., 1978), overwintering survival (Michener, 1974), and ovariole number (Leimar et al., 2012). Although body size can be driven by non-nutritional factors such as photoperiod (West-Eberhard, 1969) or genetics (Smith et al., 2008), the link between nutrition and size is direct, ancient, and the most obvious precursor to the nutritionally regulated caste switches we see in modern forms.

A seasonal gradient from poorer to richer food reserves set the stage not only for size differences but for variations in the time at which various switches activate. The later the switches activate, the more likely that endpoint differences between individuals will be physiological, not morphological, with physiological changes being of the kind that, for example, favor overwintering. By retaining enough worker-like characters, a sufficiently queen-like large female with functional ovaries can survive winter and perform all the life-history tasks necessary to sustain a species at the level of simple eusociality.

A second ancestral state subject to modification was programmed cell death (PCD) in ovarioles. PCD can be generally thought of as "normal" cell death regulated by an internal program as opposed to maladaptive cell death from injury, disease, or other external drivers. The most famous example of PCD is the cell death that happens in human fetal development that permits fingers and toes to separate from one another. PCD is important in fundamental processes ranging from removing abnormal or superfluous cells at developmental checkpoints to individualizing germ cells during gametogenesis (Baum et al., 2005; Cavaliere et al., 1998). Today, it is understood that PCD was coopted by natural selection to effect the reduced ovarioles and reproductive capacity we see in modern workers (Schmidt Capella and Hartfelder, 1998, 2002).

Although it's generally true that mother-like characters are ancestral, in the divergence of castes, we see elements of conservation and innovation in either pathway. Queens retain maternal-like ability to mate, but the workers retain maternal-like corbicula and cognitive capacities for foraging and homing. Similarly, queens innovate with increased ovariole numbers and complex socially active pheromone bouquets, whereas workers innovate with morphological specializations such as abdominal wax mirrors for forming beeswax flakes, barbed stingers for improving worker defensive performance, and reduced ovariole number.

This is a good place to bring up the *size-complexity hypothesis*—a body of thought that predicts that once group size (e.g., cells in an organism or members of a superorganism) reaches a critical number, certain group qualities will emerge, qualities that in some way describe increasing group complexity (Bonner, 1988; Bourke, 1999). These qualities are listed in table 5.3, but for now let's concentrate on one: the observation that the greater the size of a eusocial group, the greater the morphological differences between queens and workers. This is an outcome of reinforcing phenomena: greater divergence of worker-like from mother-like forms being tantamount to specialization, which improves group-level efficiencies.

In the honey bee, these specializations are not limited to physiological differences but include gross morphological divergences in form. Morphological changes cannot be switched on late in an individual's developmental sequence but must be coopted early to cause gross changes to anatomy. Thus, another prediction of the size-complexity hypothesis is an earlier developmental switch point for the commitment to mother-like or worker-like forms as group size increases (see table 5.3). We see this when we compare small-colony bumble bees with large-colony honey bees. Lacking any clear morphological differences between workers and queens, bumble bee commitment to a queen-like state is delayed until the prepupal stage, whereas full commitment in the honey bee is earlier, the fifth larval instar. The range in developmental deadlines for different worker-like characters described a few paragraphs above (Dedej et al., 1998) can be thought of as evidence for the independent cooption of genes regulating different morphological features, e.g., "abdominal sterna" genes being modified into wax mirrors, "ovipositor" genes into a barbed sting, and so on. Natural selection acting on colony efficiencies pushed developmental commitments earlier in the honey bee lineage, the result being a range of morphological specialties that improved worker and queen performance and widened the divergence between the two.

The natural selection that acted on developmental deadlines played out in the social interactions between larvae and their adult nurses. Nurses, by regulating food quantity, and later quality, and at first exploiting preexisting seasonal variation in food supplies, modified the quantities of JH, royalactin, and other royal jelly proteins responsible for regulating ovariole PCD and gene expression. Larvae, for their part, "agreed" to the nutritional conditions imposed on them and responded with variation in expression of gene sets that shifted in response to colony-level selection.

Divergence of caste forms was also partly driven by interactions between the incipient queens and workers themselves—a dynamic that seems to find antecedent in ancient forms of dominance behavior of a mother toward her daughters. However, in the social bees, direct agonism often gives way to secondary domination through inhibitory queen pheromones. In the European bumble bee *Bombus terrestris*, for example, the presence of a queen reduces JH receptivity in daughter larvae with the effect of precluding their development into queens (Röseler, 1975).

It is helpful to remember that caste dimorphisms were evolving under the shadow of simultaneous evolutionary "decisions" by daughters to remain at the nest. To the extent that a daughter's fitness was improved by casting her lot with a queen-like mother specialized for fecundity, it was adaptive for her to evolve specializations for work that improved the fitness of the group. It may prove difficult, even ultimately unnecessary, to reconstruct a chronology of cause and effect when it is likely that group stabilization and female diphenisms were so tightly integrated as to constitute a self-reinforcing evolutionary syndrome. If there were ever a place to grope for units of selection, evolution of female diphenisms is it; but it is precisely at such junctures that Dawkins's (1989) economic idea of the selfish gene is most reassuring: ultimately, the phenotype, however extended across levels, serves to transmit the gene (discussed more in the chapter 2 "Multilevel Selection" section).

In any case, it seems that intermediate forms did not persist in the evolution of honey bee female diphenisms. As long as individuals retained competence at both work and reproduction, reversibility was possible. Such cases of "secondarily solitary" transition have, in fact, happened in three lineages of halictine bees and one of allodapine, in which there are modern solitary species that descended from simple eusocial ancestors (Danforth, 2002; Wcislo and Danforth, 1997). And modern bumble bee queens, practicing an annually eusocial life cycle, have retained at least temporary competence in working and reproducing.

But in the honey bee lineage, it appears that intermediates—worker-like queens and queen-like workers—were not fit (Wheeler, 1986) and disappeared from natural history. The result was not only widely divergent castes, but widely divergent life history outcomes: queens could no longer work, and workers could no longer reproduce. Neither was any longer reproductively autonomous.

If this sounds familiar, it's because we talked about it in chapter 5 and called it the "point of no return" (Wilson and Hölldobler, 2005), a singular milestone in social evolution that marks the emergence of the superorganism—a major evolutionary transition (see table 2.1). With germ cells (the queen) segregated from the soma (the workers), it is the colony that now operates as an autonomous reproductive unit. The anatomical segregation of functions renders the decision irreversible: indeed, not one reversal is known among the more than 11,000 species of ants or 2,000 species of termites, groups that universally possess anatomically distinct worker castes (Wilson and Hölldobler, 2005).

The divergence of workers from queens is the place to talk about where the focus and strength of positive selection resides during this process. It is important because the divergence of female castes is the cutting edge of hymenopteran social evolution—the literal front where conflicting genetic interests resolve. One of the drivers of worker-queen divergence is an increase in novel tasks and the genes that control them. The task repertoire of a complex social worker can number 40–50 tasks, nearly twice that of its solitary counterpart (Oster and Wilson, 1979), and from an official honey bee gene set numbering 10,699 genes,[2] fully 696 of them were found to be taxon-unique, with 146 associated with novel eusocial traits or traits modified by

2. A later official gene set puts the number at 15,314 (Elsick et al. 2014).

social selection (Johnson and Tsutsui, 2011), including olfaction in the context of pheromone-based communication (Forêt et al., 2007), immunity (Viljakainen et al., 2009), brood care, division of labor, and glandular specialization. Moreover, Johnson and Tsutsui (2011) showed that these taxonomically unique genes bearing the stamp of social selection are overrepresented in workers compared with queens, highlighting that workers are the ones who express the majority of innovations that characterize sociality.

The idea of shifts in selection strength during caste evolution is captured in what is known as the "social ladder hypothesis" (Rehan and Toth, 2015; Toth and Rehan, 2017). Original models of molecular evolution saw eusociality arising from either regulatory shifts in old genes or rapid evolution of new ones. The social ladder model instead posits that the focus and strength of directional selection shifts during the ontogeny of eusociality and that no one model fits all social lineages. Independent work has supported this interpretation.

In a comparative analysis of the genomes of three taxa representing three grades of increasing social complexity (*Polistes* wasps<bumble bees<honey bees), Dogantzis et al. (2018) showed in the *Polistes* and bumble bees that positive selection was higher in queen-biased genes than in worker-biased genes. The reverse was true for the most socially complex taxon, the honey bees, where genes expressed mostly by workers had the strongest evidence of recent positive selection.

A honey bee population genomic analysis by Harpur et al. (2014) showed that the divergence of honey bee workers from queens is driven disproportionately by novel genes possessed by some solitary ancestors but not by others; in other words, some solitary lineages were "predisposed" to evolve eusociality. Moreover in the honey bee, selection on workers and queens is not symmetrical: positive selection is stronger for proteins coding for worker-like traits—e.g., division of labor and brood nursing—than for queen-like traits.

Results such as these show that the strength of selection is not constant during social complexification. In incipient colonies, there is an emphasis on basal mother-like characters such as independent nest founding, as retained in the modern *Polistes* and *Bombus*, but there is a shift toward worker-like traits as workers make up an increasing fraction of the nest population and their specializations render colony-level selection more important. It is during "worker-like" evolution that increasingly complex gene networks emerge that regulate the interacting battery of behaviors that typifies a complex eusocial colony (Kapheim et al., 2015).

This brings us back to an important point I raised in chapter 6 ("The Generality of Kin-Based Group Formation" section) that there are two kinds of thresholds that must be breached for eusociality to emerge: a threshold for within-nest relatedness and a threshold for ecologic benefits of group living. The monogamy hypothesis uses Hamilton's inequality (equation 1) to show that eusociality can emerge under minimal thresholds of within-nest average relatedness ($r > 0$) if even small ecologic benefits are present that favor group living ($b/c > 1$).

One of these ecologic thresholds involves nest thermoregulation, as demonstrated in the work of Bonoan et al. (2020), who focused on social fever for resisting pathogens. Honey bees are capable of elevating brood nest temperature to resist the pathogenic brood fungus *Ascosphaera apis*. Bonoan et al. (2020) showed that large colonies successfully produced elevated brood nest temperatures in response to artificial disease inoculation, whereas small colonies could not. Moreover, in their failed attempts to elevate temperature, smaller colonies spent more effort per bee. Thus, for small colonies, social fever is a cost of group living, not a benefit. The existence today of large homeothermic eusocial colonies testifies that such thresholds have been regularly breached in favor of group living, but this paper reminds us that emergent properties such as thermoregulation can express as continua, the incipient bids for which may be selectively neutral or even maladaptive.

To conclude this section, I want to talk about an interesting hypothesis that suggests how evolutionary history of female diphenism can interact with other species traits of momentous importance, in this case the differences in mode of colony reproductive swarming between the tribes Meliponini and Apini. Before these tribes diverged, their common ancestors pursued similar routes toward female diphenism, manipulating differences in seasonal body size, feeding regimens, and JH expression (Hartfelder and Rembold, 1991). The ancestral state for ovariole number was around eight (Leimar et al., 2012). In the Apini, proto-queens responded to their feeding regimes by increasing ovariole number, whereas queens in Meliponini responded by increasing ovariole length.

The strategies can be summarized as "many shorter" ovarioles versus "fewer longer." Today in *Apis*, the average divergence in ovariole number between workers and queens is 2–12 (Winston, 1987) versus 175–200 (Dedej et al., 1998), while in Meliponinae it is much narrower at four versus 4–15 (Cruz-Landim et al., 1998). Resulting colony population sizes between the Meliponini and *Apis* are quite variable, but egg size tends to be larger in *Melipona* (2.7–3.0 millimeters) (Velthuis et al., 2003) compared with 1.5 millimeters in *Apis* (DuPraw, 1961). Moreover, her emphasis on ovariole length has committed the mature meliponine queen to a permanent state of *physogastry*—a grossly distended abdomen that precludes flight (figure 8.2).

Leimar et al. (2012), in modeling the evolution of honey bee female diphenism, suggested that this divergence was the branch in corbiculate phylogeny that separated the Apini from the (Meliponini + Bombini) (see figure 1.3) and was driven by the habit in Apini of nest founding by the old queen. In modern *Apis* queens (see figure 3.3), the choice of many shorter ovarioles minimizes physogastry and permits the queen to temporarily lose weight, regain flight, and accompany a reproductive swarm. In modern Meliponini, in con-

FIG. 8.2. A queen of the stingless bee, *Melipona seminigra*. Compared with *Apis*, queens of Meliponini have fewer but longer ovarioles, the extreme condition of which results in distended abdomens as shown here, a state called physogastry. The opposite syndrome in *Apis*—more ovarioles but shorter (figure 3.3)—permits old queens in *Apis* to regain flight and found new colonies, an adaptation favoring rapid colony growth in temperate conditions. PHOTO: CRISTIANO MENEZES.

trast, nest founding is carried out by young queens who, after mating, assume physogastry and thereafter never leave the nest. It seems to me that a plausible explanation for this is the imperative for fast colony growth post-swarming in temperate-evolved *Apis*. Founding a new colony with a mature, mated queen speeds buildup of a forager force and winter food supply. A posited European origin (Kotthoff et al., 2013) for the genus supports this scenario.

WHY BEES HAVE NO MORPHOLOGICAL WORKER SUBCASTES

Ancient developmental deadlines were coopted by natural selection to produce the diphenisms we see today in eusocial queens and workers. These same coopted deadlines help explain a curious observation we see in the social Hymenoptera: among the social bees and wasps, there is not one species with morphological worker subcastes—each has only one kind of worker; whereas in ants a single colony may have numerous morpho-forms of workers such as majors (including soldiers), medias, and minors, with wide divergences in body shape and size. It is no coincidence, therefore, that in the ants, developmental switch points occur as early as the first or second instars (Robeau and Vinson, 1976). In the non-hymenopteran termites, caste diversity reaches similar levels with switch points occurring as early as the egg (Noirot, 1969).

So why do bees lack morphological worker subcastes? One investigator addressed this question by modeling the effects of a species's daily foraging mortality rates and division of labor strategy (morphological castes, age-based castes, or no division of labor) on average colony worker longevity. Average worker longevity is important because increasing values are associated with increasing colony fitness, and foraging mortality is important because foraging is the riskiest worker task class (Sakagami and Fukuda, 1968; Schmid-Hempel, 1987; Schmid-Hempel and Wolf, 1988). The model predicted that in species exposed to low foraging mortality, natural selection will optimize average worker longevity by favoring morphological subcastes over age-based subcastes (Jeanne, 1986).

I don't know of any field experiments that have followed up on this, but lower foraging mortality in ants and termites is intuitive given that these workers conduct their foraging in the shelter of tunnels and soil surface litter, in the case of termites investing in a few proven feeding sites heavily in preference to visiting many sites superficially (Delaplane and La Fage, 1989). Assuming that such strategies translate to safer foraging and overall longer worker life, then it must be adaptive for ants and termites to invest in worker polymorphisms and to link specialty to morphology. Indeed, we find such linkages in the ants: minor workers that never leave the nest, major workers that do all the foraging (Wetterer, 1999), and of course the soldiers who block entrances against aggressors with their block-shaped heads and do battle against other arthropods with their oversize mandibles (Wilson, 1971).

In contrast to worker ants and termites, the flying bees forage openly in three spatial dimensions, exposing themselves to hazards of weather and predation. With the terminal behavioral caste assured of brevity, the average worker bee life span is shortened, leaving less time to justify developmental specializations into morphological variants. Thus, it is far more efficient to invest in age-based polyethisms that optimize task proficiencies across a single morphological form. Given the comparative simplicity of honey bee caste structure, lovers of honey bees must cede the fact that this charismatic species is relatively non-remarkable on the scale of eusocial complexity on Earth.

Evolution of Worker Task Syndromes

Earlier in this chapter, we talked about how division of labor is a spontaneous outcome of groups and cannot be considered an outcome of natural selection *per se*, but rather as a scaffolding upon which natural selection can act. But once formed, logically possible only in context of a group, then division of labor became vulnerable to selection acting on the group. It's an example of something else we talked about earlier: how one thing can be selected (the group) but another thing adapted (division of labor) (Gardner and Grafen, 2009). The outcome of this selection became the two pillars of "complexed" division of labor we know today—queen-worker dimorphisms and age-based worker task syndromes, also called *age polyethism*, the subject of this section.

Examples are legion for phenotypic variation at the level of colony for some very important fitness measures including defensive behavior (Guzmán-

Novoa et al., 2004), food hoarding (Pesante et al., 1987a, 1987b), disease and parasite resistance (Spivak and Reuter, 1998, 2001), and propensity for reproductive swarming (Winston, 1980a). Virtually all colony-level variation like this ultimately accrues to variation in the behavior of workers, and that resolves to heritable differences in a worker's task specializations and the sequence in which she does her tasks. The brief sampling above of colony-level fitness metrics shows us the staggeringly diverse contexts in which natural selection has acted upon worker polyethism, resulting in the programmatic yet agile system in place today that can answer nearly any contingency a colony may face.

Workers of the social bees don't have morphological subtypes, but any individual can belong to one or more *temporal* or *behavioral* castes as she ages. The literature has long recognized an age-based progression in worker bees from in-nest tasks to out-of-nest tasks, culminating in the most hazardous task of all—foraging (Sakagami and Fukuda, 1968). Authors over the years have cataloged these tasks and attempted to discern the boundaries that mark one behavioral caste from another. The project has been complicated by competing ideas for what constitutes a behavioral caste. Most authors have defined it in terms of observed probabilities that a certain age group will perform a particular task (Calderone and Page, 1991; Jeanne, 1986; Seeley, 1982), whereas Johnson (2003) insisted that caste should discriminate physiological specializations that render a worker at least temporarily non-interchangeable with another task.

Contributing to the problem is the fact that enormous variation is built into the system, a point made obvious in figure 8.3 with the wide overlapping age ranges reported for worker tasks. This flexibility extends so far as to include the possibility of a young worker precociously jumping ahead to outdoor foraging, or an older worker reverting to more juvenile nursing physiology (Huang and Robinson, 1996), a plasticity clearly adaptive for dealing with calamitous losses to old, or young, sectors of the workforce.

But such crises excepted, a loosely age-based progression is the norm, and a worker's proclivity toward any specialization or any particular task sequence is influenced both by the genetics of her subfamily and by feedbacks, social in the case of cues from her sisters, and physical in the case of

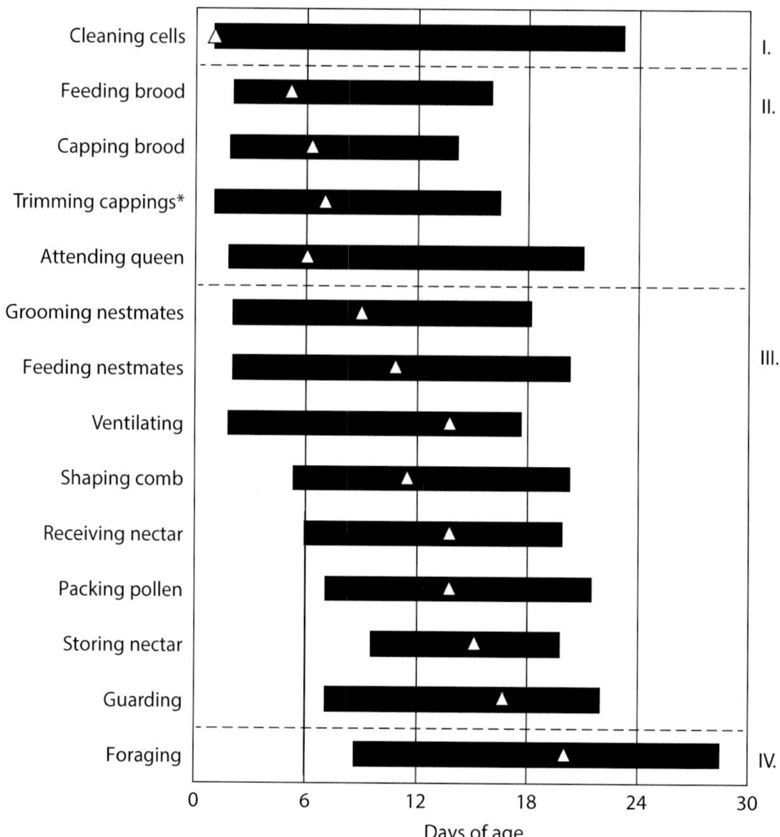

FIG. 8.3. Task groups illustrating the natural age-based progression from in-nest to out-of-nest tasks performed by honey bee workers as they age. Bars indicate age ranges from the literature, and white triangles indicate age of maximum probability of performing the task. Roman numerals indicate the four most robustly supported behavioral castes (Johnson, 2008, Seeley, 1982, Seeley and Kolmes, 1991): **I.** cell-cleaners, **II.** brood nurses, **III.** nest maintainers, and **IV.** foragers. I limit this illustration to the data for guarding behavior by Moore et al. (1987), Breed and Rogers (1991), and Huang et al. (1994) as well as the tasks studied in common by Seeley (1982) and Seeley and Kolmes (1991). I populate the graphic with data from these studies plus those of Johnson (2008) and works non-redundantly cited in Winston (1987) table 6.1. Ranges were found by calculating mean lower and upper bounds from the literature; reported maximum probability ages are means calculated from values (either maximum probability or mean age, depending on study) taken directly from the literature or from median day when most probable ages were expressed in ranges. Upper ranges are biased conservatively low because investigators sometimes set arbitrary end points for taking observations. *Trimming cappings refers to removing the shredded cappings after a baby bee emerges from its cell.

cues from the nest substrate itself (Calderone and Page, 1991). In figure 8.3, I show this age-based progression as well as the clusters of tasks that make up, in my opinion, the behavioral castes best supported by the literature, namely, (I) cell-cleaners, (II) brood nurses, (III) nest maintainers, and (IV) foragers. The fact that these castes occur as tightly associated task clusters is one reason we can also call them *task syndromes.*

As we will see below, there are practical evolutionary reasons for clustering. The four castes in figure 8.3 correspond to the behavioral castes of Seeley (1982), which were substantiated by later work (Johnson, 2008; Seeley and Kolmes, 1991). Johnson (2008), while supporting Seeley's nest-maintainer caste, was inclined to recognize it and only one other in-nest caste—brood nurses—while pointing out that other authors have tended to lump all in-nest castes into one (Amdam et al., 2004a; Amdam and Omholt, 2003).

Having established the general patterns of worker polyethism, let's shift our attention to how it evolved. We will first consider the colony-level imperatives that selected for worker specialization, then look for plausible evolutionary pathways that achieved it.

In an influential 1986 paper, Jeanne posed the matter in terms of a colony's net energy budget, pointing out that natural selection will act to optimize a colony's production of more queens. For the individual worker, this means optimizing her net energy contribution to the colony, her net energy signature being the difference between the total work she contributes in her lifetime minus the energy the colony spent to produce her. All other things equal, a colony that produces more reproductives per worker will compete better at a population level.

We thus frame the conversation in terms of efficiency. It has long been understood that a solitary can only work serially, one task at a time, whereas a division of labor among a minimally large workforce allows tasks to be performed in parallel (Robinson, 1992; Wilson, 1985) through many individuals, each contributing her share to that project or different projects: "many hands, short work." But efficiencies must resolve to the level of individual worker. Calderone and Page (1991) pointed out, for example, that multiple subfamilies of honey bee workers can, and do, cause positive synergistic improvements to task performance, but multiple subfamilies could just as easily confound tasks if workers disagreed on such behavioral components as *action thresholds*—minimally small sensory cues that trigger responses in sensitized individuals. Evidence for negative interactions between subfamilies, as well as positive ones, have been demonstrated on measures of importance to colony fitness such as seasonal weight gain and brood production (Oldroyd et al., 1992). Clearly therefore, we can predict worker efficiency to be responsive to colony-level selection.

Jeanne (1986) identified two areas in which natural selection will act on worker efficiency: the first he called *ergonomic* selection and the second *demographic* selection. With ergonomic selection, Jeanne's model predicts that natural selection will act to recover the cost of a worker's production as early as possible in the worker's life. This "early payback" selects for such things as worker cognitive capacity for gaining skill from experience and worker specialization on tasks that are spatially close to one another in the nest, thus minimizing lost time from the worker moving about searching for the next work site. With demographic selection, it proves adaptive for a colony to extend the average life span of its workers, and Jeanne's model predicts that worker life span is maximized when the riskiest tasks are delayed until late in life. Taken together, Jeanne's model shows that early payback and age-based risk allocation maximize a worker's net energy contribution to the colony.

With an imperative for worker efficiency established and theoretical hints that the imperative will involve an age-based task structure, we next look for scenarios that could lead to the patterns we see today (figure 8.3).

At the inception of groups, even small differences in an individual's experience, size, nutrient history, and genetics had huge effects on the kinds of tasks she would do once she found herself a member of a group. Ghosts of this kind of ancient variation persist to this day to affect worker polyethism. Some of these were described in the "ovarian ground plan" of West-Eberhard (1996), who proposed that ancient solitary age-related hormonal and physiological changes associated with maternal behavior were coopted and transferred onto the worker castes of social groups.

To begin understanding this, let's recall that in ancestral solitary life, one female founds one or more nests and performs all the functions of reproduction, maternal care, nest defense, and foraging. Pollen foraging is emphasized in a solitary female during the reproductive phase of her life (pollen ball formation and egg deposition) and nectar foraging in the non-reproductive phases (foraging for self and nest construction) (Dunn and Richards, 2003). It is still true in modern honey bees that those workers with comparatively high numbers of ovarioles are more likely to forage for pollen. The reverse is also true; those workers with fewer ovarioles are more likely to forage for nectar (Amdam et al., 2006). In other words, modern pollen forager honey bees seem to be recapitulating ancient maternal states, whereas nectar foragers are recapitulating non-reproductive states.

Experimental support for a genetic link between ancient reproductive state and modern division of labor came through a series of studies showing that a colony-level foraging preference for pollen hoarding can be genetically selected for, sometimes in as few as three generations (Hellmich et al., 1985; Page and Fondrk, 1995). Compared with low-pollen-hoarding bees, high-hoarding bees initiate foraging earlier in life, specialize on pollen over nectar, carry larger pollen loads, and at emergence have higher levels of juvenile hormone (JH) and the yolk precursor protein vitellogenin (Vg) (Amdam et al., 2004a). Investigators also sampled bees from ordinary non-selected colonies and found that high ovariole number in workers is associated with the individual's propensity to forage for pollen and begin foraging early. Thus, the association between a worker's reproductive state and her task specialization choice seems to be general.

What this thread of knowledge seems to suggest is that in worker honey bees, an ancient endocrine system regulating reproduction was coopted by evolution to reorient ancestral maternal care onto care for siblings (Giray and Robinson, 1996; Page et al., 2006). Natural selection pushed maternal behaviors early into the worker's life and foraging later in life, a polarity that not only maximized worker life span but established the framework for all subsequent age-based task progressions. The proximate drivers of this process were JH and Vg plus adaptive changes in the action of these molecules on the evolving worker.

In most insects, JH functions as a *gonadotropin*—one of a class of hormones that regulates growth, egg development, and Vg synthesis (Hartfelder,

2000). In the honey bee, however, JH has downgraded some of its gonadotropic functions and now regulates worker behavior transitions through its interactions with Vg. In the modern worker, JH stimulates Vg production, but only in the pupa, setting the stage for precociously maternal behaviors such as larval care and nest defense in the young adult (Barchuk et al., 2004). After JH activates Vg in pupae, the two molecules thereafter behave as "double repressors," with Vg levels dropping as brood-food production depletes Vg in nurses' bodies and JH levels rising, peaking just before onset of foraging (Amdam and Omholt, 2003; Pinto et al., 2000).

As Vg drops and JH rises, workers transition away from nest duties toward outside duties that are more representative of ancient states of nonreproductive behavior. This dual action also delays onset of foraging, and indeed, modern honey bee workers begin foraging much later in life than their solitary or eusocially simple counterparts (Michener, 1974). There is evidence that this transition is enacted in part by workers assessing their own physiology, appraising their remaining life span, and responding at critical thresholds by transitioning to foraging (Woyciechowski and Moroń, 2009). Overall, the innovations of precocious mothering, delayed foraging, and exploiting the natural age-based depletion of Vg resulted in a range of worker behaviors that effectively meets the diverse needs of the colony.

It is possible that this interlinkage between ancient maternal behaviors and age-based Vg exhaustion is the foundation for virtually all worker task specializations we see today. The shift to early expression of Vg was essential for the earliest and most basic stage of sociality—daughters staying at the nest to help their mother rear siblings. Having passed that evolutionary hurdle, subsequent task specializations can be seen as emergent properties of groups of individuals, each responding to her physiology, immediate stimuli, and genetic propensities. If we grant that Vg exhaustion is the pacemaker of this age-based progression, let's look next at some of the things that regulate individual expression of this process. Hopefully this will explain some of the range in values we see in figure 8.3.

First of all, we know that many of the tasks we see are triggered by action thresholds under genetic control, and these thresholds vary widely even within age class. Some workers respond immediately to cues at very low intensities, whereas others need repeated exposure to the same cues at high intensities before they register a response (Scheiner et al., 2004). But in general, an undertaker bee, for example, becomes an undertaker for the simple reason that she has inherited a lower tolerance threshold for corpses in the nest, and she will be among the first to jump at the chance of removing a dead bee (Robinson and Page, 1988). In a similar way, a bee specialized for hygienic behavior is more sensitive to brood anomalies (Oxley et al., 2010); a bee specialized for pollen foraging is more sensitive to the amounts of pollen stored in the nest (Page and Mitchell, 1998), and a highly defensive bee has a lower response threshold to alarm pheromone (Collins, 1979).

Such gene-based differences in task specialty often resolve to the level of subfamily, daughters of one particular male with whom the colony's queen has mated (Jones et al., 2004; Page et al., 1989; Robinson and Page, 1988, 1989). Table 1 in Mattila and Seeley (2014) gives numerous examples of how patriline membership affects worker foraging characteristics such as age of onset of foraging, time of day, preferred foraging distance, and propensity to scout and perform recruitment dances.

The exercise of an action threshold by an individual has the inescapable property of affecting the behavior of others; this can happen in either an *additive* or *non-additive* fashion—qualities that vary according to the task's objective (LeBoeuf and Grozinger, 2014). An additive effect is seen in simple *stimulus-response* behaviors in which a stimulus (such as a dead bee) presents itself, but only those individuals possessing the critical sensitivity respond to the stimulus. In the case of an acute die-off, corpses pile up because their removal rate scales directly with the number of undertakers in the nest, which does not increase in response to more corpses. Undertakers affect other undertakers by mutually removing stimuli for each other, but they do not recruit other undertakers. Evolution has not bothered to elaborate undertaking behavior because in a normal colony the mortality rate of nest bees is more or less finite and predictable.

In a non-additive effect, on the other hand, the worker's behavior is amplified through *positive feedback*. Consider a scout who is trying to recruit foragers to a profitable nectar source she has discovered. Her recruitment dance is, by definition, an attempt to amplify the response of others to her action who, if they agree with her assessment, will return to the nest and reinforce the stimulus with their own dances.

Non-additive effects also happen from stigmergy—the principle we discussed in chapter 4 in which actors leave traces in the environment upon which future actors may act. A wax-shaper deposits a bolus of wax on the nest cavity's ceiling, offering a stigmergic stimulant to other wax shapers to do the same, thus expanding the comb. One can appreciate the importance of non-additivity for optimizing those tasks that are opportunistic and fatally important, such as calorie acquisition and nest construction.

The difference between additive and non-additive effects means that individuals can vary tremendously in their impact on the group (LeBoeuf and Grozinger, 2014). Without positive feedbacks, the behavior of a group will reflect the average behavior of its members. But with positive feedbacks, the group's behavior will reflect that of a few *keystone* individuals. We have already talked about the oversized influence of a successful scout. Other examples of keystone individuals include guard bees, who amplify attack behavior in nestmates (Guzmán-Novoa et al., 2004; Paleolog, 2009), and nest-site scouts, who stimulate nestmates to depart with a reproductive swarm (Grozinger et al., 2014; Seeley, 2010).

Neither is the individual bee immune to effects from the group. Deriving from experiments in which bees of different genotypes are mixed together, we know that low-hygiene-expressing workers increase their hygienic behaviors when they are housed with high-hygiene-expressing workers (Gempe et al., 2012). Aside from genotypic differences, even group size influences behaviors of individuals. For example, large colonies perform better than small colonies in net energetic gain per forager; this is thought to be a sum effect of increased rates of parallel searching with associated increased rates of recruitment (Donaldson-Matasci et al., 2013).

The curious ability of workers to jump precociously to older tasks or revert to younger ones seems to be a self-regulating system of demographic feedbacks. The strong association of JH with foraging prompted Huang and Robinson (1992) to propose an "activator-inhibitor" model to explain how age distribution of a colony's workforce can affect individuals' transitions to new tasks. This hypothesis sees JH as an intrinsic "activator" that promotes behavioral transitions. But there is an hypothesized "inhibitor" substance that suppresses behavioral transition and is associated with older bees. Interplay between the activator and inhibitor allows task transitions to respond to up-to-the-minute changes in colony demographics.

For example, let's imagine that a colony has lost a large share of its foragers from a sudden storm. Precocious foraging in young workers is triggered in this case because the young bees encounter fewer older bees, receive less inhibitor, and respond by transitioning to foraging. The opposite happens in a colony with a disproportionately old population: the high rate of exposure to inhibitor keeps workers from transitioning out of a young state, depresses JH, and even restores nursing physiology to older workers who may abandon foraging and return to tending brood.

Twelve years after publication of the activator-inhibitor hypothesis, the existence of such an inhibitor was confirmed. The pheromone ethyl oleate was shown to be produced in the crops of older forager bees, transferred by direct contact in social food exchange, and effective at delaying onset of foraging in young bees (Leoncini et al., 2004). This is a beautiful example of a self-regulating system because the agent of change, the forager, is also the target of change, the colony's need for more or fewer of them. The old worker's presence in the nest constitutes a direct vote by existence. The feedback loop is never turned off and is therefore a force for colony-level homeostasis.

This brief treatment of regulators of individual task expression shows the extent to which high variation is engineered into the loose template we call "age-based polyethism." The wide age ranges for each task of figure 8.3 are not evidence of sloppy investigation, but rather that evolution has rewarded agility and flexibility over rigidity and specificity when it comes to colony-level adaptation to its environment.

To conclude, we can think of age-based worker task syndromes of a superorganism as analogous to the function-specific tissues and organs of an organism, each temporal caste corresponding to the comparable organ. These behavioral castes are like sandwiches of cause and effect: emergent properties of groups providing scaffolding for group-level caste selection providing self-organizing processes that promote group fitness, the sum of which rewards the gene—always the gene—with replication. It is too much for evolution alone. Perhaps Rob Page (1997) said it best: "Natural selection cannot 'see' all of the intricate and complex interactions within a colony, nor is it likely that the genome can orchestrate them. Instead, natural selection must act on colony-level organizational components that through self-organizing processes result in specific complex behavioural patterns." The age-based worker task syndromes we see today are not adaptations to any specific contingencies; they are adaptations to all contingencies. They are evolution's agile response to nearly all problems the vast range and history of *Apis mellifera* have thrown at it.

Summary

Division of labor is a pattern by which some individuals in a colony perform different functions from others, if even temporarily. This is distinct from reproductive division of labor, which distinguishes those individuals privileged with producing offspring. Being a task specialist applies to an individual who performs one task at a rate higher than the population average for that task. Division of labor is an immediate emergent property of groups and not a product of group-level selection. However, morphology and features of task execution can be altered by natural selection to improve group fitness. It follows, then, that group-level selection was operative at the inception of group formation.

The varieties of morphological castes and socially integrated task syndromes in the modern honey bee are examples of group-mediated selection, with the colony being the level selected and workers the level adapted. Caste refers to individuals who are categorical variants, based on morphology or behavior, within sex. Honey bees have two sexes, the females of which are divided into two castes—the queen and workers. Any fertilized (female) egg is bipotent for one of two developmental pathways—worker-like or queen-like. This transformation is regulated by divergent feeding regimes imposed on larvae by nurse workers. The evolutionary precursor of this divergence was season-dependent changes in food availability with the trend toward larger, more queen-like offspring emerging in late season after the simple colony had accumulated a larger workforce and food stores.

Natural selection acting on improving colony efficiencies pushed switch points earlier in honey bee immature development, the result being a range of morphological innovations that improved worker and queen performance and widened the divergence between the two. In the honey bee lineage, intermediates—worker-like queens and queen-like workers—were not fit and disappeared from natural history.

Workers of the social bees don't have morphological subcastes, but individual honey bee workers progress through an age-based sequence of tasks from in-nest to out-of-nest, culminating in foraging—a phenomenon called age polyethism. A worker's proclivity toward any task sequence is influenced by the genetics of her subfamily and by feedbacks, social in the case of cues from her sisters or physical in the case of cues from the nest substrate itself. Imperatives driving selection for polyethism are understood in terms of optimizing workers' individual net energy contribution to the colony—the difference between the total work one contributes in her lifetime minus the energy the colony spent to produce her. Efficiencies are optimized when workers pay back energetic costs early in life and when riskiest tasks are delayed until late in life.

Specialization, whether age-based or genetic, means that colony tasks can be performed in parallel—many individuals, each contributing her share to one or more projects—as opposed to tasks performed serially, one component at a time. Many tasks are triggered by action thresholds under genetic control, and these thresholds vary widely even within age class. Some workers respond immediately to cues at very low intensities, whereas others need repeated exposure to the same cues at high intensities before they register a response. There is wide variation in the expression of age polyethism, including workers capable of reverting to younger tasks or precociously jumping ahead to older tasks as colony needs dictate, a plasticity that enables a colony of *Apis mellifera* to weather a range of ecologic conditions.

CHAPTER 9

Complex Eusociality: Polyandry and Its Consequences

WE HAVE TALKED NUMEROUS TIMES in this book about polyandry—the queen's habit of mating with multiple males and storing their sperm in her spermatheca, the end result being a colony of subfamilies, each sired by a different male. The complement of this syndrome is *polygyny*—many reproductive females in one colony. This is what many ants and termites do. Their large, amorphous, subterranean colonies can host multiple secondary queens—their mother's daughters who stay in the nest, mate, and contribute progeny to the common citizenry, much like fiefdoms at the edge of empire. One can see that each strategy is a different side of the same coin—a bid for within-nest genetic diversity.

Among the honey bees, only polyandry is to be found, an exclusivity, I believe, that can be traced to their relatively small nests. It is much easier for one queen to assert her hegemony, whether through behavioral policing or inhibitory pheromones, in a finite nest the size of a couple of basketballs. Ant and termite queens, in contrast, must assert their dominance over a space the size of an acre or more. Order inevitably breaks down in the provinces, independent queenlets rise up, and polygyny creeps in and carries the day. Alternatively, a bias toward polyandry in the social bees could be the fact that their males are generally short-lived, so that the strategy of colony budding by a long-lived female pairing with a long-lived male, as practiced by the termites, is unavailable (Boomsma et al., 2005). In Hymenoptera, male life is extended only as living sperm in the spermathecae of long-lived queens.

In any case, polyandry is still the minority condition among eusocial Hymenoptera, the majority of whose species still practice monogamy (Boomsma and Ratnieks, 1996; Strassmann, 2001), but polyandry is represented widely across all eusocial Hymenoptera (Hughes et al., 2008c) which argues for its general adaptiveness across groups. Its highest expression is found in the honey bees, especially *A. dorsata*, the giant honey bee of southeast Asia (table 9.1).

Multiple mating implies a continuum of possibilities from two males to infinity, so any discussions of the matter should qualify the degrees of polyandry being talked about. The literature reports mating number (or "paternity number") a couple different ways. The most straightforward is *observed* or *actual* mating number, which is simply the number of males contributing paternity to a female's pool of daughters. This is k, which readers met in equations 6 and 7 and which makes up most of the values in table 9.2 in this chapter; k is useful when the question involves queen mating number or male mating success, as we discuss ahead. However, k assumes that all drones contribute equally to the female's brood, which is probably rarely the case. Numerous studies have been done on the extent of sperm mixing in the spermatheca and its use patterns at short time scales of the kind that matter to a colony's ecological success in the course of a season. In general, a queen uses all of her mates' sperm all the time (Haberl and Tautz, 1998; Page and Metcalf, 1982).

But constant representation is not the same thing as equal representation. Drones vary in their success at siring their mate's daughters. Haberl and Tautz (1998) showed that the proportions of a colony's workers sired by any one male can range from 3.8% to 27.3%. Franck et al. (2002) also found significant differences in male siring success and that a male's rank could change between any two seasons. Results such as these suggest the existence of sperm competition (Harbo 1990), a kind of post-mortem competition among males. Ways this could happen include differences in sperm motility or viability at fertilizing eggs. Another way it could happen is the focus of the *last-male* hypothesis. In multiple-mating insects, it is common for last male mates to be overrepresented in progeny (Gwynne, 1984), owing perhaps to a last male displacing his predecessors' sperm or, in the case of the *Apis* lineage, a last male's mating plug retaining a comparatively higher fraction of his sperm in the female's oviducts. However, only weak (Moritz, 1986) or no (Franck et al., 2002) evidence for last-male advantage has been shown in honey bees.

Perhaps variation in male fertility is driven by his age or the environmental conditions in which he is sheltered. Male fertility decreases with age and in adult males exposed to wounds and elevated colony temperatures (Stürup et al., 2013). These results suggest that a measure of a colony's fitness extends to the ability of its workers to secure nest homeothermy and sanitation for the benefit of its drones.

It is because of this variation in expression of male paternity that we need a second descriptor for mating number, and this is the *effective* mating number, m_e, which is the queen's mating number adjusted for the relative contribution of each of her mates to actual progeny. The value is derived as follows (Starr, 1985):

$$m_e = \frac{1}{\sum_{i=1}^{k} p_i^2} \qquad \text{(equation 10)}$$

where p_i = proportion of a queen's daughters sired by male i. Due to variation in male success, m_e is typically $< k$. For example, for a queen mated to $k = 5$ males whose progeny are expressed at proportions of (3:2:2:2:1), $m_e = 4.5$, whereas if males are expressed in equal proportions of (2:2:2:2:2), then $m_e = 5 = k$. Effective mating number is the parameter of choice when we're discussing paternity's effect at the colony level because m_e is a better predictor whether any two workers drawn from the nest will share the same father. Knowing such kinship structures in the colony is useful for explaining social evolution by kin selection (Nielsen et al., 2003).

Polyandry and Worker Fitness

But the point for now is that in the honey bee, polyandry has become fixed in the species (and entire Apini) for the good reason that it promotes colony fitness. I say *colony* fitness, because when it comes to the worker, it's a very different story. For the worker, her mother's switch from monandry to polyandry means that the worker's genetic reward for staying at the nest and helping rear siblings falls from $r_{sibs} = 0.5$ (equation 8) to $r_{sibs} = 0.27$ (equation 6). Rearing all those half-sisters alongside brothers and the occasional full sister means that the terms of the original "agreement" have been compromised. It is no longer equally profitable to stay at one's natal nest and help rear siblings compared with striking out on one's own.

Why would any daughter, from the perspective of natural selection, "agree" to cede such benefits? The answer is, she wouldn't. And so here, the story line tips anthropomorphically into deceit and treachery. It is untenable, theoretically speaking, that polyandry could have evolved until *after* the mother and daughters morphologically diverged into different co-dependent castes. Before that point, a reproductively autonomous daughter would quickly abandon a mother practicing polyandry in preference to solitary life and directly rearing her own progeny. But once daughters diverged into reproductively dependent sterile workers, they were committed to life with their mother and thus vulnerable to her reproductive choices. This is why this switch is called the point of no return. There is no known case of a complex eusocial species reverting to solitary life (Wilson and Hölldobler, 2005), although it is known to have happened in a few simple eusocial lineages in which the females had not yet diverged into morphological sterile castes (Wcislo and Danforth, 1997).

The math explaining these genetic stakes is covered in chapter 7, but perhaps now is a good time to reflect on the words of Ratnieks and Helanterä (2009):

> One analogy sometimes used to describe an insect society is that of a factory [inside a fortress (Oster and Wilson, 1979)]. To extend the analogy, it is a factory in which the working individuals are not as well paid as the boss or owner (the queen). But neither are they badly paid. There are few human businesses or organizations in which the highest salary is only twice the lowest, as occurring in the honeybee.

Those "salaries," of course, being the queen's gene transmittal $r = 0.5$ versus those of her helping daughters $r = 0.27$. However circumspect we may be about the diminished fortunes of workers, we will leave them for now and spend the rest of this chapter discussing the biological levels at which polyandry is adaptive.

Polyandry and Queen Fitness

Whereas workers' genetic fortunes have plummeted, the queen's direct fitness has suffered no setback at all with her move to polyandry. Every egg she lays, whether male or female, still passes on $r = 0.5$ of her genes to the next generation even though her daughters are now the progeny of numerous sires. Because her colony is an extension of her own phenotype, we can ask how natural selection has acted on the queen to increase the fitness of her colony. In this section, I foreground two hypotheses attempting to explain the evolution of polyandry from the perspective of a queen's fitness. Hypotheses explaining polyandry from the perspective of its group benefits will be covered in the "Polyandry and Colony Fitness" section.

GENETIC LOAD ON SEX LOCUS

This explanation focuses on the sex determination system of bees. It is predicted that selection for multiple mating varies positively with increasing numbers of sex alleles in a population (Page, 1980), and on this basis alone, *A. mellifera* is a good candidate for polyandry. The number of sex alleles in *A. melllifera* has been estimated from 19 to 53 in local breeding populations (Adams et al., 1977; Cho et al., 2006; Lechner et al., 2014) to over 100 at the species level (Zareba et al., 2017). Individuals that are homozygous at the sex locus become diploid males, and while still immatures, these individuals are removed by workers and thus constitute opportunity costs to the colony. A female mated to one closely related male risks burdening her progeny with homozygosity at this and other loci across their genomes, negatively affecting such individual and colony fitness measures as wing symmetry, brood nest thermoregulation, and foraging recruitment (Brückner, 1978) [but curiously not individual immune system response (Lee et al., 2013)]—the sum of which reduces the probability of her colony successfully rearing her replacement.

A female's promiscuity is expected to offset these risks (Page, 1980; Page and Metcalf, 1982; Tarpy and Page, 2001), not because it eliminates near-kin matings (indeed, it may increase them for some colonies), but because the benefits of increased genetic variation compensate for the presence of some diploid males (Crozier and Fjerdingstad, 2001). One example of this offsetting variation is the presence of subfamilies that can detect and remove diploid eggs early enough that they can be replaced with fresh diploid eggs, hopefully female (Ratnieks, 1990). This process may happen more efficiently in a colony with genetically diverse workers.

But more broadly, this hypothesis foregrounds a feature of polyandry that applies to other explanations we will cover ahead. Rather than a means of directly increasing population, reducing disease, or in the present case reducing homozygosity at the sex locus, polyandry is understood as a "bet-hedging" strategy that reduces variance around the mean of these characters; it is, in other words, a moderator of extremes (Yasui, 2001). A monandrous colony risks expressing an extreme maladaptive phenotype, and without the compensating benefits of additional subfamilies, the colony is uncompetitive and selectively handicapped. Of course, it's possible that a single-mating queen may hit the jackpot and mate with an exceptionally viable male who provides complete heterozygosity at the sex locus and other loci, but in geologic time, this approach has not competed well because monandry is rare in modern *Apis*.

This feature of convergence on a population mean has been modeled by Page and Metcalf (1982) for the sex load hypothesis and visualized by Palmer and Oldroyd (2000) (figure 9.1). Assuming (1) random mating with a large drone population, (2) random mixing of sperm in the spermatheca, (3) all queens mate k times with equal amounts of sperm at each mating, and (4) all sex alleles are present at equal frequency in the population, the expected average viability of diploid brood in the population given k matings and n sex alleles is

$$1-\left(\frac{1}{n}\right) \qquad \text{(equation 11)}$$

The variance about this figure is

$$\frac{1}{2k}\left(\frac{1}{n}\right)\left(1-\frac{2}{n}\right) \qquad \text{(equation 12)}$$

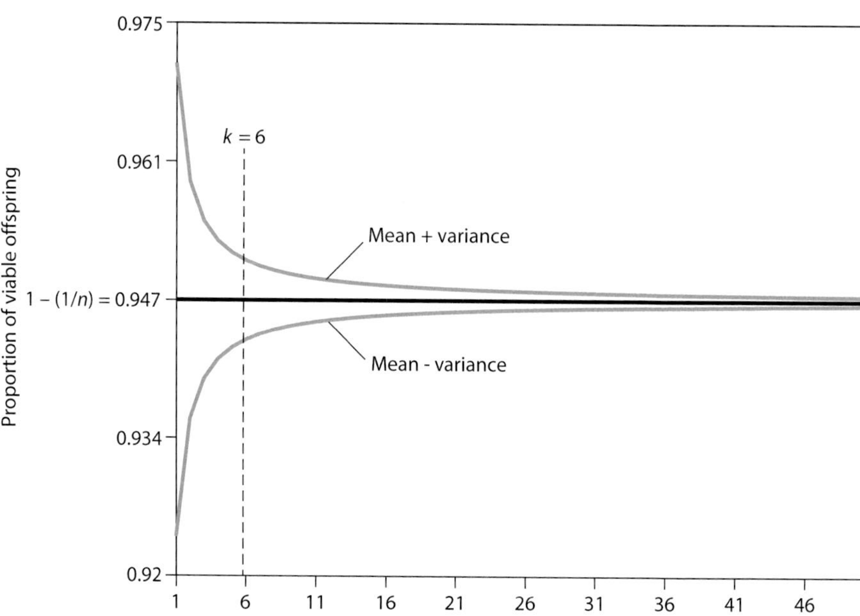

FIG. 9.1. A plot showing that extreme expressions of a character, in this example colony brood viability, converge toward the population average as the queen mother increases her mating number with local males. This simulation assumes a local breeding population with 19 sex alleles and mean expected brood viability of 0.947. The rapid convergence shows that early queens were rewarded with more environmentally stable colonies by practicing even low levels of polyandry.

Figure 9.1 plots this relationship assuming a local breeding population with an early estimate of 19 sex alleles (Adams et al., 1977) and mean expected brood viability of 0.947. The change in variance about the mean is rapid. Over 90% is accomplished by the tenth mating, 75% by the first. This empirical demonstration suggests that early queens were rapidly rewarded with more environmentally stable colonies, at least to the extent influenced by brood viability, by practicing even very low levels of polyandry.

However, let's not forget that the transition from monandry to polyandry constitutes nothing less than the transition from simple to complex eusociality (table 5.3). As such, it is fraught with potential conflict; any gains in colony fitness are purchased at the cost of worker fitness. There is evidence that within the sisterhood of corbiculate bees, this contest has not always tipped in favor of group fitness. Baer and Schmid-Hempel (2001) instrumentally inseminated queens of *Bombus terrestris* with sperm from one male, two males, or four males. Colony infection rates with the parasite *Crithidia bombi* decreased as polyandry numbers increased from one to two or four mates. However, colony fitness as measured by the number of new reproductive males and queens produced at season's end was highest in colonies whose queens were mated to either one male or four. For queens in the middle, inseminated with two males, there was a pronounced and significant drop in colony fitness.

Baer and Schmid-Hempel (2001) suggested that this hints at an adaptive valley over which early innovators in polyandry had to cross. The sharp reduction in worker kinship represented by the first excursions into polyandry, i.e., $k = 2$ matings, may have set up subfamily conflicts over rearing the next generation of queens. The rarity of polyandry in *B. terrestris* and most other social Hymenoptera suggests that this adaptive valley proved insurmountable for most lineages. But this was not so for the Apini. For reasons unknown, with Apini the costs of intra-nest conflict from early polyandry were offset with mitigating colony-level benefits, the leading candidates being marginal gains in parasite resistance as indicated by their corbiculate cousin *B. terrestris*, and immediate benefits to brood viability and environmental stability from avoiding diploid males.

SPERM LIMITATION

The *sperm limitation hypothesis* places the level of adaptation squarely on the queen by focusing on the details of her mating behavior. It is based on the assumption that there are positive relationships among the number of sperm a queen collects, the queen's reproductive longevity, the number of fertilized worker eggs she produces, the resulting population size of her colony, and the colony's performance at rearing her sons and reproductive daughters (Bourke, 2011; Cole, 1983; Starr, 1985). It is closely aligned to the size-complexity hypothesis, which is the idea that increasing size of social groups, whether cells in an organism or workers in a superorganism, is associated with increasing social complexity and environmental stability of the group (Bourke, 2011, p. 163). A queen who fails to secure enough sperm suffers terrible fitness costs in terms of a shortage of working daughters and a surplus of non-working sons, adding up to a short-lived colony with low likelihood of producing new reproductives. Conversely, colonies with large worker populations issue large reproductive swarms with improved probabilities for overwintering and reproducing (Seeley, 1985). Thus, a queen's fertility is strongly correlated to her fitness. The sperm limitation hypothesis predicts positive selection for queen promiscuity to ensure enough sperm for her to sustain a large colony and to compensate for the occasional infertile male or brother she mates with.[1]

Against these positive selection vectors for polyandry must be balanced the risks of promiscuity. Multiple mating involves one or more mating flights per queen of about 3–57 minutes each (Heidinger et al., 2014) with hazard from birds and other predators, as well as multiple copulations, which expose the queen to sexually-transmitted diseases (De Miranda and Fries, 2008; Roberts et al., 2015; Schlüns et al., 2005). These risks are real and documented. Queen loss on mating flights ranges from 4.85% to 14% (Ratnieks, 1990; Schlüns et al., 2005; Tarpy and Page, 2000), or put another way, the probability of fatality is 0.0026 per minute of flight (Tarpy and Page, 2000). Therefore, we can predict that queens will be selected to minimize flights and maximize sperm acquisition, all under some kind of optimizing threshold mechanism. The fact that queen mating frequency is heritable, responsive to genetic selection (Kraus et al., 2005), reinforces the idea that mating behaviors are evolvable.

So, what is the minimum necessary quantity of sperm? From an ecologic perspective, one could say that this threshold is the number required to fertilize the eggs for a queen's expected lifetime worker retinue. Using data from managed colonies, Seeley (1985) estimated that a colony produces 150,000–200,000 workers annually, which, given an expected queen life span of three years, adds up to a functional minimum requirement of 600,000 eggs. However, sperm depletion rates behave logarithmically not linearly, which means that much is wasted, especially early in the queen's tenure. Using John

1. As far as we know, a queen cannot judge a male's quality before mating with him, and encounters with infertile drones (copulations without insemination) are known to occur.

Harbo's (1979) logarithmic equation, I show in figure 9.2 that a queen receiving 5 million sperm at mating, a typical spermathecal load for *A. mellifera* (table 9.1), will have enough sperm to produce a lifetime cohort of 600,000 workers—but with only about 99,000 sperm to spare. Harbo (1979) pointed out that the logarithmic flattening of the depletion rate in late life is an adaptive measure with the effect of prolonging the queen's productivity. Baer et al. (2016), making direct sperm counts on freshly laid eggs, found a similar logarithmic depletion rate, with higher sperm loss early in the queen's life and more conservative usage later.

Both of these studies support a scenario where a queen releases a uniform volume of spermathecal fluid with each fertilization, fluid that is regularly replenished. But having a finite sperm supply, the spermathecal fluid contains progressively fewer sperm as the queen ages. Baer et al. (2016) posited that as the queen's sperm concentration decreases, eventually she begins laying unfertilized eggs, an "honest signal" to her daughters that she is reproductively senescing and a stimulant for them to rear her replacement. It is in the queen's selfish interest that she be replaced by one of her own daughters before she goes completely infertile. Baer et al. (2016) calculated that a well-mated queen can be expected to fertilize over 1.5 million eggs, consistent with an earlier estimate of 1.7 million eggs (Bozina, 1961).

But given a typical initial demand for 5 million sperm, we arrive at an apparent problem with the sperm limitation hypothesis in the case of *Apis*. Among the features of modern honey bee mating biology—the wide range of mating numbers achieved by queens, the copious amounts of sperm produced per drone, the comparatively small holding capacity of a spermatheca, and the extravagant wastage of sperm (table 9.1)—none of them seems to be an adaptation driven by sperm shortage. It is for these reasons that the sperm limitation hypothesis has met with resistance over the years (Crozier and Page, 1985). Moreover, a comparative analysis concluded that polyandry, surplus sperm production by males, and excretion of excess semen by females are plesiomorphs—ancestral states shared across the genus (Oldroyd et al., 1998). In other words, however important overcoming sperm limitation may have been for queens early in social evolution, the problem was solved by the time we get to the common ancestor of *Apis*.

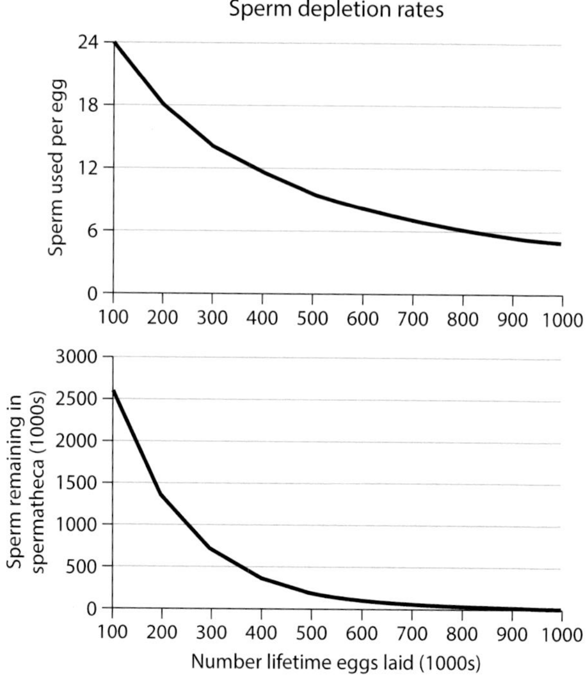

FIG. 9.2. Logarithmic rates of sperm depletion after the model of Harbo (1979). Sperm used per egg and sperm remaining in the spermatheca are derived from $\ln a = \ln a_o - t$ (1/153,000) where a = number of sperm remaining in the spermatheca after a queen has laid t eggs, and a_o = original number of sperm in the spermatheca after mating, which for these lines I have set at 5 million.

So how was the problem of sperm limitation solved? Aside from polyandry itself, to me, the most plausible early adaption would have been an enlargement of the spermatheca in a bid to improve sperm retention, which is famously inefficient in the Hymenoptera and bees in particular. And in fact, a comparative analysis did show that spermathecal volumes tend to increase from solitary to social bee taxa, with spermathecae of *Apis mellifera* conspicuously large (Pabalan et al., 1996). Nevertheless, these gains in volume seem to have been functionally modest because sperm retention in modern *Apis* is still quite low (table 9.1). Maybe there were physiological or energetic constraints against evolving a large spermatheca.

I think a more satisfying explanation for the mating behaviors of modern *Apis* is that selection imperatives for queens overcoming sperm limitation were quickly met and superseded by imperatives of a different order—colony-level benefits—which I take up in the next section.

TABLE 9.1. Some data on mating biology across *Apis*

SPECIES	SPERM PER DRONE (MILLIONS)	CAPACITY OF SPERMATHECA (MILLIONS OF SPERM)	SPERM RETAINED (OLDROYD ET AL., 1998) (%)	MALE MORPHOLOGY (KOENIGER ET AL., 1991)	EFFECTIVE MATING NUMBER, M_E	QUEEN AGE AT MATING (DAYS)	DURATION OF FLIGHT (MIN)
florea	0.44 ± 0.04 (Koeniger et al., 1989b)	0.48–1.35 (Koeniger et al., 1989b)	28.1	Bulb pointed, no mucus plug	5.6 ± 1.0 (Oldroyd et al., 1995) 7.9 ± 3.3 (Tarpy et al., 2004b)	6–8 (Koeniger et al., 1989b)	18–21 (Koeniger et al., 1989b)
andreniformis	0.13 ± 0.01 (Koeniger et al., 1990)	0.33–1.26 (Koeniger et al., 2000)	74.1	Bulb pointed, no mucus plug	9.1 ± 0.83 (Oldroyd et al., 1997) 10.5 ± 1.9 (Tarpy et al., 2004b)	3 (Koeniger et al., 2000)	19–23 (Koeniger et al., 2000)
laboriosa	—	—	—	Bulb intermediate, little mucus	28.4 ± 3.2 (Tarpy et al., 2004b)	—	—
dorsata	2.5 ± 0.15 (Koeniger et al., 1990) 1.24 ± 0.39 (Tan et al., 1999)	3.64–3.69 (Koeniger et al., 1990) 4.6–7.0 (Tan et al., 1999)	5.5	Bulb intermediate, little mucus	20.0 ± 6.6 (Oldroyd et al., 1996) 25.6 ± 1.1 (Moritz et al., 1995) 44.2 ± 27.2 (Tarpy et al., 2004b)	6 ± 1 (Tan et al., 1999)	15.4 ± 4.3 (Tan et al., 1999)
mellifera	4.6 ± 0.9 (Rinderer et al., 1985) 5.7 ± 0.9 (Rinderer et al., 1985)	4.7 (Woyke, 1966) 5.0 (Harbo, 1979) 0.057–5.6 (Baer et al., 2016) 1.05 ± 0.17 (Simone-Finstrom and Tarpy, 2018)	3.4–6.8	Bulb large, mucus plug	20.1 ± 1.7 (Kraus et al., 2005) 11.6 ± 7.9 (Tarpy et al., 2004b) 11.0 ± 5.7 (Simone-Finstrom and Tarpy, 2018) 13.2 ± 4.9 (El-Niweiri and Moritz, 2011)	10.4 ± 2.8 (Yoshida et al., 1994) 7–17 (Yoshida et al., 1994)	3.1–57.1 (Heidinger et al., 2014) 17.7 ± 13.2 (Heidinger et al., 2014) 11.7 ± 5.0 (Yoshida et al., 1994) 12–47 (Simone-Finstrom and Tarpy, 2018)

(*continued*)

TABLE 9.1. (Continued)

SPECIES	SPERM PER DRONE (MILLIONS)	CAPACITY OF SPERMATHECA (MILLIONS OF SPERM)	SPERM RETAINED (OLDROYD ET AL., 1998) (%)	MALE MORPHOLOGY (KOENIGER ET AL., 1991)	EFFECTIVE MATING NUMBER, M_E	QUEEN AGE AT MATING (DAYS)	DURATION OF FLIGHT (MIN)
koschevnikovi	1.7 ± 0.2 (Koeniger et al., 1994)	1.4–3 (Koeniger et al., 1994)	<11.2	Bulb large, mucus plug	13.3 ± 10.4 (Tarpy et al., 2004b)	5–6 (Koeniger et al., 1994)	12–27 (Koeniger et al., 1994)
cerana	1.5 (Ruttner et al., 1973)	0.66–2.0 (Woyke, 1975) 2.665 (Ruttner et al., 1973)	4.8–7.2	Bulb large, mucus plug	12.0 ± 1.6 (Oldroyd et al., 1998) 14.1 ± 3.9 (Tarpy et al., 2004b)	3.6 (Sharma, 1960) 12.4 ± 5.2 (Yoshida et al., 1994) 6–24 (Yoshida et al., 1994)	20–38 (Woyke, 1975) 17–46 (Ruttner et al., 1972) 26.3 ± 6.5 (Yoshida et al., 1994)
nigrocincta	—	—	—	—	40.3 ± 23.4 (Tarpy et al., 2004b)	—	—

Polyandry and Colony Fitness

The persistent poor efficiency of the sperm retention process seems to beg for an explanation. Rob Page, cited as a personal communication by Crozier and Fjerdingstad (2001), suggested that "the queen's anatomy is adapted to excluding most sperm so as to maximize genetic diversity." This gets us to an important point that will stay with us for the rest of our discussion on polyandry: that the hypotheses attempting to explain its evolution are dominated by the so-called *genetic variation* (GV) model and its variants—the idea that polyandry increases within-nest genetic diversity with fitness advantages for the colony. The sperm limitation hypothesis discussed in the previous section is really the only viable exception.

Early in this chapter, we established that polyandry trades worker fitness for colony fitness. The queen's multiple mating and ensuing drop in average nestmate relatedness appear to be affronts to group cohesion and the very basis of eusociality itself. But if these costs and risks borne by the workers and mother are compensated for by fitness benefits to the colony, this could be an example of balancing selection pitting costs of the individual against benefits of the group. As recently as the beginning of the 21st century, it was still possible for Kraus et al. (2005) to argue that group-level benefits of polyandry are rare or equivocal, but a spate of newer papers has tipped the weight of evidence in favor of a positive relationship among degree of polyandry (mating number), high intra-colony genetic variation, and measures of colony fitness (table 9.2).

An influential paper came out in 2007 that addressed the question in direct Darwinian terms: Does within-colony genetic variation promote colony fitness? Mattila and Seeley (2007) compared performance of colonies whose queens had each been instrumentally inseminated with either one or 15 males. Queens of each type and one kilogram of their worker progeny were caged and fed sugar syrup to simulate pre-swarm engorgement of honey. Then, each simulated swarm was housed in a comb-less hive during local swarming season and left to grow in the normal fashion of a natural swarm. The investigators measured rates of comb construction, brood production, foraging activity, food storage, worker population, and colony weight

TABLE 9.2. Some colony-level fitness effects of increasing polyandry in *Apis mellifera*

FITNESS CHARACTER	LEVELS OF POLYANDRY COMPARED
Positive effects	
Presence or intensity of brood diseases	1, 10 (Seeley and Tarpy, 2007) 1, 24 (Tarpy, 2003) 1, 10 (Tarpy and Seeley, 2006)
Prevalence of acute bee paralysis virus and *Nosema ceranae*	1, 12 (Desai and Currie, 2015)
Concentration of deformed wing virus	1, 12 (Desai and Currie, 2015)
Brood production or efficiency	15, 30, 60 (Delaplane et al., 2015) 1, 15 (Mattila and Seeley, 2007) 1, 10 (Seeley and Tarpy, 2007) 1, 2, 4, 8, 16, 32 (Delaplane et al., 2024)
Brood survival	9, 54 (Delaplane et al., 2021)
Varroa mite incidence	15, 30, 60 (Delaplane et al., 2015) 9, 54 (Delaplane et al., 2021) 1, 2, 4, 8, 16, 32 (Delaplane et al., 2024)
Varroa mite mortality rate	1, 12 (Desai and Currie, 2015)
Comb construction rate	1, 15 (Mattila and Seeley, 2007)
Foraging rate	1, 15 (Mattila and Seeley, 2007)
Food hoarding rate	1, 15 (Mattila and Seeley, 2007)
Colony weight gain	1, 15 (Mattila and Seeley, 2007)
Drone production	1, 15 (Mattila and Seeley, 2007)
Colony worker population	1, 12 (Desai and Currie, 2015) 1, 15 (Mattila and Seeley, 2007)
Nest temperature stability	Modeled 1, 15 (Graham et al., 2006) 1, 5, open-mated (Jones et al., 2004)
Recruitment dance efficiency and information transfer	1, 14–15 (Girard et al., 2011) 1, 15 (Mattila et al., 2008) 1, 14–16 (Mattila and Seeley, 2011) 1, 15 (Mattila and Seeley, 2010)
Number of inspectors (foragers reassessing a known food site)	1, 18–20 (Mattila and Seeley, 2014)
Pollen hoarding rate	1, 20 (Eckholm et al., 2011)
Nurse bee protein investment in larvae	1, 20 (Eckholm et al., 2015)
Colony mortality rate	$m_e \leq 7$, $m_e > 7$ (Tarpy et al., 2013)
Worker/queen conflict over sex ratio	Modeled (Moritz, 1985b)

(*continued*)

TABLE 9.2. (Continued)

FITNESS CHARACTER	LEVELS OF POLYANDRY COMPARED
No significant effects	
Brood area	9, 54 (Delaplane et al., 2021)
Colony worker population	1, 10 (Seeley and Tarpy, 2007) 9, 54 (Delaplane et al., 2021)
Colony weight gain	1, 10 (Seeley and Tarpy, 2007)
Comb construction rate	15, 30, 60 (Delaplane et al., 2015)
Return rate of non-scout foragers	1, 14–16 (Mattila and Seeley, 2011)
Exit rate of dance-following foragers	1, 14–15 (Girard et al., 2011)
Nest temperature stability	$8.72 \leq m_e \leq 37.75$ (Simone-Finstrom et al., 2014)
Average measures of worker physiological immune function	$8 \leq m_e \leq 29$ (Wilson-Rich et al., 2012)
Negative effects	
Taxonomic diversity of pollen collected	1, 20 (Eckholm et al., 2011)

Note: Unless indicated otherwise, all values are actual mating numbers (k) derived from instrumental insemination. Effective mating numbers (m_e) are derived from genotyping worker progeny of focal queen. "Modeled" indicates results of theoretical modeling and may not include discrete comparison of levels.

gain. By every measure, genetically diverse "swarms" out-performed their genetically uniform counterparts. Moreover, 50% of the genetically uniform colonies perished during a late August cold snap, and the remainder were all dead by December, whereas 25% of the genetically diverse group survived until the following May. Because large colonies, as measured by brood, bees, and food stores, produce more drones, survive winter better, and produce more swarms (Free and Racey, 1968; Lee and Winston, 1987; Seeley and Visscher, 1985), the results of this study are evidence that polyandry-enabled genetic diversity is adaptive at the level of colony.

Further evidence for the overarching relevance of a GV explanation for complex groups is the fact that a polyandrous honey bee queen uses all of her mates' sperm all of the time (Haberl and Tautz, 1998; Page and Metcalf, 1982). Any evidence that sperm were layered by mate within the spermatheca and metered out sequentially would undermine the premise that subfamily-based genetic variation works in ecologic time to promote colony stability.

So, we find the GV model taking its place alongside sperm limitation and genetic load on sex locus as imperatives for the evolution of polyandry.

And although compensating for diploid males technically belongs under the cloud of GV hypotheses, I treated it in the previous section because of its ancient importance to queen fitness and incipient group formation. But with that exception, the hypotheses under the GV umbrella speak forcefully to group complexification and environmental stabilization. For these reasons, I expect to see an oversized importance for sperm limitation and avoiding male diploidy early in the formation of hymenopteran social groups, whereas GV dynamics will predominate later, once groups are competing with other groups.

There is, to me, however, a shortcoming in our understanding of the GV model, and that is a general inattention to the processes by which multiple mating could have evolved. What explains the range of effective mating (m_e) values in table 9.1? Did these values arrive gradualistically, first one mate, then two, and so forth, with adaptive benefits accumulating along the way? Or did they arrive saltationally, by leaps? Gradualism has held the high ground for most of the history of evolutionary biology. "*La nature ne fait jamais des sauts*," Gottfried Leibniz famously wrote in 1704,[2] "nature never jumps."

But the persistent appearance of abrupt changes in the fossil record has renewed interest in saltational explanations for evolutionary transitions. Katsnelson et al. (2019) showed mathematically that gradualism is sufficient for explaining most evolutionary process but that saltational ("multi-mutational leap") changes may apply under conditions of stress-induced high mutagenesis. But a later model by Bakhtin et al. (2021) suggested that saltational processes happen more generally than thought, even under conditions expected to favor slow, gradualistic mutational changes.

Whether gradual or by saltations, the drivers of increasing mating numbers are poorly understood, a matter I take up in more detail in the "Capturing Unique Gene Compositions and Rare Specialist Alleles" section.

However relevant the GV model is for explaining fitness in complex eusocial honey bees (table 9.2), the jury is still out over how exactly genetic

2. Although Leibniz's manuscript was finished in 1704, it was not published until 1765 and is available today in an English 1996 edition; see references.

variation translates into colony-level benefits. Having discussed hypotheses on genetic load and sperm limitation in the "Queen Fitness and Polyandry" section, I spend the rest of this section summarizing five explanations put forward for polyandry's fitness benefits at the colony level. In my overview here, I focus on hypotheses considered plausible in the critical reviews of Crozier and Page (1985) and Crozier and Fjerdingstad (2001) or that seem particularly relevant in the case of social Hymenoptera. I add two that I think are warranted by the literature and deserve more experimental attention, one I call sociophysiological synergies and another that invokes rare specialist alleles and unique genotype compositions. Because the following are all variants of one GV concept, it's not surprising that some of them share points of overlap.

RESOLVING SEX ALLOCATION CONFLICT

Important changes in colony sex ratios were happening as polyandry and genetic diversity enlarged and complexified the colony, changes that led to the demographics we see in modern *Apis*—numbers dominated by female workers and very high ratios of drones to queens. In modern *A. mellifera*, roughly 20,000 males are reared for every queen (Page and Metcalf, 1984).

Early changes in colony sex ratios were driven by selection balancing the competing sex allocation interests of queens (preferring 1:1 females to males) against those of workers (preferring 3:1). The queen, passing along 50% of her genes with every egg she lays, is expected to comply with R. A. Fisher's (1930) classical prediction that sexually reproducing populations approach a 1:1 ratio of females: males. But the queen's daughters, as explained elsewhere, are predicted to favor rearing their supersisters over brothers by a ratio of 3:1. Thus, the sex allocation interests of queens and workers are at odds, with potentially negative consequences for colony cohesion and efficiency.

By practicing polyandry, a queen narrows the distance between genetic interests of herself and her workers, thus reducing conflict. As shown in equation 6, average relatedness of workers in a colony drops from $r_{\text{sisters}} = 0.75$ for a singly mated queen mother ($k = 1$) to $r_{\text{sisters}} = 0.29$ for a queen mated to $k = 12$ males. From the workers' perspective, this shifts the relatedness ratio with their mother's progeny from 3:1 to something much more Fisherian—0.29:0.25.

Robin Moritz (1985b) modeled a queen's risk of increasing mating flights against the benefits of reducing sex allocation conflict in her colony and arrived at near 1:1 agreement with polyandry levels approaching natural ranges of $k = 7–10$. So when it comes to the efficacy of polyandry at reducing sex allocation conflict, the evidence seems strong; however, the importance of eliminating sex allocation conflict in the early formation of social groups remains a theoretical point of argument (discussed in the "Sex Ratios, Split Sex Ratios, and Reproductive Value" section).

GENETIC POLYETHISM

Although I have alluded to it many times, the genetic polyethism hypothesis makes explicit that benefits of polyandry accrue from the actions of a more efficient workforce. The existence of specialization in workers, both genetic- and age-based, has been known for decades. The benefits of serial, as opposed to parallel, task performance; the existence of threshold-variant specialists; and the social feedbacks by which workers multiply or repress reactions in their sisters are all vehicles by which tasks are parsed in a colony and their execution optimized. The "Evolution of Worker Task Syndromes" section in chapter 8 provides a review of worker polyethism.

The hypothesis at hand here is that polyandry increases colony fitness by the inclusion of one or more subfamily-based genotypes whose task specializations increase colony fitness over that of colonies with lower levels of polyandry. Central to this is the idea of *behavioral dominance*—defined as the action of one or more genotypes that exerts colony-level fitness benefits beyond its proportional representation in the workforce (Crozier and Fjerdingstad, 2001). Included under behavioral dominance are non-additive benefits from subfamily-based specialists improving colony performance over colonies in the population lacking such specialists.

The colony-level fitness benefits of a mother's polyandry have been the focus of a string of papers summarized in table 9.2. But only some of them have characterized the degree of worker polyethism under a range of polyandrous conditions and associated that polyethism with colony fitness. Importantly, the general assumption that there is a positive correlation between mating number and worker specialisms is empirically supported. Delaplane et al. (2024) found a significantly higher number of practicing worker specialists in an experimental colony whose queen was inseminated with 32 males compared with a colony whose queen was mated to 16 males.

A few groups of investigators have focused on the relationship between a mother's mating number and expression of worker specialisms in a notable trait of immense colony-level importance—the honey bee's recruitment dance language (von Frisch and Chadwick, 1967). The dance language is not only attached to a critical fitness parameter—a colony's efficiency at acquiring food—but it's parceled into at least seven levels of participant: novice forager, scout, recruit, employed forager, unemployed experienced forager, inspector (temporary non-forager who spontaneously revisits a previous known site), and reactivated forager (Biesmeijer and de Vries, 2001). There are at least two levels of information—information going "out," which is information about the resource being communicated by a scout, and information coming "in," how that information is being received and processed by recruits. Finally, there may be related signals that have nothing directly to do with communicating information but instead alert potential recruits to the presence of a dance or stimulate the workforce to forage.

Exploiting this rich gold mine for specialization, Mattila et al. (2008) found that high-polyandry colonies, compared with their low-polyandry counterparts, hosted 36% more waggle dances daily with 62% more waggle circuits per dance, made food discoveries further from the nest, and produced 91% more stimulatory shaking signals, all of which added up to an overall 87% increase in colony foraging rate.

Mattila and Seeley (2010) focused on behavioral components of workers engaged in recruitment and foraging and found that "high-participating" subfamilies in high-polyandry colonies did not express lower action thresholds for dancing, such lower thresholds often being the criterion for identifying specialists. Rather, these high-participating patrilines simply engaged more in foraging behaviors, but curiously, once engaged, these workers were neither better foragers nor more inclined to engage in recruitment dances than other patrilines.

For these reasons, those authors resisted calling these patrilines specialists. Instead, they pointed out that in the case of a syndrome like foraging recruitment in which group-oriented behavior-amplifying non-additive dynamics are at work, it may be enough to simply have high participation. Large numbers of scouts, dancers, inspectors, and recruits cause geometrically increasing positive feedbacks that lead to higher overall colony foraging rates. At the other end of the spectrum, workers from the least-participating single-patriline test colony were "shockingly miserable dancers and mediocre foragers at best," underscoring the hazards for a queen who mates only once and unfortunately acquires an under-performing father for her offspring.

Another study by the same authors (Mattila and Seeley, 2011) gave more evidence that ecologically important foraging propensities parse at the level of subfamily. Among six high-polyandry colonies set up for this experiment, 21 of their combined patrilines qualified as either a forager-rich patriline or a scout-rich patriline, but only four qualified for both, suggesting that optimization of the syndrome depends on multiple patrilines.

If high levels of polyandry are associated with increased waggle dance signaling (Mattila et al., 2008; Mattila and Seeley, 2010), then Girard et al. (2011) have shown that the same is true for increased signal receiving. In multi-patriline colonies, 33% more workers followed a recruitment dance and for almost two additional waggle circuits beyond that expressed by workers in single-patriline colonies. Dance followers in multi-patriline colonies were 2.7 times more likely to leave the nest after watching a dance than were dance followers in single-patriline colonies. Those authors were able to show that the increase in dance following was not simply an artefact of the longer duration of dance signaling by other workers in the colony but rather its own independent behavioral cohort. A multi-patriline colony thus has a better chance of possessing the full repertoire of actors necessary to execute a complicated behavioral syndrome such as foraging recruitment.

Girard et al. (2011) speculated on the mechanisms that could drive a cohort of workers to excel at following dances and considered genetic predispositions, lower action thresholds for initiating dance following, faster response times to dances, greater attention spans, or higher probability of responding once a threshold is breached, most of which seem to me consistent with formal categories of behavioral specialization and could apply generally to other components of the recruitment syndrome discussed here.

The results of Eckholm et al. (2011) offer nuance to what otherwise seems a straightforward narrative. Focusing on pollen collecting and scouting, this team, predictably, found more pollen foraging in high-polyandry colonies. But curiously, they found more diverse plant taxa represented in the pollen of low-polyandry colonies. In other words, scouts of high-polyandry colonies were recruiting to a more taxonomically narrow range of food plants. Eckholm et al. (2011) invoked subfamily-based genetic specialization as a possible explanation, noting that in a genetically diverse colony, those patrilines with lower response thresholds will respond to cues faster, one result being that patrilines with higher thresholds may never perform a particular task. If "vacancies" for pollen scouting are filled up by one specialist subfamily in a diverse colony, that colony may actually field fewer pollen scouts, and given that each pollen scout is an independent actor, that colony by extension may discover a narrower taxonomic range of food plants.

It is worth noting that this scenario seems to contradict the conclusions of Mattila and Seeley (2011), who found quantitatively more scouts in high-polyandry colonies; however, Mattila and Seeley were focusing on nectar scouting. In any case, the argument of Eckholm et al. (2011) suggests more broadly that by monopolizing threshold-based tasks, specialist subfamilies in a genetically diverse colony may push other subfamilies into more diverse roles, the sum effects being higher colony-wide range of task performance and greater ecologic stability.

SOCIOPHYSIOLOGICAL SYNERGIES

Support for this hypothesis comes from Eckholm et al. (2015), who compared pollen foraging and nutrient dynamics in colonies whose queens were inseminated with either one male or 20. As known from earlier studies, high-polyandry colonies collect more pollen, but Eckholm et al. (2015) showed that nurse bees from high-polyandry colonies consume more pollen and have lower levels of midgut protease (an indicator of higher and more frequent protein consumption). This effect was mirrored in cage studies with same-aged cohorts: workers from high-polyandry brood cohorts consumed more pollen and had lower protease concentrations than workers from low-polyandry cohorts even though the two cohorts had access to equal amounts of pollen. This suggests an independent behavioral effect that stimulated pollen consumption in the polyandrous cages; as the cages equally lacked brood or incoming pollen foragers, the effect cannot be attributed to external cues normally associated with pollen consumption.

Eckholm et al. (2015) also directly sampled newly emerged bees from colonies of the two types and found, once again, that adults from polyandrous cohorts had lower protease concentrations. These results hint at legacy effects from different rearing conditions on the nutrient physiology of subsequent adults. Either scenario leans toward positive independent effects of polyandry on nutrient status of cohort members. There are numerous candidate interacting elements at work here—presence of high-participating pollen foragers, high propensity to consume pollen thus elevating protein demand and stimulating foragers, and effective distribution of protein to larvae resulting in healthier "hungrier" adults, to name a few.

HERD IMMUNITY

The term *herd immunity* describes the collateral protections afforded some members of a population when a critical mass of others in the population expresses resistance or immunity to a transmissible disease or parasite. It's almost effortless to extend this understanding to the teeming nestmates of a complex eusocial colony. But the themes we've covered in this book on evolutionary transitions along with the topic of this section, within-colony (intra-superorganismal) genetic diversity, should alert us that intra-genomic heterogeneity, or "hybrid vigor," may be a better metaphor. In any case, a large and growing base of literature has emerged showing that parasites and pathogens are among the most powerful selectors favoring polyandry in social insects (Liersch and Schmid-Hempel, 1998; Palmer and Oldroyd, 2000; Sherman et al., 1988; Shykoff and Schmid-Hempel, 1991), and a string of papers has shown a positive relationship between polyandry and a colony's resistance to brood diseases and parasitic *Varroa destructor* mites (table 9.2).

For those disease and parasite resistance phenotypes with a strong behavioral component, mechanisms for polyandry's benefit on herd immunity are likely indistinguishable from those we covered in the "Genetic Polyethism" subsection. This would, for example, be the case for so-called Varroa

Sensitive Hygiene—the heritable phenotype by which honey bees selectively detect and open brood cells and remove pupae infested by parasitic *Varroa destructor* mites (Harbo and Harris, 2009; Harris et al., 2010). This character is controlled by no fewer than five genes associated with olfaction, learning, and social behavior and one affiliated with circadian movement patterns that influence the probability of threshold-sensitive bees encountering infested brood cells (Oxley et al., 2010; Scannapieco et al., 2017; Spötter et al., 2016).

Tarpy (2003), however, foregrounded another mode of action that contributes to the adaptive benefits of polyandry in colony parasite resistance, and that's the principle we encountered in the "Genetic Load on Sex Locus" section—reducing variance around colony phenotypes (Gillespie, 1977). Rather than directly reducing parasite populations, polyandry reduces extreme expressions of parasite resistance, making a workforce that is more "average" relative to the local population (Page et al., 1995). Polyandry spreads in evolutionary time because colonies headed by monandrous queens are more likely to express maladaptive extremes, unmitigated by other compensating phenotypes, and thus die and remove their genes from the population. Counterintuitively, population saturation with extremely "good" phenotypes, i.e., zero parasite load, is not selected for because such phenotypes are too ecologically specialized or energetically costly to sustain.

CAPTURING UNIQUE GENE COMPOSITIONS AND RARE SPECIALIST ALLELES

This hypothesis pays attention to levels of polyandry seen in *A. mellifera* that appear both routinely high and inexplicable by other hypotheses. With average mating numbers up to $k = 44$ (table 9.1), the genus *Apis* expresses the highest levels of polyandry known in the Animal Kingdom.

The term "genetic diversity" when applied to polyandry implies a finite number of alleles in a breeding population and a range in their rate of capture across colonies. The number of *Apis mellifera* genes (the official gene set, or OGS) is estimated (Elsik et al., 2014) at 15,314, with the number of alleles, or variants, of studied loci ranging between seven and 30 per locus (Estoup et al., 1995). The distribution of those alleles, however, varies across populations according to local histories, founder effects, genetic drift, and inbreeding. The GV body of explanations for polyandry leads us to predict that natural selection will optimize a queen's chances of capturing as many alleles as possible in her local breeding population. It becomes a question of interest to ask how much diversity is "enough," or more precisely, toward what target fraction of allelic representation will selection drive polyandry?

Part of the answer comes from examining the mating biology of queens and drones, specifically the makeup of drone congregation areas (DCAs), which I cover in more detail in the next section. Baudry et al. (1998), working in Germany, showed that males participating in a DCA derive from virtually every colony within that DCA's recruitment perimeter. This means that a queen flying in a DCA has an equal probability of mating with a male from any colony in that population. In a separate study, Estoup et al. (1994), working in France, found that the distribution of paternal alleles among workers in a colony is statistically indistinguishable from that of the local population. Thus, it appears that natural selection has shaped polyandry (and male mating behavior) to achieve in any focal colony an unbiased representative sample of the local alleles. Honey bees, in fact, express among the highest rates of population allelic mixture known in nature—a state called, appropriately enough, *panmixia*.

For the rest of the answer, however, we need to focus on the variability in polyandry expression across colonies. For polyandry to be evolvable at the colony level, it must be heritable, variant in its expression across the population, and the target of positive selection. We discussed some of this in the "Sperm Limitation" subsection, where we covered the risks and costs constraining polyandry. But what about positive selectors? How many mates is "enough" under a GV perspective of explanations?

The alleles available to any focal colony are those established by their mother's genomic heterogeneity plus the allelic heterogeneity of the sperm in her spermatheca. These two reservoirs of alleles give us a handle for comparing within-nest genetic diversity among colonies—equation (6) which we met in Chapter 7 and which yields average percentage of alleles shared in common (r) between sisters and which I reproduce here:

$$r_{\text{sisters}} = 0.25 + \left(\frac{0.5}{k}\right)$$

where k = observed or actual mating number of their common mother, assuming that each male contributes equally to paternity. While yielding $r = 0.75$ for supersisters of one sire, the formula asymptotes to $r = 0.26$ after the 50th mating, showing that average r among workers in a colony can never drop below 0.25, a value fixed among sisters of one mother owing to rules of diploid inheritance.

Now bear in mind that equation 6 is very good at showing the decline in average relatedness among workers as their common mother increases her polyandry. But "decline in relatedness" is simply the flip side to "increase in diversity," for which I here introduce the term relative genetic diversity (RD):

$$\text{RD} = 1 - r_{\text{sisters}} \qquad \text{(equation 13)}$$

If r is the average fraction of alleles shared by familial inheritance between any two workers in a colony, then its complement $1 - r$ is the average fraction of alleles unique to one or the other—that is, their allelic divergence, or diversity. It is *relative* diversity because it is based (the "1" in equation 13) on the alleles present in their mother's genome and her spermatheca. But, importantly, RD scales positively with the queen's mating number (figure 9.3), which is intuitive given that a queen who mates with more males will capture more population alleles than one who mates with fewer. If genetic diversity is adaptive, then the GV hypotheses would lead us to predict increasing colony fitness with increasing RD.

Even though RD increases positively with mating number, the manner of increase is non-linear, with most increases happening in the first few matings and additional gains thereafter rapidly falling off. In figure 9.3, we see that over 90% of the increase in genetic diversity is accomplished by the tenth mating, and even 75% of it by the first. In short, any hypothesis for polyandry based on genetic diversity *per se*, runs out of explanatory power after the tenth mating.

Yet natural mating rates in excess of this GV asymptote happen all the time—a condition that I believe warrants its own qualitative label,

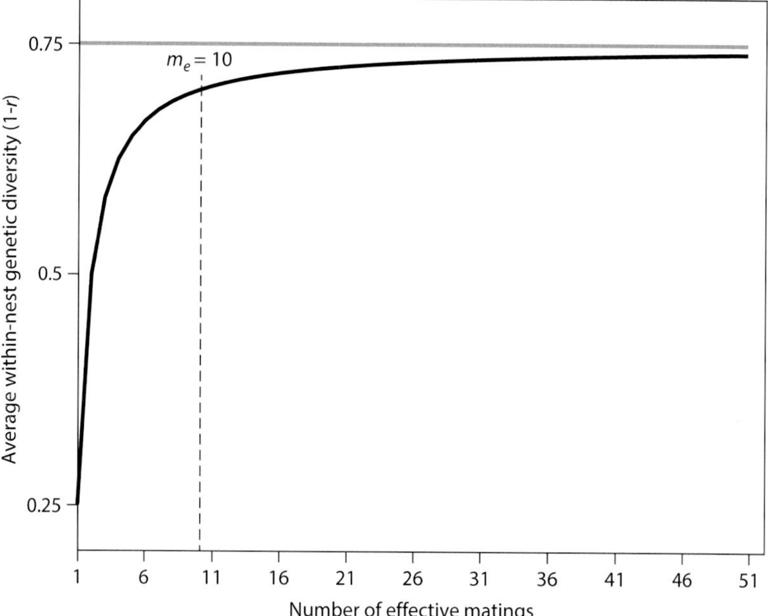

FIG. 9.3. A plot showing that over 90% of the increase in within-nest genetic diversity is accomplished by the queen's tenth mating. However, queens of *Apis mellifera* routinely breach this asymptote. This suggests that simple increase in genetic diversity cannot explain rates of polyandry as high as $k = 77$ (Withrow and Tarpy, 2018) reported in the literature.

hyperpolyandry. Queens of *Apis mellifera* routinely breach this asymptote with reports as high as $k = 77$ (Withrow and Tarpy, 2018). Given that mating flights are the riskiest interval in a queen's life, how does evolution produce such high levels of polyandry when marginal increases in benefit seems so small?

One answer could be that hyperpolyandry is not under positive selection but is an outcome of chance, a possibility put forward by Tarpy and Page (2000). The activity at drone congregation areas is chaotic and intense, with hundreds of drones jockeying for opportunity to mate with a flying queen. A queen at the right DCA at the right time may encounter and mate in rapid succession with tens or even scores of males within seconds—an outcome over which she has little control—whereas most other queens achieve no more than average mating numbers. But the idea that hyperpolyandry is selectively neutral is challenged by the many studies that seem to show colony-level adaptive improvements with increasing mating numbers (see table 9.2).

Another answer could be that favorable interactions among patrilines continue increasing at mating frequencies beyond the GV asymptote. Individual honey bee workers progress from interior brood- and nest-oriented tasks toward exterior guarding and foraging tasks in a predictable age-based sequence (discussed in chapter 8, "Evolution of Worker Task Syndromes" section). Subfamily-based specialists have been shown for several life-critical common tasks including guarding, undertaking, pollen collecting, nectar foraging, scouting, regulating nest temperature, and nursing larvae (Calderone and Page, 1991; Chapman et al., 2007; Jones et al., 2007). Optimizing ratios of these subfamily-based proclivities may be one adaptive outcome of hyperpolyandry according to a model by Fuchs and Moritz (1999).

Although average within-nest allele diversity as a function of mating number is capped by rules of hymenopteran inheritance (figure 9.3), the number of unique genotypic compositions is potentially infinite, with many compositions possible at any given value of average relatedness. Optimal compositions may continue increasing at mating frequencies above $k = 10$, and indeed it is in this zone that the model of Fuchs and Moritz (1999) predicts the highest fitness returns.

A second consideration of the model of Fuchs and Moritz (1999) applies to the inheritance of phenotypes that are rare and beneficial. Rarity in a highly potent beneficial allele is selected for because a colony dominated by specialists would be short on workers who perform common tasks. Rarity may also be an outcome of selection that balances costs of maintaining the special character against benefits realized by sustaining it (Boots and Begon, 1993; van Baalen, 1998). As a queen enlarges her sampling of drones, the allele frequency of her colony approaches that of the local breeding population. This averaging effect reduces the likelihood of environmental failures from phenotypic extremes (Page et al., 1995; Rueppell et al., 2008) and applies especially to rare beneficial alleles vulnerable to *cliff edge effects* (Boyce and Perrins, 1987; Mountford, 1968). A member of a population, a bee colony in our case, "falls off the cliff" if specialisms are so rare that it fails to procure any specialists at all (Fuchs and Moritz, 1999). A modern application of this scenario with *Apis mellifera* is the mite *Varroa destructor*, a parasitic mite of the eastern honey bee *A. cerana* moved by humans onto its non-natural host *A. mellifera* in the twentieth century. During its brief history with this parasite, few populations of *A. mellifera* have had time to evolve or acquire gene sets able to resist this novel threat. Hyperpolyandry can be seen as a ploy for avoiding such contingent ecologic catastrophes.

In 2021, my colleagues and I published a paper showing that colony fitness measures continue increasing with mating levels above the GV asymptote (Delaplane et al., 2021). We compared fitness measures in colonies whose queens were instrumentally inseminated with semen of nine or 54 drones. Drone sources derived equally from either a non-selected wild line or from a Varroa Sensitive Hygiene (VSH) line selected to resist *V. destructor*. Hygienic behavior directed at *V. destructor* is known to be uncommon, typically with no more than 3.5% of workers expressing some component of the syndrome (Spötter et al., 2016). We showed that survival rate of young brood increased in colonies whose queens were inseminated with 54 males regardless of genetic source. This result is consistent with the idea that novel genotype compositions for common task suites, brood nursing in this case, continue increasing above the GV asymptote with adaptive benefit to the colony. We also showed reduced colony parasite loads with queens inseminated with VSH drones, but only at the highest mating level tested, $k = 54$, in the hyperpolyandry zone.

We consider this result consistent with the model put forward by Fuchs and Moritz (1999): by practicing hyperpolyandry, queens capture rare specialist allele combinations, thus resisting contingent ecologic catastrophes. Such a source of genetic novelty may help *Apis mellifera* weather other human-induced stressors associated with intensified agriculture such as new pesticide chemistries (Tsvetkov et al., 2023) and nutritional disruptions associated with deteriorating foraging ranges (Naug, 2009).

In a follow-up study, Delaplane et al. (2024) were interested in showing how, or whether, inheritance dynamics change below and above the GV asymptote featured in figure 9.3. We hypothesized that if polyandry evolved in a gradualistic manner from two mates, to three and so forth, then heritable changes in ancient character states fundamental to eusociality would be sensitive to polyandry changes in this early range and mirror the curved line in figure 9.3 with benefits leveling off after about the tenth mating. We called this the population allele capture (PAC) model. Mating numbers above this asymptote, we reasoned from the model of Fuchs and Moritz (1999), would be the domain for resistance to rare, contingent environmental events responsive to rare specialized alleles or unique genotype compositions. Heritable changes in these phenotypes would vary with mating number in a linear manner unconstrained by the strictures of hymenopteran haplodiploid inheritance. We called this the genotype composition (GC) model.

We set up field colonies whose queens had been instrumentally inseminated with 1, 2, 4, 8, 16, or 32 males (constituting *observed* mating numbers or m_o), bracketing the qualitative divide of 10. Over five months of active season we recorded total colony brood area, a proxy for cooperative brood care, an ancient and fundamental character of eusociality, and colony levels of *Varroa destructor* mites, a proxy for an unevolved-for environmental contingency. Consistent with our predictions, increases in brood area increased with mating number mirroring figure 9.3, with increases leveling off at around $m_o = 16$. In contrast, colony growth rates of *V. destructor* declined with increasing mating number in a linear manner with no leveling off effect, consistent with the idea that a queen keeps capturing rare resistance alleles and unique, beneficial, and potentially infinite gene combinations even after she has captured most of the available alleles in her breeding population. Together, the PAC and GC models may explain the full range of mating numbers seen in *Apis mellifera*.

In the end, it is unlikely that the full picture will be this tidy. I point out that we have also shown improvement in the hyperpolyandry zone for another proxy of cooperative brood care, brood survival (Delaplane et al., 2021). Clearly, many gene sets or individual specialisms contribute to the ancestral syndrome "cooperative brood care," some of which are responsive to mating levels in the hyperpolyandrous zone. In contrast, our results for the contingent environmental stressor *Varroa destructor* have consistently shown increased mite mitigation with increasing mating numbers in the hyperpolyandrous zone (Delaplane et al., 2024, 2021, 2015).

Overall, I believe that the sum of evidence confirms that hyperpolyandry confers colony-level benefits and is adaptive. Alternative explanations are still viable: extreme mating levels may be responsive to sperm limitation (Simmons, 2002) or chance (Tarpy and Page, 2000). But it appears that hyperpolyandry increases acquisition of favorable genotype combinations for tasks both rare and common, ancient and derived, in a process that is both independent of within-colony genetic diversity and possibly even continuous.

Polyandry and Evolution of Modern Mating Behaviors

Whether considered from queen-focused or colony-focused explanations, selection imperatives for polyandry have left their indelible mark on modern mating syndromes—descriptions of which I take up now.

The theoretical arguments for the size complexity-hypothesis are compelling (chapter 10), and we cannot dismiss the importance of large sperm supplies in the social evolution of the honey bee. But the biology of modern *Apis* suggests that this problem was solved rather early: recall from a previous section that it's difficult to interpret modern *Apis* mating behaviors as adaptations for sperm scarcity. One could argue that avoiding diploid males was an even more primitive and compelling imperative selecting for early polyandry, and as I showed earlier, this problem was also resolved quickly with even very low expressions of polyandry. In either case, one can imagine a self-reinforcing iterative process of selection: polyandry selecting for increasing brood viability and spermathecal volumes leading to higher sperm retention in turn leading to larger populations, at which point spontaneous division of labor leads to improved task efficiency, which leads to greater colony stability and fitness. The point here is that relatively quickly, the colony becomes a viable level of selection in its own right with the level adapted being the mating behaviors of males and females.

The following events in the evolution of mating biology would have been contemporaneous with Apini's divergence from the other corbiculate tribes about 70–90 million years ago (see figure 1.3), while other features distinguishing simple from complex eusociality were being sorted out. For example, individual males can mate multiple times in simple eusocial taxa such as the Bombini, but a rising male skew in the complex eusocial Apini ("Resolving Sex Allocation Conflict" section) diminished opportunities for male multiple mating until selection veered away altogether from male traits that promoted multiple mating to male traits that promoted any mating at all (Boomsma et al., 2005). The result was short-lived males with fixed complements of sperm (in contrast, for example, to long-lived termite males who produce sperm continuously) and suicidal means for optimizing sperm delivery such as the explosive ejaculation of *Apis* males. Thus, the evolution of mating behavior occurs as an interacting syndrome with other basic features of social life history.

To a human observer, the earliest mating behavior adaptations under GV descriptions would have been indistinguishable from adaptations under sex load or sperm limitation. Mating efficiency, specifically sperm retention in the spermatheca, is famously inefficient in the Hymenoptera and bees in particular. Once caste divergence rendered mothers (technically, now "queens") at liberty to practice polyandry, there would have been rapid selection for increasing spermathecal volumes (Pabalan et al., 1996) to offset low sperm loads of males and morphological inefficiencies in sperm retention. At the same time, increasing polyandry and spermathecal volumes would have selected for male adaptations like increasing body size (Berg et al., 1997), robust flight morphology (Jaffé and Moritz, 2010), and increasing sperm loads (Crozier and Page, 1985, Kraus et al., 2004, Schlüns et al., 2003) coming together to improve performance at aerial mating and to increase one's representation in the female's spermatheca.

Aerial copulation (Koeniger and Koeniger, 1991), polyandry, surplus sperm production, and excretion of excess semen (Oldroyd et al., 1998) were all in place by the common ancestor of the *Apis*. But within the genus, members of subgenera have pursued divergent pathways of sexual selection (Oldroyd et al., 1997, 1995, 1996). There are striking differences in male morphology and sperm delivery mechanisms (Koeniger and Koeniger, 1991) (see table 9.1). The endophallus bulb of males of the small open nesters *Micrapis* is pointed, which lets males deliver sperm directly into the spermatheca, whereas the bulb of males of the cavity nesters *Apis* is large, and semen is delivered to the median oviducts from where it migrates to the spermatheca over the course of hours. The bulb of the large open nesters *Megapis* is intermediate, and sperm delivery is accordingly intermediate, with almost all sperm found in the spermathecas of newly mated queens and very few in the oviducts (Tan et al., 1999).

Moreover, a pronounced mucus plug is present in *Apis*, absent in *Micrapis*, and intermediate in *Megapis*. And finally, attachment *in copulo* is accomplished in males of the open-nesting *Micrapis* and *Megapis* with the help of forceps-like "thumbs" (*A. florea*) or adhesive hairs (*A. dorsata*) on the hind legs, whereas attachment in the cavity-nesting *Apis* is accomplished almost solely by the enlarged endophallus (Koeniger and Koeniger, 1991).

Although the "standard" *Apis* phylogeny of [*Micrapis*(*Megapis* + *Apis*)] (figure 9.4) is virtually unchallenged (Koeniger et al., 2011), investigators have differed in their interpretations of the evolutionary polarity of male characters that have gone into its construction. There's consensus that the leg clasping features of the *Micrapis* and *Megapis* males are derived (Alex-

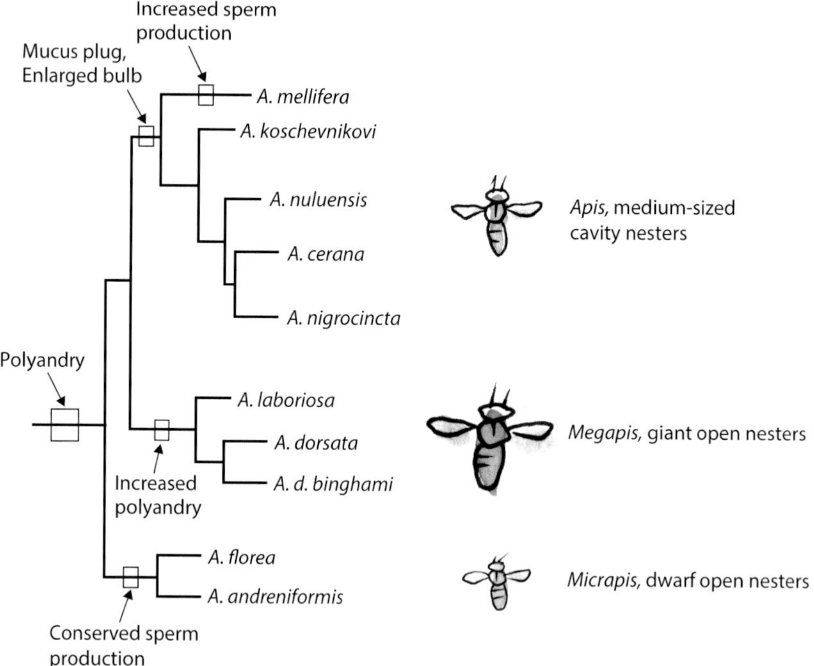

FIG. 9.4. *Apis* phylogeny of Raffiudin and Crozier (2007) superimposed with the reproductive character states of Oldroyd et al. (1998).

ander, 1991; Koeniger and Koeniger, 1991; Oldroyd et al., 1998), but when it comes to the enlarged penile bulb and mating sign, Oldroyd et al. (1998) interpreted them as unique innovations (autapomorphies) of subgenus *Apis*, whereas Koeniger and Koeniger (1991) saw them as ancestral. An overview of table 9.1 further suggests that extreme polyandry is an autapomorphy of *Megapis*, and high male sperm counts an autapomorphy of *A. mellifera*.

Oldroyd et al. (1998), in addressing this difficult set of evidence, summarized by saying that mating behavior in *Apis* has been subject to divergent sexual selection. Drones in cavity-nesting species have been selected to produce large amounts of sperm to increase their representation in their mate's progeny, whereas females have been selected to expel most of their mates' semen to gain genetically diverse colonies. In contrast, drones in open-nesting species increase their paternity by directly delivering sperm to the spermatheca, thus saving energetic costs of producing massive amounts of sperm.

Understanding such patterns of divergent sexual selection in *Apis* can be aided by considering the genus's biogeography, one scenario of which considers central Europe the geographic origin of the genus, with the Asian species products of separate expansions southeast from Europe and the *A. mellifera* line a uniquely western branch centered in Africa (Kotthoff et al., 2013). The geographic separation of the *A. mellifera* lineage from its nearest congeners (see figures 9.4 and 22.3) is consistent with its possession of autapomorphic high sperm loads in males. When it comes to the sympatric Asian species, divergent sexual selection may have fostered divergence of ecologic traits—an adaptive process that can help closely related populations coexist in similar habitats (Bonduriansky, 2011). In this manner, sexual selection may help explain the seasonal resource-based colony migrations common in *Micrapis* (Wongsiri et al., 1997) and *Megapis* (Dyer and Seeley, 1994) but rare in the cavity-nesting *Apis* (*Apis*). Flight energetics may have favored migratory drones invested in lightweight leg claspers over drones with heavy mucus glands (Rinderer et al., 1985) and enlarged bulbs.

Let's turn next to mode of copulation, which in all four corbiculate tribes is widely divergent with no discernable polarity for the character. In the Euglossini, basal with respect to all other corbiculates, we see mating on foot (Dressler, 1982); in the Apini, basal relative to the remainder, we see mating on wing (Winston, 1987), and within the following sister group we see mating on foot [Meliponini (de Camargo, 1972; Engels and Engels, 1988)] and wing and foot [(Bombini (Alford, 1975)] (figure 1.3). However difficult the polarity of this trait, mating on wing is a synapomorphy of the Apini (Koeniger and Koeniger, 1991).

Moreover in *A. mellifera*, aerial mating seems to take on forms adapted for exaggerating out mating and polyandry. In advance of the presence of flying females, an average of over 11,000 males (Koeniger et al., 2005) from distances ranging up to five kilometers (Ruttner and Ruttner, 1972) form mobile aggregations called drone congregation areas (DCAs) that patrol a circuit at a landscape scale of square kilometers. Because these aggregations tend to fly along permanent linear landmarks such as streams, roads, or forest margins, DCAs can be persistent over many years.

A genetic analysis of 13 DCAs across the natural range of *A. mellifera* found that the median number of colonies contributing to a DCA ranged from 2 to 26 (Jaffé et al., 2010), although numbers as high as 200–240 have been found in areas with high densities of managed colonies (Baudry et al., 1998). Such extremes of panmixia means that a queen has an equal chance of mating with a male from every colony in her recruitment perimeter. This perimeter includes her own brothers, but the exhaustive population sampling afforded by a DCA combined with her polyandrous habit means that the probability of a queen mating with her brother is no more than $p = 0.061$ (6.1%) for a queen mated 15 times. By extension, the average relatedness between a queen's mates one with another is predicted at $p = 0.0021$ (0.21%). Clearly, a queen's polyandry coupled with DCA genetic structure are strong antidotes to inbreeding depression.

Virgin queens fly into DCAs during their mating flights,[3] and when males detect a flying queen, around 30 of them will hotly pursue her in a comet-shaped formation, attracted by her silhouette against the sky (Menzel et al., 1991; Strang, 1970) and pheromone plume (Gary, 1962). It is males in these comets that accomplish mating, especially those in a space of about 2,000 cm^3 behind the queen (Koeniger et al., 2005). Mating lasts no longer

3. Although drones are responsive to visual and chemical cues of queens, there is no corresponding evidence that queens actively seek out DCAs.

than a few seconds. As described by Koeniger and Koeniger (1991), the male approaches the female from behind, grasping her abdomen. When she opens her sting chamber, he inserts his endophallus and with its first-stage eversion, he is paralyzed, stops flying, and falls backward. His mucus-engorged endophallus swells to fill the queen's sting chamber, at this point constituting the only connection between the mating pair. The queen, still flying, then squeezes the endophallus, an action which releases sperm into her median oviducts. In fast succession, the progressing eversion breaks the contact between the pair, and the male falls to the ground and dies. Parts of his endophallus, including congealed mucus, chitin plates, and a conspicuous orange coloration produced by cornual glands (Koeniger, 1988), remain in the queen's sting chamber.

This "mating sign" poses no obstacle to subsequent inseminations; Woyke (2011) has described in detail how the endophallus of a succeeding male enters the queen's sting chamber beneath the mating sign of the previous male, pulling it out in the process of its own expansion.

In fact, far from an impediment, a previous male's mating sign may be an active attractant and facilitator for subsequent mates. Gudrun Koeniger (1990), using physical queen models and live drones, found evidence that the mating sign provides visual cues that help subsequent males identify and mount flying queens. If so, then this must be interpreted as a derived form of a more primitive state. Ancestrally, mating plugs are obstructions to subsequent males—a ploy for monopolizing mating (Boomsma et al., 2005) as known for fire ants (Mikheyev, 2003) and bumble bees (Baer et al., 2001; Sauter et al., 2001). The mating plug's curious reversal in *Apis* from deterrent to facilitator of subsequent matings is a male adaptation for both exploiting and facilitating benefits of within-colony genetic variation. To the extent that within-colony genetic diversity increases colony ecologic fitness (discussed in the previous section), it increases a drone's fitness for his progeny to be nurtured in such a colony.

This may explain the apparent lack of male-on-male aggression (Koeniger, 1990; Koeniger et al., 2005) in *Apis* DCAs, although males do jockey for prime positions in the mating comet where advantage goes to the fastest and strongest flyers (Jaffé and Moritz, 2010; Koeniger et al., 2005). It is also a ready explanation for the fact that no last-male paternity advantage has been shown in honey bees (Franck et al., 2002; Moritz, 1986) as is known in other insects. Gudrun Koeniger (1990) summed it up this way: ". . . after competing with other drones to grasp and mate with the queen, a drone seems to cooperate 'post mortem' with his successor by marking the queen with a conspicuous colour in order to reduce the mating flight time and to ensure the production of a large colony."

In general, the DCA is a brilliant adaptation for maximizing polyandry and minimizing risks of multiple matings. The valuable queen must undergo several minutes of risky flight, exposing herself to predators and weather. The DCA mitigates these risks by concentrating in time and place a maximum pool of drones from virtually all colonies within the local breeding population. Mating can be rapid, intense, and chaotic. During my graduate school days while working with suspended pheromone traps at a DCA near the Baton Rouge bee lab, I could hear the air above me crackling with the sound of ejaculating drones. The mysterious permanence and charisma of DCAs led Karl Showler, a prominent beekeeper and author, to make a life hobby of tracking DCAs in the English and Welsh countrysides. Karl once suggested to me that the ephemeral flash and audible woosh of DCAs swirling around dolmens and Neolithic landmarks of the British landscape were the sensate inspirations for leprechauns, fairies, sprites, and other creatures of folklore.

When it comes to females, there is reason to think that mating evolution will involve some kind of optimization threshold informing a queen when she has acquired "enough" sperm. A queen experiences strong fitness gains if she can collect a large supply of sperm to sustain a large colony capable of producing new reproductives. Balancing this, however, is substantial risk from mating flights. The queen has voluntary control over this process on two fronts: the number and duration of her mating flights and the number of her copulations.[4] Of the two, flights are the riskier behavior because once a queen is airborne, any subsequent copulations will be comparatively low-risk. Therefore, we predict that natural selection will act to reduce flight numbers while maximizing the number of copulations. Numerous field investigators have studied the relationships among the number of

4. A flying queen receptive to mating voluntarily opens her sting chamber as she flies through a drone congregation area.

a queen's mating flights, number of matings, and number of sperm in her oviducts or spermatheca. These studies require hours of observation at entrances of hives fitted with devices that let investigators observe the queen exiting or entering or to trap her or restrict her as needed. Mating number is determined by genotyping her subsequent worker brood.

Schlüns et al. (2005), working in Berlin, monitored queens returning from a single mating flight and restrained some of them when they were attempting subsequent flights. Other queens were permitted to take as many flights as they wanted. Queens that were restricted to one flight but tried to take more had significantly fewer inseminations than free-flying queens. This sounds predictable, but the interesting detail is that some queens never attempted a second flight, vying instead to immediately begin laying eggs. These early egg-layers also had more inseminations than the queens that were flight-restrained, even though both groups had only one mating flight. These results suggest that queens are responding to some in-flight signal, not a programmed number of mating flights. As soon as the threshold is achieved, queens cease mating flights and begin laying eggs.

Support for this idea came from Koeniger and Koeniger (2007), who found a negative correlation between flight duration and number of sperm in the spermatheca. The longer the flight, the fewer the sperm. This seems counterintuitive unless queens are indeed self-monitoring their mating success and stop flying once they've collected enough sperm. Efficient flights would, by extension, be brief, and inefficient flights long. But either pole implies a criterion and its associated cue for what constitutes "enough" matings. During mating flights of *A. mellifera*, drones deposit semen into the queen's oviducts, and once the queen is back at the nest, it takes more than 24 hours for sperm to migrate to the spermatheca (Woyke, 1988). Thus, it is not likely that the in-flight cue is fullness of the spermatheca, but rather the number of copulations or fullness of the oviducts.

Across bees of the genus *Apis*, up to 97% of a drone's semen is lost after mating (Koeniger and Koeniger, 2000). After her numerous mating flights, a typical *Apis mellifera* queen's spermatheca contains 4.7–5.0 million sperm (see table 9.1), which is enough to supply her with a lifetime's worth of workers yet virtually no more sperm than that delivered by even one male (4.6–5.7 million, table 9.1). Jerzy Woyke (1960), moreover, showed that if a queen is instrumentally inseminated with only one male, her spermatheca will realize no more than 20% of its holding capacity. Thus, it appears that a queen must accumulate a certain number of (inefficient) matings in order to achieve a full spermatheca. A rough estimate with Woyke's figure would suggest that that number is at least five.

The cues that inform a queen that she has met this threshold are most likely the number of matings she has counted or the fullness of her oviducts. Evidence exists that queens do (Schlüns et al., 2005) or do not (Simone-Finstrom and Tarpy, 2018) adjust their number of flights to their mating success, so I lean toward stretch receptors in the oviducts, given that this cue would be direct and persistent at a scale of minutes or hours and not demand that queens have the cognitive capacity for counting their mates. Moreover, injecting a queen with 10-μL volumes of either semen or saline in her oviducts causes her to cease mating flights, suggesting that queens are indeed responsive to stretch receptors as a behavioral trigger (Kocher et al., 2010).

Whatever the threshold signals, there is evidence that queens adjust their threshold expectations according to drone availability. When Koeniger and Koeniger (2007) restricted the number of drones available for mating in an isolated valley in the Austrian Alps, they found longer average queen flight times compared with an earlier study in the same valley when drone numbers were nearly four times larger (Koeniger et al., 1989a). Perhaps early encounters with plentiful drones, or shortages thereof, calibrate a queen's expectations and help her judge her number and duration of flights. Learned pessimism would be consistent with the negative correlative relationship described above between flight time and sperm number. A queen in a drone-impoverished habitat will be compelled to take more and longer flights; apprehending that her sperm acquisition rate is unlikely to improve she adjusts her threshold downward and upon meeting it, she stops flying. What does *not* happen is a positive correlative relationship—an open-ended pursuit of more males. Clearly, the queen is constrained against infinite mating flights. The subtitle of Koeniger's and Koeniger's (2007) study on mating flight duration seems to capture the strategy: "as short as possible, as long as necessary."

In summary, we see in both males and females complementary adaptations that accrue toward minimizing mating flights, maximizing copulations, and encouraging within-nest genetic diversity. The male DCA concentrates in time and place the number and genetic diversity of males available for mating; the persistent inefficiency of sperm retention in the *A. mellifera* queen leads to diverse genetic representation in her spermatheca, and her threshold-sensitive mating strategies optimize sperm acquisition per flight. For a genus that must handle environmental extremes—whether as a migrator or perennial occupant of a stationary nest—polyandry and its complementary traits has been a winning ecologic strategy. More broadly, it has helped propel the honey bee lineage into complex eusociality.

Summary

Polyandry is the queen's habit of mating with multiple males during a brief interval in her young life, storing their sperm, and using it to fertilize her life's retinue of daughter workers. It's the minority condition across the eusocial Hymenoptera, yet it is widely represented across social groups and universal in the honey bees Apini, where it reaches its highest known expression. Polyandry, along with its complement polygyny, practiced among certain ants, is a strategy for increasing within-nest genetic diversity of the workforce, the result being improved ecological competitiveness at the level of colony. For the worker, however, her mother's switch from ancestral monandry to polyandry means that the worker's genetic reward for staying at the nest and helping rear siblings falls from $r_{sibs} = 0.5$ to $r_{sibs} = 0.27$. For this reason, polyandry could not have evolved prior to the morphological divergence of worker and queen castes, a state that rendered the two mutually co-dependent—an event signaling an evolutionary transition from organism to superorganism. While workers' genetic fortunes have plummeted, the queen's direct fitness has suffered no setback because every egg she lays still passes on $r = 0.5$ of her genes even though her daughters are now the progeny of numerous sires.

In this chapter, we discussed seven hypotheses that attempt to explain the adaptive benefits of polyandry. The first two focus on adaptive benefits to the queen. The genetic load hypothesis notes that individuals that are homozygous at the sex locus become diploid males, and these individuals are killed by workers and constitute opportunity costs to the colony. A female mated to one closely related male risks burdening her progeny with homozygosity at this and other loci across their genomes. Her promiscuity is predicted to offset these risks. The sperm limitation hypothesis notes that there are positive relationships among the number of sperm a queen collects, the queen's reproductive longevity, the number of fertilized worker eggs she produces, the resulting population size of her colony, and the colony's performance at rearing her sons and reproductive daughters. The rest of the hypotheses emphasize colony-level benefits.

The sex allocation resolution hypothesis notes that whereas the queen is predicted to prefer a 1:1 population ratio of females to males, her supersister daughters will prefer a 3:1 female to male ratio. By practicing polyandry, a queen narrows the distance between genetic interests of herself and her workers, thus reducing conflict. The genetic polyethism hypothesis proposes that polyandry increases colony fitness by the inclusion of one or more subfamily-based genotypes whose task specializations increase colony fitness over that of colonies with lower levels of polyandry. A hypothesis focusing on sociophysiological synergies focuses on evidence that highly polyandrous worker cohorts have social interactions that alter worker nutrient physiologies and promote task efficiencies. The herd immunity hypothesis seeks to describe collateral protections afforded some members of a population when a critical mass of others in the population expresses subfamily-based resistance or immunity to a transmissible disease or parasite. A final hypothesis interprets polyandry as a female's bid to capture a limited number of unique gene compositions and rare specialist alleles that confer colony fitness.

The mating behaviors of males and females show strong signs of adaptation to promote high within-colony genetic diversity. Large flying cohorts of males, called drone congregation areas (DCAs), concentrate in time and place males from virtually all colonies within a queen's mating perimeter. A young queen flies through a DCA and rapidly mates with 10 or more drones. There is evidence that queens self-monitor their mating success and take more or fewer flights to achieve a local mating optimum.

PART II

Emergent Properties and Outcomes of Social Evolution

CHAPTER 10

Emergent Properties and the Honey Bee Superorganism

FOR SCIENTISTS, THE SUPERORGANISM concept is not a theory; it is a place-keeper—a milepost along a gradient of increasing biological complexity. There is a vector in evolution toward lower levels of biological organization being subsumed, clumped, and incorporated into higher levels of organization. Millions of years ago, a daughter bee rejected independent life in preference to staying at the nest and helping her mother, thereby subsuming her genetic interests to the group and setting in motion the events that led to the teeming colonies of social bees we see today. Far more anciently, a single-celled microbe became incorporated into another microbe, accommodating its own genetic interests to its host and setting in motion the events that led to the modern-day mitochondria, quasi-alien entities that still live in the cells of eukaryotes such as you and me.

Superorganisms mirror these clumping processes, subsuming individuals and their individual genetic interests into a higher level of organization—the colony—that does the sort of things that organisms do. It grows, it acquires resources from the habitat, it thinks, it digests, it defends, it reproduces, and it dies. It was put this way by Queller and Strassmann (2009): "The evolution of organismality is a social process. All organisms originated from groups of simpler units that now show high cooperation among the parts and are nearly free of conflicts. We suggest that this near-unanimous cooperation be taken as the defining trait of organism."

There is another vector in biological history I've mentioned a few times in Part 1, and that's a tendency for complexification to march in tandem with increasing group size, with "size" meaning cell numbers in an organism or worker numbers in a superorganism. Bourke (2011, pp. 162–197) devoted an entire chapter to what he calls the *size-complexity hypothesis*, which essentially states that (1) external ecological and evolutionary drivers push social groups toward increasing size; (2) increasing size selects for social traits that distinguish simple from complex sociality (table 5.3); and (3) these evolved traits further increase group size via positive feedbacks of self-reinforcing social evolution (figure 10.1).

One of these positive feedbacks is *emergent properties*—a phenomenon long recognized by information theorists, physicists, and evolutionary biologists. A product of increasing group size, emergent properties describe the emergence of order out of pre-existing physical constraints and the action of independent actors, each responding to its immediate local conditions (Holland, 2014). A feature of emergence is that it describes properties of a group that cannot be explained by the sum of the group's parts. Holland

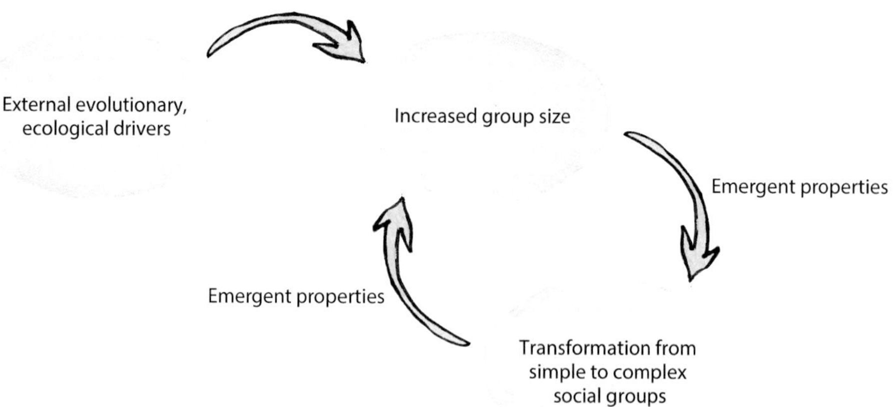

FIG. 10.1. The size-complexity hypothesis. Redrawn from figure 6.1 in Bourke (2011).

(2014, p. 49) noted that the "wetness" of water is not obtained by summing the wetness of H_2O molecules; it is obtained by the *interactions* of the H_2O molecules. Mayr (1982, p. 62) stated, "Systems almost always have the peculiarity that the characteristics of the whole cannot (not even in theory) be deduced from the most complete knowledge of the components, taken separately or in other partial combinations. This appearance of new characteristics in wholes has been designated as *emergence*."

The pessimism Mayr expressed in 1982 is not unfounded, but neither is it indomitable against advances in knowledge and computing power. Predictability is conceivable if one has defined a set of variables and captured enough information on their non-linear relationships across multiple scales. But it is to be granted that capturing those dynamics above or below the scale of interest is daunting. Nevertheless, successes have been won in predicting emergent properties, especially at the microbial scale (Gilbert and Henry, 2015).

It remains that it is easier to statistically describe emergence and its components than to predict it. To that end, I want to familiarize readers with some of the leading descriptive models on emergent properties seen in biological systems and the honey bee superorganism. In this, I draw heavily from Ponge (2005) and Holland (2014).

Two conditions must be met in order for group properties to emerge that eclipse the sum of the components: (1) the components themselves must share common properties and connectedness, and (2) sufficient matter must be concentrated in time and space to effect a permanent outcome. Space- or time-restricted phenomena are amplified to the point that chaos passes into order. Consider an infection. The pathogen is the agent acting at the scale of individual host cell. Disease is the property that emerges once a critical threshold of pathogen is reached in the body of the host. Or to use a honey bee example, consider cold-weather social thermoregulation. Clustering bees, each seeking her physiological optimum temperature zone in the cluster, are the agents acting at the scale of nest cavity. Survivable group temperature is the property that emerges once a critical density of clustering bees is reached.

Note that both of these examples involve non-linear effects. The host doesn't get sicker with each additional bacterium; the cluster doesn't survive better with each additional bee. Both emergent outcomes are either/or states—the host gets sick, the cluster survives only once a critical threshold of the agent is breached. This is why additive effects are not emergent. Many group properties—volume, size, population, weight—*are* equal to the sum of the parts. It is the non-additive interaction of parts that gives us the emergence of group properties that cannot be achieved by summation.

Non-linear processes that produce emergence involve a tension between forces that amplify the material phenomena and forces that cohere, contain, and stabilize them. In biology, these processes fall into three descriptive models, as summarized by Ponge (2005).

The "bubble" model describes phenomena in which the most important property is a strong boundary or "skin." This boundary delimits the system and is the point of interface with other systems. In organisms, the boundary is, in fact, the organism's physical skin. But bubbles exist also as territories in which boundaries are maintained by signals, whether pheromone, audible, optical, tactical, or electrical; or as ecologic zones with boundaries maintained by forest canopies, margins, or topographic features. The common feature of a bubble boundary is that it acts as a filter with other systems and its integrity is essential for the persistence of the system.

In essence, a bubble system expresses material amplification from internal expansion/reaction forces, cohesion via strong connectedness between components of the boundary, and action/reaction forces between the inside and outside mediated by the boundary. If bubbles of compatible types come in contact, the result is fusion. We have encountered this in our earlier discussions of evolutionary transitions, such as when early prokaryotes fused to result in the modern mitochondria residing in eukaryotic cells (table 2.1). A bubble system may describe an entire honey bee superorganism, nest and all, but its features are also discernible in finer scale emergent properties such as size of the honey bee colony worker population (chapter 14) and the size, coherence, and occasional fusion of reproductive swarms (chapter 16).

The "wave" model describes patterns in biology that result from periodic, cyclic processes that provoke a chain reaction in connected components. Unlike bubbles, wave phenomena are not delimited by a boundary and therefore propagate themselves in time and space. A good example of a wave phenomenon is movement of cars in a traffic jam. When the car in front of you moves forward, there is a delay before your car moves forward. This delay is the sum of time required to accommodate your sense organs, your own learned caution, nerve transmissions from your brain to muscles, and the mechanical inertia of your car. Viewed from above, what one sees is a wave of alternately moving and stopping cars.

The movement of a column of foraging ants can be described by wave dynamics. So can the propagation of a signal of virtually any kind that passes through an organism or animal group. An example from the honey bee superorganism is the dance seen in open-nesting *Apis dorsata* by which bees flip their abdomens in a concentric wave that wards off predatory hornets (Kastberger et al., 2008). Another example is the concentric banding patterns of nectar, pollen, and brood seen on beeswax combs regulated by the variance in rates of pollen acquisition and consumption (chapter 12). In all these examples, the wave is the emergent property; the interaction of components (cars or bees flipping abdomens) is the cohesion force that sustains the wave, and the amplifying agent is the changing density of matter in the system (density of cars, frequency of wasp threat, or rate of incoming pollen).

The "crystal" model is similar to the wave model in that repetitiveness builds the system and there is no boundary, but unlike waves, the connections between components are rigid, not flexible. An initiating action creates a template upon which synergizing reactions build up more matter, rapidly changing chaos into order. A feature of a crystal system is that any part exhibits the same properties as the whole. When crystals come in contact, attraction/repulsion forces may generate either fusion or regular spacing patterns.

Examples in the honey bee superorganism are the regular hexagonal cells that together comprise contiguous combs (chapter 12). Another example is the tight assembly of identical components such as cells in a membrane or skin. Another less intuitive example comes from the trial-and-error processes by which individuals optimize their position in a group, resulting in regular spatial distributions, whether it's individual bees flying in a swarm or natural colonies across a landscape. In crystal dynamics, the resulting matrix (comb, skin, swarm, and colony spatial distribution) is the emerging property; the cohesion forces are those that sustain the individual component, and the amplifying agent is imported matter (additional units) that is built upon the template (the worker's first beeswax deposit, the first dividing cell, or the first pioneering colony).

It is perhaps obvious at this point that in nature these three models describe co-occurring, non-mutually-excluding, and nested phenomena. One honey bee colony/superorganism constitutes a bubble; its combs are crystalline, but their contents are organized by waves. Flying bees in a reproductive swarm orient themselves in a crystalline pattern, and the colony they found complies with a crystalline pattern with other colonies across the landscape. And when winter comes, the clustering bees contract and expand according to waves imposed by ambient temperatures and the changing physiological temperature optima of individuals.

In all these examples, one sees that emergent properties cannot be obtained from unities. There is a pervading need for a critical number of independent *actors*, which most often in the honey bee means worker bees. Hence, the positive feedbacks between emergent properties, large group size, and increasing social and superorganismal complexity (see figure 10.1). It can become difficult to distinguish a *cause* of complex order from an *effect* of complex order.

In the honey bee, the constraints of cavity size and type, seasonal availability of food, worker physiological state, temperature, gravity, and tens of thousands of independent decisions made by workers every minute of every day have produced numerous examples of emergent properties, some of which are hallmarks of honey bee life. These emergent properties are not direct products of natural selection as much as they are outcomes of evolved adaptations that in turn become scaffolding for natural selection to continue adapting the system.

In Part 2, I highlight several examples of emergence which have become icons of complex eusociality as expressed in the honey bee. The research stories behind these topical chapters often highlight the elegance of *parsimony*—a scientific bias toward explanations that require the fewest assumptions and agents. We humans are by no means inclined to arrive at the simplest explanation for anything, even if we think our answer is the simplest and most direct. The human mind, always hungry to detect order and quick to see human properties like volition, intentionality, and forward planning everywhere in the natural world, is slow to accept answers based on explanations as mundane as gravity, temperature, and the random actions of individuals. Part of the scientific mindset is recognizing this bias and its tendency to blind us to more direct and testable explanations. An understanding of emergence is a healthy antidote to this vulnerability. Perhaps it was something like this that the twentieth-century satirist H. L. Mencken (1982, p. 443) had in mind when he wrote, “Explanations exist; they have existed for all times, for there is always an easy solution to every human problem—neat, plausible, and wrong.”

Summary

The size-complexity hypothesis in biology states that (1) external ecological and evolutionary drivers push social groups toward increasing size; (2) increasing size selects for social traits that distinguish simple from complex sociality; and (3) these evolved traits further increase group size via positive feedbacks. One of these positive feedbacks is emergent properties—the emergence of order out of pre-existing physical constraints and the action of independent actors, each responding to its immediate local conditions. A feature of emergence is that it describes properties of a group that cannot be explained by the sum of the group’s parts. Two conditions are necessary: (1) the components themselves must share common properties and connectedness, and (2) sufficient matter must be concentrated in time and space to effect a permanent outcome.

Consider an infection. The pathogen is the agent acting at the scale of individual host cell. Disease is the property that emerges once a critical threshold of pathogen is reached in the body of the host. Crucially, this example involves non-linear effects. The host doesn’t get sicker with each additional bacterium. The emergent outcome “disease” is an either/or state. The host gets sick only after a critical threshold of the agent is breached. This is why additive effects are not emergent. Many group properties—volume, size, population, weight—are equal to the sum of the parts. It is the non-additive interaction of parts which gives the emergence of group properties that cannot be achieved by summation.

Non-linear processes that produce emergence involve a tension between forces that amplify the material phenomena and forces that cohere, contain, and stabilize them. In biology, these processes fall into three models. The "bubble" model describes phenomena in which the most important property is a strong boundary. This boundary delimits the system and is the point of interface with other systems. The "wave" model describes patterns in biology that result from periodic, cyclic processes that provoke a chain reaction in connected components. Unlike bubbles, wave phenomena are not delimited by a boundary and therefore propagate themselves in time and space. A good example is movement of cars in a traffic jam. The "crystal" model is similar to the wave model in that repetitiveness builds the system and there is no boundary, but unlike waves, the connections between components are rigid not flexible. An initiating action creates a template upon which synergizing reactions build up more matter, rapidly changing chaos into order. A feature of a crystal system is that any part exhibits the same properties as the whole. An example in the honey bee superorganism are the regular hexagonal cells.

These three models describe co-occurring, non-mutually-excluding, and nested phenomena. One honey bee colony/superorganism constitutes a bubble; its combs are crystalline, but their contents are organized by waves. In all these examples, one sees that emergent properties cannot be obtained from unities.

CHAPTER 11

Thermoregulation

THE VAST NATURAL RANGE of *Apis mellifera* extending from southern Africa to the subarctic would not be possible without the species's extraordinary resilience to temperature extremes. This resilience must be achieved as a group owing to the commitment of *A. mellifera* to year-round eusociality. Honey bees can no longer "opt out" of group living and retreat to individual winter hibernation or summer estivation of the kind practiced by solitary insects. Given that solitary cold and heat resistances are ancestral, we can expect to see elements of them conserved in the honey bee superorganism along with more recent innovations bearing the stamp of social selection.

Different authors have placed the geographic origin of the genus *Apis* in Europe (Kotthoff et al., 2013) or in Asia (Han et al., 2012; Ji, 2021; Ruttner, 1988) of the late Oligocene to middle Miocene, a period in Earth's history coinciding with a shift from generally warm conditions to glacial (Liu et al., 2009). The geographic origin of the modern species *Apis mellifera* is variously defended as either tropical Africa (Whitfield et al., 2006), western Asia (Dogantzis et al., 2021), northern Africa (Tihelka et al., 2020), or northern Europe (Carr, 2023). The divergence dates of numerous European sister subspecies follow the onset of glaciation in the Northern Hemisphere around 3 million years ago (Dogantzis et al., 2021; Livina et al., 2011). The general drift of the evidence points to a tropically evolved bee adapting for cold-weather survivability.

An organism whose body temperature is more or less dependent on ambient conditions is called cold-blooded or *poikilothermic* ("*poikilos*" in Greek = "changing"), whereas an organism capable of maintaining its core temperature across a range of conditions is called warm-blooded or *homeothermic* ("*homois*" = "same"). Additional terms describe special conditions or where the heat comes from: a *heterothermic* animal can go from an unregulated (cold-blooded) state to a regulated (warm-blooded) state as activity requires; an *ectothermic* animal gets its heat from external sources, as does a turtle basking in the sun, and an *endothermic* animal gets its heat from internal sources like shivering muscles. In truth, most animals move across these categories with a fluidity that defies labels, and none more so than the honey bee, which at the level of individual is a cold-blooded ectotherm that nevertheless shivers, making it an endotherm, whereas at the level of superorganism, it is an outright homeotherm.

The Importance of Nest Site

No innovation is more important than cavity nesting when it comes to the expansion of *Apis* outside of the tropics. The most recent common ancestor of *Apis* was already a superorganism by the time the genus emerged *circa* 30 million years ago in Southeast Asia (figure 1.3). The oldest lineages of *Apis* built single exposed combs (figure 4.6) as do the modern Asian *A. florea*, *A. dorsata*, *A. binghami*, *A. laboriosa*, and *A. andreniformis* today. However, some ancient *Apis* adapted to the novelty of nesting inside cavities, diverging from their open-nesting sister group about 7–25 million years ago (Dogantzis et al., 2021; Smith, 2021). It seems no great surprise that it was cavity-nesting *Apis* that expanded in temperate zones and eventually gave rise to the western honey bee *A. mellifera*. In other words, cavity nesting was a pre-adaptation for temperature extremes. By the time *A. mellifera* successfully pioneered the cold latitudes of Europe, it had evolved other behavioral adaptations beyond nest choice that rendered it as homeothermic as the fire-wielding hominins (Roebroeks and Villa, 2011) who followed it a few hundreds of thousands of years later out of Africa.

Bees use their well-known symbolic dance language (chapter 15) not only for group foraging decisions but for what is arguably the most important decision of all—choice of nest site (Seeley, 1977; Seeley and Morse, 1978). Tree hollows are the preferred choice, and it is not necessary for the tree to be alive as long as the cavity walls are thick and sturdy. As part of the annual reproductive swarming cycle, scout bees inspect candidate cavities for sufficient volume and reject those smaller than 20 liters. An adequately large volume is necessary for storing food and housing a critical mass of workers sufficient for group clustering and heat generation. Bees prefer cavities whose entrances are small, defensible, and (in the northern hemisphere) south-facing. Entrances facing south are probably warmer than alternative orientations, and an entrance below the clustering bees minimizes the loss of rising warm air. Once a new swarm occupies a cavity, the bees immediately set to work plastering shut any unnecessary openings with *propolis*—a resinous material collected from tree sap. Then they begin constructing their parallel interior beeswax combs, which contribute to the insulating properties of the total nest.

Although a tree hollow's insulating properties can be advantageous for either temperature extreme, the benefits of cavity nesting in warm climates are mainly limited to protection from parasites, predators both vertebrate and invertebrate, and nectar-thieving honey bees from neighboring colonies. Clearly, the effect size of a good cavity is stronger in cold climate conditions, and this single adaptation was the crucial springboard allowing *Apis* to spread into temperate zones.

Warming the Nest

In insects generally, overwintering species are lumped into one of two groups—those that avoid freezing and those that tolerate it.

Freeze avoidance is expressed on a behavioral scale as complex as seasonal migrations (Urquhart and Urquhart, 1978) or as simple as movement by an individual along a temperature gradient. Group migrations of any kind, whether temperature- or resource-driven, are unknown in temperate-evolved

Apis mellifera, but we will see below that freeze avoidance at the scale of an individual's local movement plays a crucial role in group survival.

Freeze tolerance in insects is driven by physiological processes involving synthesis of ice nucleating agents, cryoprotectants, antifreeze proteins, and changes in membrane lipids (Bale and Hayward, 2010). In the case of temperate honey bees, two physiologically distinctive cohorts of workers are produced over the course of a year: a cohort of short-lived (25–40 days) summer bees and a cohort of long-lived (> 250 days) winter bees [reviewed in Knoll et al. (2020)]. Notably, there is no such bimodal longevity distribution in tropical-adapted *Apis mellifera scutellata* (Winston, 1980b; Winston and Katz, 1981).

Winter bees have lower titers of juvenile hormone, larger fat bodies, and higher blood proteins, particularly of vitellogenin, a molecule associated with increasing winter bee longevity (Knoll et al., 2020). Vitellogenin increases winter longevity by its effects on storage protein concentration and antioxidant functions (Amdam and Omholt, 2002; Omholt and Amdam, 2004) and on cellular immunity (Amdam et al., 2004b), and through the cryoprotectant properties of its breakdown products (Qin et al., 2019). A comparative study showed that European honey bee workers have higher blood vitellogenin accumulation during brood-free periods than do workers of African-derived *A. m. scutellata* (Amdam et al., 2005). Together, these studies suggest that winter bees evolved as an adaptation favoring cold-latitude penetration, an innovation driven by their increasing capacity for accumulating vitellogenin.

The iconic overwintering behavior of *Apis mellifera* is group clustering, but the cluster's formation and maintenance is a socially mediated property emerging from the interactions of both ancient strategies—freeze avoidance and tolerance. Elements of these ancestral building blocks have been modified by social selection and retained in the honey bee superorganism.

Year-round sociality demands a certain minimum population biomass for ensuring group homeothermy. For this reason, a temperate honey bee colony's population is normally in the range of tens of thousands, achieving peaks of 50,000–60,000 workers just before reproductive colony fission. In autumn, all brood rearing ceases, and bees allow the core nest temperature to drop to as low as 18°C (64.4°F). This number is significant because it represents the minimum temperature at which adult bees can shiver and generate heat (Allen, 1959); it also illustrates the razor's edge of efficiency by which bees consume their precious reserves. Such levels of efficiency help temperate honey bee colonies routinely survive temperatures of −30°C (−22°F) or colder (Gates, 1914).

But as impressive as this efficiency is, what bees do next in the middle of winter once days begin lengthening borders on the unbelievable. This is when the wintering bees resume rearing brood. Energy demands skyrocket, not only from worker bees producing brood food to feed developing larvae, but from generating the extra heat required to incubate them. Core nest temperatures now jump to 34.5°C (94°F), rarely deviating thereafter by more than one degree [reviewed by Seeley (1985)]. All the while, this surge in energy consumption is underwritten by the same finite food supply held in reserve from the previous summer.

This profligate energy expenditure at the most vulnerable time of year can only be understood as a strategy for improving reproductive success of the future swarm. Although overwintered colonies, especially populous ones, may produce two or more swarms during spring, the first or prime swarm is the largest and the only one accompanied by the tried-and-proven mother queen. There are strong positive correlations among earliness of swarming, population size of the swarm, and subsequent measures of its growth rate and survivorship as a colony [reviewed in Winston (1987)].

Bees assume a clustering configuration in the center of the nest as ambient temperatures begin dropping (figure 11.1). Bees remain active at the core of the cluster with space between them, whereas bees at the cluster's edge are packed together tightly. In the layers between, bees are packed together in varying stages of density. Necessary to this clustering behavior is the presence of empty cells into which bees enter head-first. In this manner the contiguous nature of the cluster is maintained in spite of the combs separating them (figure 11.2). The cluster tightens as temperatures drop and expands as temperatures rise.

Since the early twentieth century, it has been known that temperatures at the center of the cluster begin increasing as ambient temperatures begin

FIG. 11.1. Sagittal view of a winter cluster. The top oval encircles younger bees more loosely clustered in the center whereas the bottom yellow oval encircles older bees more tightly packed near the cluster's edge. PHOTO COPYRIGHT: WYATT MANGUM, USED BY PERMISSION.

dropping (Phillips and Demuth, 1914). This formed the basis for an early model of honey bee winter thermoregulation—that core bees "know" when ambient temperatures drop and respond by shivering to generate heat and warm the cluster. More fanciful additions became attached to this foundation myth, including acts of altruism with mid-level bees forcing their way to the surface, pushing their cold-immobilized sisters inward to get warm, and bravely taking their place for a tour of duty at the surface. This model has been replaced with a much simpler one that takes into account nothing more than the actions of individual bees as they respond to temperature conditions in their immediate vicinity.

Heinrich (1993) synthesized the results of numerous field studies into an elegant, parsimonious explanation for cluster-facilitated group winter survival in the honey bee. In early autumn, as temperatures begin dropping, bees do what any of us would do if we were in a crowded, freezing room; we

FIG. 11.2. A winter cluster viewed from the comb edges. Individual bees enter empty cells head-first, thus minimizing the distance between clustering bees in spite of the combs that separate them. PHOTO COPYRIGHT: WYATT MANGUM, USED BY PERMISSION.

begin moving closer together to share and conserve body heat. In the case of bees, young adults cannot shiver to generate heat and they naturally have a low cold tolerance, so these individuals aggregate by choice toward the center where it is warmer. Older bees, in contrast, are able to shiver and generate heat (Stabentheiner et al., 2010), plus they have a higher cold tolerance, so these individuals naturally aggregate away from the center, forming the middle layers and outer mantle.

As temperatures continue dropping, the bees, young and old, respond by clustering tighter, the mature bees shivering to keep warm, with their heat, not that of the young bees at the core, responsible for warming the cluster. Eventually, the problem at the core is no longer excess cold but excess heat from over insulation, and the young bees respond by moving out from the center, forcing open channels that allow exchange of inner air that warms the mantle bees and outer air that cools the interior bees. It is this excess heat in the middle that tricked earlier investigators into thinking that the core was the heat-generating furnace for the cluster. Over the months of winter, aging workers not only increase their innate cold tolerance, but they are able to metabolize antifreeze proteins with an efficiency that mirrors environmental demand so that antifreeze properties are optimized in January, the coldest month in the northern hemisphere (Qin et al., 2019). The result is a set of conditions that permits a high degree of individual low-temperature tolerance.

There are two numbers that regulate much of what's going on here: The first is 35°C (95°F), and the other 15°C (59°F). Young bees naturally seek out 35°C and move further in or out of the cluster to find it. For older bees, 15°C is the minimum temperature at which an individual can shiver its flight muscles and generate heat. Bees at the very surface of the cluster may fall below this threshold and enter a state of cold coma, but they continue serving as insulation even in their state of torpor, passively relying on the shivering and contracting bees beneath them to stay above lethal hypothermia, waiting to resume shivering until their bodies exceed 15°C again. Mature adults can survive cold coma for several hours, with death from hypothermia a real risk only if they sustain temperatures of −2.0°C to −6.0°C (28.4°F to 21.2°F) (Free and Spencer-Booth, 1960). But if that lower ambient limit is realized, and if bees deeper in the cluster cannot keep surface bees warm enough, then a deadly cascade of events ensues—surface bees die from hypothermia and drop off the cluster with a corresponding loss of insulation, which imperils bees deeper in the cluster until they too fall off the cluster, and so on, until the whole colony perishes.

It is not superfluous to state that winter death from a cold extreme is more likely than winter death from a hot extreme. It is easier for young bees to push their way out of a hot interior and force the cluster to cool than it is for cold mantle bees to stay above the lethal cold threshold of −2.0°C to −6.0°C. Hence the value of insulation—the original pre-adaptive benefit of cavity nesting that *Apis mellifera*'s ancestor brought with it out of the tropics. Although open-air exposed combs of *A. mellifera* are not unknown in temperate zones (figure 11.3), their overwintering survival is not good (Byers, 1959).

A comparative genomic analysis of the tropically-evolved *A. mellifera scutellata* and the recently recognized subspecies *A. mellifera sinisxinyuan* of Central Asia gave insight into the kinds of gene families that were selected

FIG. 11.3. An exposed nest of *Apis mellifera* in Georgia. Such nests are presumably a result of cavity scarcity in the local habitat and rarely survive winter in temperate latitudes. PHOTO: BOBBY CHAISSON.

for as the species advanced into temperate latitudes (Chen et al., 2016). In *A. m. sinisxinyuan*, the investigators found evidence for positive selection for genes associated with metabolism of fat body and vitellogenin, which are in turn related to longevity of the queen and winter bees. Similar evidence was shown for genes associated with control of flight muscles, which may have improved cool temperature flight capacity or increased efficiency of thoracic shivering and heat generation. A third overrepresented gene family in *A. m. sinisxinyuan* was related to neurotransmission, which may hint at positive selection for a wider range of social behaviors under conditions of seasonal climate fluctuations. A fourth set of enriched genes in *A. m. sinisxinyuan* was associated with the Hippo signaling pathway, which regulates organ growth and in this case may underlie differences in fat body deposition between winter and summer bees.

In summary, winter thermoregulation in temperate *Apis mellifera* is an ebb and flow—a dynamic of bee age, individuals' preferred temperature strata, temperature limits imposed on shivering, and the constant pulse of individuals in the cluster, each seeking her optimum temperature. Group survival is not the result of individual altruism nor top-down command: it is the sum effect of pre-existing conditions and individual actions—each bee responding to its immediate situation, each shifting in or out of the role of ectotherm or endotherm as its near conditions dictate. Even the empty cells that permit contiguous clustering are unplanned, simply a pre-existing result of reduced brood rearing in autumn. There is evidence for socially mediated genetic adaptation as the species extended its range out of the tropics into cold-weather latitudes. The sum effect of these dynamics is something that looks very much like homeothermy at a scale practiced by mammals.

Cooling the Nest

Heinrich (1993) summarized the mechanics of nest cooling in the honey bee colony—a process driven by wing fanning and evaporative cooling, processes necessary to the desiccation of nectar and its transformation into honey. Although honey ripening no doubt provides a measure of collateral nest cooling, it is also clear that bees actively import water for this purpose.

Raw nectar accumulates in cells above the brood area (figure 3.11), a configuration that itself is an emergent property of the collective actions of bees rearing brood and depositing and withdrawing pollen and nectar from the combs (chapter 12). Rising warm air from the brood area helps evaporate water from nectar, but only to the extent that relative humidity is low enough in the cavity for evaporation to continue. This relative humidity differential is maintained by bees circulating air in and out of the cavity through active fanning. A fanning bee remains firmly grounded and “flies in place,” forcing air to flow behind her in the direction her abdomen is pointing (figure 11.4). Fanners position themselves along existing airflows, and fanning increases dramatically at 36°C (96.8°F) or higher, at which point bees must not only rid the nest of CO_2 and water vapor but also excess heat.

In the normal age-based worker polyethism schedule, fanning peaks at middle age (figure 8.3); however, the probability of a worker engaging in fanning is also influenced by her subfamily-based genetic proclivities (Su et al., 2007), the colony’s need for more, or fewer, fanners (Egley and Breed, 2013), and social contexts (Cook and Breed, 2013). For example, Kaspar et al. (2018)

FIG. 11.4. A fanning worker bee. Air flow is directed behind the bee in the direction its abdomen is pointing. This bee is simultaneously exposing her Nasonov gland, seen as the fleshy area between her terminal two abdominal segments. The pheromone helps bees orient to the nest entrance and is particularly important when a swarm is entering a new domicile. PHOTO: BENJAMIN ROUSE.

noted that nurse bees (young bees focused on brood care) normally initiate fanning at lower temperatures than do older fanners collected from a hive entrance. This threshold difference seems to ensure that nurse bees, who do not normally engage in fanning, will do so in the event colony homeostasis deteriorates and brood are in danger of overheating.

But when Kaspar et al. (2018) experimentally placed one experienced fanner with a group of four young nurses, the fanner had the effect of raising the temperature fanning threshold of the whole group. In other words, the single experienced fanner was a group influencer. Those authors holistically interpreted the evidence to show that fanning behavior, normally prescribed by individual age and colony genetic structure, is nevertheless a labile behavior responsive to changing colony needs; through social learning individuals can voluntarily initiate fanning outside of one's normal schedule. It may be an example of the kinds of novel cognitive capacities selected for as *Apis mellifera* expanded into latitudes with more variable climates (Chen et al., 2016).

Groups of fanners form alongside existing airflows inside the nest, the result being detectable unidirectional currents. Some of this hot air escapes through cavity entrances and gaps by paths of least resistance. But most of it is actively expelled by gangs of fanners aggregated at the nest entrance, who are easily visible to any beekeeper inspecting hives on a hot summer day.

Without these entrance fanners, there would be very little air exchange between nest atmosphere and the exterior.

As it is, a colony experiences a tidal respiratory cycle following a pronounced day/night rhythm (Southwick and Moritz, 1987). Air leaving the entrance has higher temperature, higher CO_2 content, and lower O_2 content, and air entering has vice versa. Entrance fanners are the critical regulator of this colonial "breathing." With their abdomens pointed outward, bees force air outward, effecting active expiration, whereas inspiration happens passively not from fanners reversing positions but by ceasing fanning in a coordinated fashion. During daytime, a colony "breathes" about three times per minute, whereas rates at night are much lower, about 0.4 breaths per minute.

Forced ventilation is augmented by active evaporative cooling. Beginning at about 29°C to 30°C (84.2°F to 86°F), foragers collect water and spread droplets on comb surfaces (Stabentheiner et al., 2021). The evaporative action that follows causes a temporary and localized temperature decrease. Water is, in fact, one of the nest assets (along with nectar) actively appraised by nest monitor bees who "reward" a forager collecting the limited resource by rapidly relieving her of her load (Kühnholz and Seeley, 1997). Successful water foragers may even perform dances to recruit helpers (Lindauer, 1954).

Another form of evaporative cooling—and perhaps the evolutionary precursor for the water painting behavior described above—is "tongue lashing." Beginning at temperatures exceeding 31°C (87.8°F) (Lindauer, 1954), tongue lashing is expressed when a bee regurgitates a droplet of foregut contents—dilute nectar or pure water—and spreads it as a film over her expanded proboscis, enlarging its surface area and exposing it to evaporative water loss. This is a normal step for converting watery nectar into honey, but it is also used by foragers away from the nest for cooling their thoracic flight muscles. Tongue lashing inside the nest serves a similar function for cooling the immediate vicinity.

A final mode of nest cooling involves the active reduction of adult bee density from the hot nest interior. The immatures in their cells are utterly incapable of moderating their own temperatures. It is the adults who must maintain air temperature between combs at a steady 35.5°C (96.9°F), a differential that is sufficient for allowing heat to flow away from the combs. At temperatures between 34°C and 36°C (93.2°F and 96.8°F), adult bee densities in the brood area are at the minimum necessary for brood care and cooling (Stabentheiner et al., 2021). If heat stress increases, adult bees are known to move toward the sunward side of the cavity, acting as a kind of heat shield (Starks and Gilley, 1999). At its most extreme, adult bees will vacate the cavity *en masse*, a phenomenon known to beekeepers as "bearding" (figure 11.5).

We thus find ample fodder for parsimonious explanations for nest cooling—explanations demanding nothing more than the actions of individuals, each responding to her local conditions. There arc properties that are emergent, not the direct effects of natural selection at all. Fanning behavior is thought to have evolved from precursors as simple as escape reaction to

FIG. 11.5. Managed colonies showing "bearding" behavior on a hot afternoon. Adult bees vacate the nest interior *en masse* to reduce bee density in the temperature-sensitive brood rearing area. PHOTO: GEORGIA P. ZUMWALT.

noxious odors, and organized gangs of fanners inside the nest evolved from individuals simply seeking airways for cooling their own hot bodies (Heinrich, 1993). The fact that tongue lashing helps cool a social nest distracts in no way from the assumption that the behavior first evolved as a means for solitary foragers to cool themselves. And the mass exodus of adult bees bearding a hive front on a hot evening are simply a signal that the bees are escaping uncomfortable temperatures.

The elements of nest cooling behavior that show the stamp of social selection are, like we saw for nest warming, modified from simpler, solitary precursors. Marks of social evolutionary tinkering are evident when we see fanning behavior forced into an age-based polyethism schedule (chapter 8) or expressed as a genetic subfamily specialism (Su et al., 2007), or water collecting rewarded in a socially regulated decision-making process (chapter 15), or evaporative cooling expanded from tongue lashing to water painting on substrate surfaces. All of these are candidates for the kinds of innovative cognitive capacities *Apis mellifera* would have needed for emigrating into climatically diverse latitudes (Chen et al., 2016). The sum of these inputs, evolved or individually enacted, adds up to the emergent property of colony-level resistance to heat extremes.

Summary

In insects generally, overwintering species are lumped into one of two groups—those that avoid freezing and those that tolerate it. Elements of both are seen in the honey bee. The bulk of evidence suggests that *Apis* evolved in hot tropical conditions and that cavity nesting was a pre-adaptation for temperate expansion.

Choice of nest site is the most fundamental adaptation for temperature extremes. The cavity must have walls thick enough to provide insulation, a volume sufficient for storing food and housing a critical mass of workers, and entrances small, defensible, and positioned below the clustering bees to minimize heat loss from rising air. Bees assume a clustering configuration in the center of the nest as ambient temperatures begin dropping. The cluster tightens as temperatures drop and expands as temperatures rise. The cluster's survival is an emergent outcome of the age-based attempts by individual bees to find their temperature optimum. Young adults cannot shiver to generate heat and they naturally have a low cold tolerance, so these individuals aggregate by choice toward the center where it is warmer. Older bees, in contrast, are able to shiver and generate heat, plus they have a higher cold tolerance, so these individuals naturally aggregate away from the center, forming the middle layers and outer mantle. Winter thermoregulation in temperate *Apis mellifera* is a dynamic of bee age, individuals' preferred temperature strata, temperature limits imposed on shivering, and the constant pulse of individuals in the cluster each seeking her optimum temperature.

Nest cooling is similarly an outcome of individual actions. Raw nectar accumulates in cells above the brood area, a configuration itself an emergent property of the collective actions of bees rearing brood and depositing and withdrawing pollen and nectar from the combs. Rising warm air from the brood area helps evaporate water from nectar, but only to the extent that relative humidity is low enough in the cavity for evaporation to continue. This relative humidity differential is maintained by bees circulating air in and out of the cavity through active fanning. Air leaving the entrance has higher temperature, higher CO_2 content, and lower O_2 content, compared with vice versa for air entering.

Another form of evaporative cooling is "tongue lashing" which is expressed when a bee regurgitates a droplet of foregut contents—dilute nectar or pure water—and spreads it as a film over her expanded proboscis, exposing it to evaporative water loss. If heat stress increases, adult bees move toward the sunward side of the cavity, acting as a kind of heat shield. At its most extreme, adult bees will vacate the cavity *en masse*, a phenomenon known to beekeepers as "bearding." Fanning behavior is thought to have evolved from precursors as simple as escape reaction to noxious odors.

CHAPTER 12

Comb Construction and Use Patterns

BEESWAX COMB—the stimulants and regulators of its construction by bees, the design of its constituent cells, and the patterns of its contents—is another example of emergence, or biological self-organization. It is not an outcome of hierarchical top-down regulation imagined by early writers but instead emerges from the actions of numerous independent decision-makers (Camazine et al., 2001). The superorganism with its hordes of inmates constitutes a gold mine of opportunity for observing biological self-organization in action.

Comb, Stimulants, and Regulators of Its Construction

If colonial *Apis mellifera* can be thought of as a superorganism, then it is no stretch to imagine the combs as the superorganism's skeleton. The material entity of the comb is so integral to the colony's life that it almost acts as a colony member itself. The comb is the substrate upon which everything happens—brood rearing, food storage, pheromone deposition, dance recruitment, and transmission of vibration signals. Nothing can happen without it, and building it is the highest priority for a new swarm occupying a new cavity. And as a metabolic product of consumed honey, it is expensive. As early as the 1960s, Weiss (1965) calculated that it costs a colony 6.3 kilograms of honey to produce 1 kilogram of comb. Using data from Seeley (1985) from New York, we can extrapolate from Weiss (1965) that a normal colony consumes 7.5 kilograms of honey, roughly 12.5% of its annual honey income, just to build its comb. It is for this reason that a new swarm gains a significant advantage if it has the good fortune to find a cavity with pre-existing combs. Such colonies can end up collecting twice as much honey as new colonies founded without combs (Szabo, 1983).

To put the cost/benefit ratio into sharper relief, let us consider the time of year when all this is happening—for temperate-evolved honey bees, that means late winter/early spring. In other words, this comb-building campaign is underwritten on stored honey—and in the case of a new swarm, that stored honey goes no further than the honey in the honey stomachs, or foreguts, of individual bees. This is why swarms always occur on days with good nectar flows; bees immediately need the new honey to supplement the honey in their foreguts. If the flow shuts off, the swarm is in peril of starvation. This is one reason why the 12-month survival rate of newly founded swarm colonies is only 24% (Seeley, 1985).

But the cost/benefit calculation applies as well to mature colonies, albeit with lower stakes. Consider: a colony benefits if it has plenty of empty comb on hand as soon as a strong nectar flow begins. Foragers can immediately unload their nectar to "receiver" bees (chapter 15), who in turn immediately deposit it into empty cells, the combined effect of which encourages more foraging and more honey storage. If, on the other hand, the colony builds "too" much comb too early, then it runs the risk of exhausting its precious honey stores. If the colony/superorganism really does constitute a Darwinian unit of selection, then we should be able to make scientific predictions about how it behaves to optimize its survival chances between these two extremes.

There are at least three primer signals necessary for the initiation of comb construction. These are colony states, some of which are months in the making: (1) bees prefer to initiate comb construction in the dark (Morse, 1965), (2) building ceases in the absence of a queen (Ledoux et al., 2001), and (3) workers must have adequate protein nutrients early in adult life if they are to activate their wax glands, produce beeswax, and serve as comb builders (Goetze and Bessling, 1959). These three conditions are static and act rather like on/off switches. However, there are other important regulators that act dynamically, changing constantly. We can think of them as releaser signals initiating comb construction any particular day: (1) the rate of nectar intake from the field, (2) the amount of filled comb in the colony, and (3) the amount of empty comb in the colony. Managing these dynamics requires of individual bees the ability to gather information, appraise it, and respond appropriately.

Field experiments and mathematical simulations have shed light on how colonies respond to these dynamics, at least in temperate zones. For his model, Pratt (2004) collapsed the three releaser signals into two: (1) there must be nectar coming in from the field, and (2) there must be a certain "threshold of comb fullness" or percentage of available comb filled. If a colony has plenty of empty comb, bees will not metabolize nectar into beeswax to build more until a certain fullness threshold is reached. Once that threshold is breached with incoming nectar, pollen, or brood, comb construction begins, and only the nectar flow is necessary thereafter to keep it going. But once the nectar flow stops, bees stop building more comb—even if the quantity of stored products now exceeds the former fullness threshold. Thus, it appears that the nectar flow is more important than comb fullness in regulating comb construction (figure 12.1).

The fullness threshold is itself a dynamic that changes relative to the quality of nectar flow. Simulation models (Pratt, 2004) predict that bees will have a very low fullness threshold when nectar flows are strong and a very high threshold when flows are weak. In other words, they should be conservative about initiating comb construction when nectar flows are weak. But what bees do in fact is stop building altogether when nectar flows are weak. This is an example where reality does not match up to a computed optimum. Evolution, in this case, has erred on a slightly suboptimal outcome—the only loss being wasted effort by bees who spend time appraising comb and field conditions and never put it to use.

The point about "wasted effort" leads us to the level of activity that makes all this possible—and that is the decisions made by individual

FIG. 12.1. Producers of award-winning comb honey know intuitively the power of incoming field nectar to stimulate comb construction by bees. It is so powerful that bees will keep secreting wax and adding to the comb even if storage space is nearly used up. The result is visible in two dimensions: thicker combs (longer cells) and wider combs (filled to the edges). Commercial comb honey production is thus a practical application of evolutionarily-driven optimization choices made collectively by the colony. It is adaptive for bees to invest liberally in building storage space while nectar flows are strong. PHOTO: GEORGIA P. ZUMWALT.

bees that culminate in the colony's near-optimal exploitation of field nectar and nest storage space. As early as the 1950s, Ribbands (1953) put forward an hypothesis that individual bees, most likely nest bees who receive fresh nectar loads from foragers, are stimulated to secrete wax and build comb by the physical sensation of a distended crop or honey stomach. As attractive as this hypothesis is, it has not withstood experimental challenge. Our latest knowledge now suggests that it is not nectar receivers who build comb, but instead a separate behavioral caste of workers drawn from labor reserves otherwise un-preoccupied with other duties (Pratt, 2004).

These "resting" bees are numerous in a colony, especially at night (Kaiser, 1988), and it appears to be from their ranks that comb builders are drawn. Given the two dynamics that regulate comb construction—comb fullness and richness of the nectar flow—these comb builders must be in a position to appraise each, presumably by directly inspecting combs and by appraising the rate at which they are offered nectar loads. Experimental evidence suggests that comb builders do in fact engage in both these behaviors at higher than average rates (Pratt, 2004).

Comb Design and Construction

Cell construction is an outcome of emergent properties interacting with instinctive behavior of individuals. A behavior is traditionally considered *instinctive* if it is genetically "hardwired," universal to a species, and not learned (Blumberg, 2005). It is sometimes called innate behavior, and the sequence of events in an instinctive response is sometimes called a *fixed action pattern*. Examples include language acquisition in humans (Pinker, 1994), the movement of newly hatched sea turtles from the beach toward the surf, and the suckling urge of newborn mammals. The instinctive behavior of bees we're concerned about here is their propensity to build individualized cells.

Readers may recall from chapter 4 that the earliest ancestors of the honey bee nested in simple excavated tunnels, which over time were elaborated into individualized cells for holding food or rearing young, in many cases the cells being lined with water-resistant glandular secretions. The individualization of cells is both ancient and persistent across all bee families; hence, we can assume that it was strongly selected for as a means to protect food from spoilage and the young from pathogens, predators, and trampling by nestmates. I am considering it an instinctive behavior, not learned, and even though the social life of colonial bees offers many opportunities for learning such a behavior, the antiquity and ubiquity of individualized cell making argue against it being a modern innovation.

We established in the previous section that building the comb is high priority for a swarm newly settled into a cavity. The initiation of comb construction is always associated with sheets of bees hanging together in a posture called *festooning* (figure 3.2). Inside these festoons are found wax-bearing workers and cell builders. In a newly occupied cavity there can be many of these festoons—a behavior obscured in the manufactured Langstroth hive of beekeeping with its interspersing frames—and the physical location of these festoons doubtless imparts a sort of template onto the ultimate outcome. In a new cavity, there will numerous independent starts of comb construction, with each festooned work crew working without regard to an overarching design. Comb initiation inside these festoons has been difficult to observe without disruption in spite of centuries' worth of effort, owing naturally enough to the dark cavities and density of the festoons themselves.

For this reason, our best observations come from the open-nesting Asian honey bee *Apis florea* as described by Hepburn at al. (2014). As with their western cousin *A. mellifera*, the initiation of comb construction is immediate upon setting onto a nest site—in *A. florea*'s case, the underside of a limb. Within hours, a first row of cells appears. These cells are not hexagonal nor even polygonal, but plainly circular, strong evidence for an imprinted design inherited from their solitary tunneling ancestors. The deposition of wax happens in the following sequence: a cell builder removes a wax scale from her ventral abdominal wax glands, moves it to her mouth, chews it while adding salivary secretions to render the scale malleable, then applies it to the cavity substrate or previous cell, shaping it with her mandibles. The worker can lengthen a cell wall by depositing and shaping new wax or by shaving off wax from an existing section and transferring it to the edge, thinning the wall in the process.

Although each worker is acting out this fixed action pattern that culminates in a cylindrical cell, the worker does not fix her attentions on only one cell. Rather, the whole group works on all available cells at the same time, all without any appearance of coordination, focus, or order. We get a sense of this from Smith (1959), who observed one marked bee for more than one hour while the bee was capping brood cells. During the first 11 minutes, the bee began and finished capping one cell; she started another, and eight minutes later had "nearly finished it" when she went exploring and taking honey offered by another bee; she started a third cell, and 16 minutes later left it unfinished to return to cell 2 and finish it. Then she returned to cell 3, and two minutes later finished it. She then went exploring, and three minutes later began cell 4. I could go on, but perhaps you get the gist. But I can't resist the last entry: "appeared to have lost interest—walking around."

What I'm illustrating here is that fusion of innate behavior and emergent group properties I mentioned at the beginning. Cell builders may apply and shape their wax boluses in a random fashion within the work zone, but each bee's action is nevertheless aimed at her imprinted template of a "cylindrical individualized cell."

But, remembering that a social insect colony is a field of complexity within which self-organization and spontaneous order emerge, from this point on, we'll see that the collective actions of these workers, each responding to her innate template, give rise to a higher order of complexity—the combs, their posture, and their position relative to each other, and all this without any hierarchical coordination imposed from above.

To continue, let's recall that multiple combs is an evolutionary innovation deriving from the more primitive state of single-comb open nesting (Ji, 2021). Readers may remember from chapter 4 that open-nesting *Apis* commit a large fraction of their adult population to forming a living curtain over the comb as a sort of proxy shelter. Cavity nesting not only freed up this cohort of adults to more useful activities but permitted the enlargement of the nest from one comb to several. This also created new problems, not the least of which was comb spacing.

L. L. Langstroth popularized the idea of "bee space," the observation that bees maintain a more-or-less fixed distance around combs to accommodate their movement within the nest (Johansson and Johansson, 1967). As much as beekeepers have ever since celebrated bee space as an example of animal intelligence, it is in fact only another emergent property—a natural outcome of the bees moving around the nest (figure 12.2). French scientist Roger Darchen, in an extraordinary string of papers between 1952–1968 [reviewed by Hepburn et al. (2014)], described the bees' tendency to make combs parallel to one another. Darchen's "rule of parallelism" edges out other considerations such as cell depth.

Darchen's rule of parallelism also applies to the early stages of comb construction, picking up where we left off a few paragraphs above. Whereas each festoon of bees in a new cavity begins its wax deposition on the cavity ceiling without regard to other festoons, eventually the products of their labors collide, at which point the working bees simply bend their comb to accommodate the other, establishing comb parallelism and bee space as outcomes of their own need to move about.

Another emergent property is the regularity of cells in a comb. Once an initial row of cells is deposited, then that row constitutes a template for the next row, and so forth. Hence, the appearance of regularity and order. Cells in the earliest rows tend to be simple circles at first, but a curious thing happens with them and subsequent cells as the comb gets larger—the cells begin to assume their famous hexagonal configuration. Perhaps no other feature

FIG. 12.2. Bee space is not an imposed order on the bee nest, but rather an emergent property of a colonial insect living in a volume-constrained cavity with multiple combs. Bees build neighboring combs no thicker than what permits the bees' free movement in between them. Moreover, bee space is not nearly as fixed and consistent in natural nests as the beekeeping literature would suggest, as a casual look at this natural colony makes apparent. PHOTO: BOBBY CHAISSON.

in the honey bee nest has been more misrepresented as a triumph of animal engineering—nor a greater violation of the principle of parsimony—than the hexagonal beeswax cell. One of the most delightful tropes about it goes back to the Greek mathematician Pappus of Alexandria (*circa* 290–350 CE), who wrote that the bees construct the hexagon, among all geometric possibilities, with awareness that the hexagon represents the most economical number of units per surface, a conjecture that was not mathematically proven until the twenty-first century (Hales, 2001).

Alas, the hexagon is yet another emergent property, this time an outcome of a physics principle that when pressure is applied equally to all sides of an elastic cylinder, a hexagon results (Karihaloo et al., 2013). It has been pointed out, however, that this spontaneous geometry obtains only if an isodiametric cylinder is previously surrounded by six other cylinders of the same size.

This necessary condition is, in turn, regulated by behavioral rules that cause bees to initiate new cells in the groove between two pre-existing cells and to erect the cell's walls only after the cell's base reaches a certain threshold size (Nazzi, 2016).

A final feature worth mentioning is the verticality of combs, long understood to be an important engineering feature that permits optimum load bearing without breakage. Both *A. mellifera* and *A. cerana* prefer to build cells with opposing peaks of the hexagon in true vertical orientation (Yang et al., 2022). Similarly, the aggregate combs bees build in the darkness of their cavities are, in the main, plumb to a true vertical, but there are competing explanations jostling for prominence in the scientific literature.

The first model proposes that bees impose a verticality on combs with the use of gravity-sensitive organs in their necks. This research was done in a series of laborious experiments in Germany in the 1960s. Martin and Lindauer [cited in Hepburn et al. (2014)] searched for gravity-sensitive organs in the honey bee by trial and error—systematically covering different parts of the bees' bodies with glue and seeing which affected comb construction. They found that comb construction ceased in bees whose heads had been immobilized, presumably by deactivating a sensory hair plate in the neck. The subsequent discovery of magnetic oxide particles in the bee's abdomen (Gould et al., 1978) supported the idea that bees actively appraise gravity cues as they construct their combs.

But more parsimonious explanations are available that take into account only existing properties of the bees and the nest. A honey bee experiment on a NASA space shuttle in the early 1980s conclusively showed that bees are capable of building normal combs in zero gravity (Vandenberg et al., 1985), albeit without any reference to a true vertical, further inviting the consideration of alternate hypotheses. A so-called substrate-dependent model foregrounds the fact that incipient comb midribs are always perpendicular to the ceiling in a cavity nesting bee primitively conditioned to make hanging combs. Considering that each row of cells uses the previous row as a template, one sees how a vertical comb can result (Pratt, 2000).

It seems to me that another, even more parsimonious explanation should be considered—and that is the simple fact that the comb-making festoons themselves, free-hanging and plumb, are the most direct and simple explanation for a vertical comb. But this scenario invokes the agency of gravity, which the results of the NASA experiment seem to marginalize.

Patterns of Comb Content Distribution

The stereotypic spatial pattern of brood comb contents is a concentric series with brood in the middle, an arc of pollen above it, and the rest of the cells filled with honey (see figure 3.11). Cells of brood are close to other cells of brood nearly the same stage (see figure 3.8). The center-to-edge pattern of brood-pollen-honey is so regular and fixed that it constitutes one of the descriptive constants of beekeeping books and biology texts. Moreover, the arrangement makes evolutionary sense. By positioning brood centrally, the colony optimizes incubation temperatures and minimizes temperature swings for the developing immatures. By positioning pollen immediately next to brood the colony facilitates its rapid access by protein-hungry nurse bees who synthesize and secrete worker-jelly for developing larvae. By positioning honey above and to the sides of the brood area, workers provide additional insulation to the critical nest interior.

Because of these obvious adaptive benefits, it's tempting to imagine that cell allocation and use patterns in combs is a fixed template imprinted on honey bee instinct: that bees "know" to put brood in the middle, pollen immediately above it, and honey immediately above that. The trouble with that approach is that it too-quickly defaults to a vague and ultimately complex solution before it has exhausted simpler explanations based on inherent properties and existing conditions. In the case of comb content patterns, there is evidence that instinctive templates are at least partly involved, but otherwise the stereotyped brood-pollen-honey concentric pattern can be explained as a natural outcome of the actions of independent bees.

In a landmark paper, Camazine (1991) laid out the conditions necessary for the stereotyped brood-pollen-honey pattern to emerge spontaneously. Camazine realized that the stereotyped pattern could emerge if the following four behavior rules were met: (1) the queen preferentially deposits eggs centrally and in cells near other cells containing brood, (2) foragers returning from the field deposit nectar and pollen indiscriminately on the comb,

(3) nurse bees preferentially consume cells of nectar and pollen occurring near cells of brood, and (4) on average, colony deposition of nectar exceeds deposition of pollen.

Camazine (1991) then challenged each of these rules to experiments and literature searches, and finding supporting evidence, proceeded to design a computer simulation that used the rules to mimic development of one side of an empty Langstroth comb located in the center of the nest. The model proved capable of producing the stereotyped patterns, and interested readers can look up the original article and see illustrations of the kinds of "errors" that occur when one or more of the rules is compromised.

Camazine's 1991 model has stood the test of time, albeit with modification. Johnson (2009) argued that self-organization interacts with instinctive behavioral responses to physical templates, suggesting that bees respond to gravity and preferentially deposit nectar near the top of combs. Montovan et al. (2013), noting that Camazine's model was limited to describing only the formation of the stereotyped pattern, made modifications to the queen's movement and workers' consumption of pollen and honey (rules 1 and 3, respectively) that permit the model to sustain the pattern for up to 60 days of brood rearing.

Now, taking into account these papers, here's a summary of my best understanding of how the stereotyped brood-pollen-honey pattern emerges in the honey bee comb. It can best be described as a blend of self-organizing emergent principles and instinctive behavioral responses to physical templates.

The queen deposits eggs in the center of the nest, following an instinctive template of temperature preference; this explanation coheres with the observation that the queen avoids laying eggs at the bottoms and edges of combs. She moves from cell to cell in a disorderly manner, yet on average tends to deposit eggs next to other brood cells and only a few cell lengths from her previous deposition. Thus, in spite of her random visitation pattern, her relative economy of motion means that brood of similar age will be situated together (figure 12.3). While all this is happening, nectar-handling bees and pollen foragers are depositing their loads throughout the brood nest. There is a gravity-based template that biases some of this behavior toward the top of combs, but this effect is variable and much honey and pollen still end up in the middle of the brood.

The nurse bees tending brood are concentrated, as a matter of course, in the brood area. As the "lactating" members of the superorganism, these bees consume large amounts of stored pollen and honey, and energy conservation dictates that they move no further than necessary to find it. Thus, the probability of a nectar or pollen cell being emptied is proportional to its proximity to brood, with nectar or pollen cells in the middle or edge of brood being consumed more quickly than nectar or pollen further away. The fourth behavioral rule now gets involved—and that's the colony-wide case that foraging income and effort are higher for nectar than pollen. Citing data by Seeley (1985), Camazine (1991) pointed out that a typical natural colony collects about 60 kilograms of honey, of which about 40% is not immediately consumed and therefore constitutes "stored" honey. In contrast, a colony collects about 20 kilograms of pollen, all but maybe 10% of which is immediately consumed.

This storage disproportionality is what causes the segregation of pollen into its own band nearest the brood. Because the rate of nectar foraging is much higher than pollen, any cells of pollen in the periphery that are emptied are likely to be refilled with nectar that has a higher chance of being "stored," i.e., permanent—hence the growing concentration of honey in the nest edges. Increasingly, the only places remaining to store pollen are those areas of the nest not in permanent storage mode—and that's the edge of the brood where young bees are emerging and randomly-deposited foraging loads are rapidly emptied by hungry nurse bees. The queen's focus on depositing eggs in the brood center, and the fact that pollen is consumed even more rapidly in the brood center, explain why pollen doesn't accumulate there.

There are feedback loops that encourage the preservation of the stereotypic pattern. The queen of course is the source of brood, and the presence of brood stimulates pollen foraging, while hungry nurse bees are a check on its encroachment into the queen's egg-laying arena. Beekeepers sometimes note irregularities in the stereotypic pattern, and these irregularities are clues to colony and queen status. A so-called "honey bound" condition is seen

FIG. 12.3. The queen is instinctively biased to deposit eggs in the warmest parts of the nest and to deposit eggs near other cells with brood. She does not necessarily move sequentially from cell to cell but jumps around. In spite of this haphazard deposition pattern, her temperature bias and relative economy of motion mean that brood cells normally occur near other cells of similar age.

when honey is encroaching on the brood production area. This is usually associated with a dead or failing queen who is not able to sustain the brood/nurse bee/food cell removal cascade that regulates the pattern. Conversely, a pollen-bound condition, although more rare, may occur if a huge burst of pollen availability temporarily outpaces its consumption or if a sudden drop in nurse bees causes pollen to accumulate in the brood area.

Summary

Beeswax comb, the design of its cells, and the position of its contents are classic examples of biological self-organization. Its construction constitutes one of the costliest priorities of a new colony. There are three primer conditions, i.e., colony states, some of which are months in the making, necessary for comb construction: (1) the cavity must be dark, (2) a queen must be present, and (3) workers must have adequate protein nutrients early in adult life to activate their wax glands and serve as comb builders. Two dynamics serve as releaser signals initiating comb construction on any particular day: (1) there must be nectar coming in from the field, and (2) there must be a certain "threshold of comb fullness" or percentage of available comb filled. If a colony has plenty of empty comb, bees will not spend stored honey building more until a certain fullness threshold is reached. Once that fullness threshold is reached and comb construction begins, only the nectar flow is necessary thereafter to keep it going. But once the nectar flow ceases, so does comb construction. Comb builders constitute a separate behavioral caste of workers drawn from labor reserves otherwise un-preoccupied with other duties.

Cell construction is an outcome of emergent properties interacting with instinctive behavior of individuals, the imprinted behavior in this case being the propensity of bees to build individualized cells. Initiation of comb construction is always associated with sheets of bees hanging together in a posture called festooning, inside which are found wax-bearing workers and cell builders. In a newly occupied cavity, there can be many of these festoons, each festooned crew working without regard to an overarching design. Each crew deposits an initial row of circular wax cells, then that row constitutes a template for the next row, and so forth. Hence, there is the appearance of regularity and order. Although each festoon works without regard to other festoons, the products of their labors eventually collide, at which point the working bees bend their combs to accommodate comb parallelism and bee space as simple outcomes of the need to move about.

The stereotyped hexagonal shape of cells emerges from the physical principle that when pressure is applied equally to all sides of an elastic cylinder, a hexagon results. The verticality of combs emerges from the fact that comb-making festoons are themselves free-hanging and plumb. The stereotypic spatial pattern of brood comb contents is a concentric series with brood in the middle, an arc of pollen above it, and the rest of the cells filled with honey. A landmark 1991 study showed that this pattern can emerge spontaneously if four behavior rules are met: (1) the queen preferentially deposits eggs centrally and in cells near other cells containing brood, (2) foragers returning from the field deposit nectar and pollen indiscriminately on the comb, (3) nurse bees preferentially consume cells of nectar and pollen occurring near cells of brood, and (4) on average, colony deposition of nectar exceeds deposition of pollen. With a few modifications, this model has stood the test of time.

CHAPTER 13

Regulators of Colony and Body Size in the Honey Bee Superorganism

I INTUITED BERGMANN'S RULE before I knew of it. It happened in the context of quitting my boyhood home in northern Indiana (40th latitude) for graduate school in subtropical Louisiana (30th latitude). The epiphany came on my first trip home during summertime, the previous trips having been limited to winter holidays when squirrels were not to be seen. Squirrels, that is, because my epiphany had to do with squirrels, and summertime because that meant it had been a long time since I'd been home. The first time I looked at an Indiana squirrel after years of absence, my newly naïve eyes dilated at their size. Indiana squirrels are bigger than Louisiana squirrels. And beekeepers who have had the opportunity to work bees in different parts of the world know that the same applies to honey bees: worker bees are bigger in more extreme latitudes. Only later did I learn that this observation had been noted before: animal body size increases as latitude increases. It can apply both within species and across species within a genus. It is general enough to be called a rule of biogeography, named after Carl Bergmann, the nineteenth-century German biologist who first wrote about it in 1847 (Bergmann, 1847).

But it is far from universal, and in the case of insects may even be weak. Among the Hymenoptera, the order making up the ants, wasps, and bees, only 25 of 62 studies (40%) have confirmed Bergmann's rule (Shelomi, 2012), and among the bumble bees in particular, the opposite seems to hold—that body size gets smaller as latitude increases (called "converse Bergmann") (Ramírez-Delgado et al., 2016). I raise the issue because I think Bergmann's rule has something to say in our evolutionary history of the honey bee, and what little information exists on the matter suggests that *Apis mellifera*, along with the other cavity-nesting *Apis*, do indeed follow it (Ruttner et al., 2000).

We will return to the idea of body size in the worker bee later in the chapter, but for now, let us think about honey bee body size at the superorganismal level—how many workers make up a colony? Among members of its genus, *A. mellifera* has a relatively high worker population. A comparison of natural colony populations in the genus *Apis* is shown in table 13.1 along with some other measures of interest.

A US Department of Agriculture (USDA) scientist, C. L. Farrar, showed in an applied context the importance of large colony populations for honey production. His landmark 1937 paper showed that honey hoarding efficiency increases as colony population increases. One colony of 50,000 bees can be expected to make more honey than the sum of two colonies of 25,000 bees (Farrar, 1937). The application of this insight led directly to today's standard practices for honey production, but more broadly, it underscored the ecologic benefits of large colony populations. Large colony size in *A. mellifera* can be thought of as another optimized outcome of selection by natural forces

TABLE 13.1. Some measures of body/soma size in *Apis*

		COLONY TRAITS		
SPECIES	BODY MASS (MG)	WORKER POPULATION	NUMBER OF COMBS	NEST SITE
A. florea	22.6	6,000	1	Open
A. dorsata	118.0	>35,000	1	Open
A. cerana	43.8	8,000	5–6	Cavity
A. mellifera ligustica	77.2	15,000–40,000	5–8	Cavity

Note: Measurements are based on those of Seeley (1985) and Dyer and Seeley (1987).

acting on physical constraints placed on the bee and positive feedback loops that reinforce colony growth.

We can reconstruct plausible pathways for this evolution from Southeast Asia, a leading candidate for the ancestral home of adaptive radiation for the genus *Apis* (chapter 4 section "The Cavity-Nesting Modern Forms"). *Apis* can be divided into two camps, namely, those that nest on single open combs and those that nest on multiple combs in cavities. It has been resolved beyond reasonable doubt that the ancestral condition is single open combs, and the later derived condition is cavity nesting (Fouks et al., 2021). All modern single open-comb nesters, of which *A. florea* and *A. dorsata* are shown in the table, express a primitive behavior unknown in the cavity nesters *A. cerana* and *A. mellifera*. The open nesters have a "curtain" of living bees that hangs over the comb covering the queen, the brood, and the nurse bees tending it (figures 4.6 and 4.7). This curtain provides a defensive barrier against enemies such as predatory hornets and a measure of temperature insulation. It approaches the function of a physical shelter.

It is believed that the shift from open nesting to cavity nesting was a watershed moment in the evolution of *Apis*, beginning with the obsolescence of the living bee curtain. With the onset of cavity nesting, the worker cohorts formerly engaged passively as living insulation were now freed to contribute more directly to the colony's economy.

The implications of this evolutionary fork in the road were hinted at by a pair of studies that compared metabolism and foraging rates of modern open nesters against cavity nesters (Dyer and Seeley, 1987, 1991). Those authors showed that cavity-nesting species have a higher "tempo," by which they meant metabolic rate and foraging rate. This pattern did not track with body size as one might expect—metabolism usually increases as body size gets smaller (Calder, 1984)—because workers of the low-tempo open nesters included both the largest and smallest workers in the study, and workers of the two cavity nesters are intermediate (table 13.1). Instead, nesting habit, not body size, explained colony metabolism.

By considering Dyer and Seeley's work alongside other studies, we can infer a plausible scenario for the evolution of worker population size in *A. mellifera*:

1. As workers were freed from passive curtain duty, more of them were available for nursing, foraging, and other tasks in support of brood.
2. As the colony's foraging economy improved, this encouraged more brood rearing and the innovation of multiple combs to support it.
3. The increasing ratio of brood to workers had a stimulating effect on foraging, thus increasing selection for high worker metabolism and productivity.
4. The positive feedbacks of more brood and increased foraging reinforced larger colony populations.
5. Larger worker populations had a stimulatory effect on worker division of labor and task specialization, further increasing efficiency and social complexity, constituting yet more positive feedbacks on large populations.
6. Increasing social complexity biased the *A. mellifera* lineage toward its temperate expansion, providing additional positive feedbacks toward large worker populations capable of buffering temperature extremes.

At this point, it is worth stating that large populations are not only an outcome of complex eusocial colony life, but they are *necessary* to complex eusocial life. If this sounds like a chicken-or-egg paradox, we only need recall *worker coercion*—the act of workers eating each other's eggs and how this behavior reinforces altruistic group living (chapter 7's "Restraining Genetic Conflict in the Honey Bee Superorganism" section). In small populations, it's easier for one worker to physically dominate her sisters and monopolize egg laying, but this license fades in larger populations where aspirants to dominance have more competitors. Large populations therefore have a selecting effect for social cohesion (Bourke, 1999).

With all these positive feedback loops, one must wonder what prevented even larger worker populations than the 15,000–40,000 we see today—and the answer is seasonal and physical constraints imposed upon early *Apis mellifera* and its ancestors. First, the seasonal: For a vegetarian insect committed to living off stored nectar and pollen, there is a finite periodicity to the availability of these resources. The few weeks of nectar flow in much of the temperate world are too brief to allow unlimited population growth. Second, the physical cavities: It seems to me that there is a finite range of cavity sizes that could meet the conditions necessary for the natural history I propose above. Too small a cavity would not permit the extra combs and extra brood that initiate the positive feedback loops toward large populations; too large a cavity—a cave for example—would limit climate control and daytime flight orientation.

Moreover, in the case of African and European *A. mellifera* that nest in hollow trees, there is a limit to the range of volumes available in tree hollows. It so happens that we know a little bit about the volumes available in tree hollows, and for European bees in temperate latitudes, that figure ranges between 15 and 80 liters (Seeley, 1985). I have always thought it was an interesting case of serendipity that L. L. Langstroth and his colleague, the beekeeping equipment innovator A. I. Root, in the nineteenth century settled on an American hive body volume of 43 liters, which, with additional hive bodies and supers, generously accommodates the evolutionary constraints imposed upon *Apis mellifera*.

Within the *Apis* cavity nesters, Bergmann's rule seems to apply at both the level of colony population and worker body size. *A. mellifera* has conspicuously larger populations and worker body size than its sister cavity nester *A. cerana* (table 13.1). Bergmann's rule emphasizes thermoregulation—that larger bodies confer greater homeostasis, resistance to temperature extremes, and a measure of buffering against food dearths. In mammals, this means fat reserves, whereas in temperate *A. mellifera* workers, it means storage of the egg yolk protein vitellogenin (Chen et al., 2016; Wallberg et al., 2014).

But this principle applies as well at the superorganismal level. In the ant *Solenopsis invicta*, it has been shown that large colonies buffer the queen against a prolonged food dearth better than small colonies (Kaspari and Vargo, 1995), and in managed honey bees, high autumn colony weight (a proxy for food store quantity) and population size are the strongest predictors of overwintering survival (Döke et al., 2019). Larger individuals and larger worker numbers in *A. mellifera* are consistent with its history of pioneering into cold latitudes with variable forage availability.

And finally, we should stress that the colony populations we've been talking about are colony populations in nature. In managed colonies, it's not uncommon for populations to achieve 50,000–60,000 bees or even higher. This is an outcome of generous hive volumes and swarm-prevention management practices precipitated by the discoveries of C. L. Farrar. In this, modern beekeeping is in step with all other sectors of agriculture—a human enterprise of coaxing levels of productivity in plants and animals far in excess of their natural optima for survival. Whether this is a good thing or not is an important question that invites reflection on the kinds of production paradigms we modern humans have created. Whether these paradigms can be sustained will be among the most important questions facing the next generation.

Summary

Bergmann's rule from the nineteenth century states that animal body size increases as latitude increases. It can apply both within species and across species within a genus. In the context of honey bee social evolution, the question of "body" size and its evolution is properly aimed at the size of the superorganism's soma—colony worker numbers, and a case can be made that cavity-nesting *Apis* do comply with Bergmann's rule in the classical sense. Among members of its genus, *A. mellifera* has a high worker population. The more ancestral subgenera *Micrapis* (dwarf open nesters) and *Megapis* (giant open nesters) maintain a "curtain" of living bees that hangs over the comb covering the queen, the brood, and the nurse bees tending it, which provides a defensive barrier against enemies and a measure of temperature insulation. With the onset of cavity nesting, the worker cohorts formerly engaged passively as living insulation were now freed to contribute more directly to the colony's economy.

Compared with open nesters, cavity-nesting honey bee colonies have higher metabolic and foraging rates, a pattern that does not track with a traditional expectation that metabolism increases as body size gets smaller; workers of low-tempo open nesters include both the largest and smallest workers in the genus. Instead, nesting habit, not body size, explained colony metabolism.

With this background, a plausible scenario for the evolution of worker population size in *A. mellifera* can be inferred: (1) as workers were freed from passive curtain duty, more of them were available for nursing, foraging, and other tasks in support of brood; (2) as the colony's foraging economy improved, this encouraged more brood rearing and the innovation of multiple combs to support it; (3) the increasing ratio of brood to workers had a stimulating effect on foraging, thus increasing selection for high worker metabolism and productivity; (4) the positive feedbacks of more brood and increased foraging reinforced larger colony populations, (5) larger worker populations had a stimulatory effect on worker division of labor and task specialization, further increasing efficiency and social complexity, constituting yet more positive feedbacks on large populations, and (6) increasing social complexity biased the *A. mellifera* lineage toward its temperate expansion, providing additional positive feedbacks toward large worker populations capable of buffering temperature extremes.

Constraints on upper worker populations in *A. mellifera* are seasonal availability of forage resources and physical volume limits of available nest cavities. Within the cavity nesters, colony population and individual body size both seem to comply with Bergmann's rule because *A. mellifera* has conspicuously larger populations and body size than its sister subspecies *A. cerana*. This is consistent with *A. mellifera*'s history of pioneering into high latitudes.

CHAPTER 14

Mechanisms and Evolution of Dance Language

THE WAGGLE DANCE IS ONE of the most celebrated examples of animal communication and a window into foundational inquiry on the honey bee's evolution and modern functioning as a superorganism. Indeed, the waggle dance figures prominently in any discussion of group decision-making, a topic I'll cover in chapter 15. The waggle dance is a true symbolic language, by which we mean its users employ *symbols*—models or representations of real things. For analyzing the honey bee dance and recognizing what it is, the Austrian scientist Karl von Frisch won the Nobel Prize in Physiology or Medicine in 1973 (von Frisch and Chadwick, 1967), and since then, scores of scientists have deepened our understanding of the intricacies and layers of this behavior.

Mechanisms of Dance Language

The waggle dance is a means by which scout foragers communicate to potential recruits the distance, direction, and quality of a potential resource. The resource can be a nest site in the context of a swarm or a foraging asset such as nectar, pollen, water, or propolis in the case of an established colony. A recruit may then use the information to fly out and inspect the resource herself, and if she agrees about its desirability, she will return to the nest and reinforce the campaign with a dance of her own. In this manner, the foraging force can rapidly mobilize around the most profitable sites and, with voting by direct participation, steer the group toward a decision. Inversely, non-profitable sites receive fewer follow-up dances and, by extension, interest in them decays.

Think for a moment how many moving parts there are in this machine. There are at least two levels of participant—the scout/dancer and the recruit/dance-follower; there are at least two levels of information—the information going "out," that is, information about the resource being communicated by the scout, and the information coming "in," that is, how that information is being received and processed by the recruits. And finally, there are related signals that have nothing directly to do with communication but rather serve to alert nestmates to the availability of new information. For a human observer, it can be difficult to know the difference between an announcement or real information. I'll try to explain the waggle dance taking into consideration these moving parts. In what follows, I draw heavily upon the excellent review by Fred Dyer (2002) and sources cited therein.

The scout/dancer performs the waggle dance in a figure-eight pattern (figure 14.1), during the straight run of which she waggles her abdomen left and right. After completing a straight-run waggle, she circles back and does it again, each return constituting one circuit. She alternates her return trips left and right, hence the figure-eight pattern. The straight run encodes for directional information, and it works like this: the angle of the straight run relative to gravity, in our case the top of the comb, is correlated to the angle of the resource to the sun relative to the nest ($\alpha°$ in figure 14.1). It's as if the bees have agreed to pretend that the top of the comb corresponds to the direction of the sun when viewed from the nest. The degrees of the straight run to the left or right of true vertical correspond to the degrees left or right of the sun as viewed from the colony. True vertical has become a *symbol* for the direction of the sun, and the angle of the straight run relative to vertical has become a *symbol* for the resource's direction relative to the sun.

The distance to the resource is symbolized by the tempo of the dance, the number of waggles performed in a straight run, and the duration of

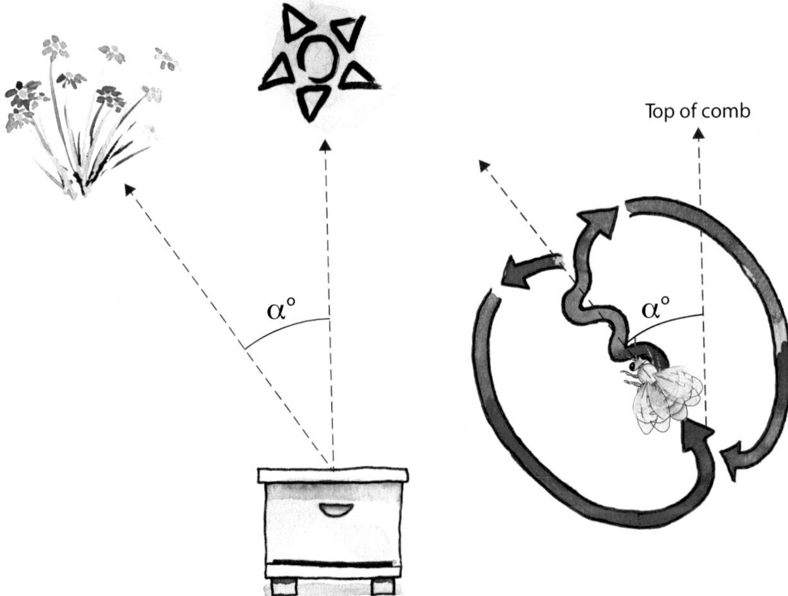

FIG. 14.1. The waggle dance is a true symbolic language that communicates to potential recruits the direction and distance of a resource. The top of the comb, or true vertical, corresponds to the direction of the sun as viewed from the nest. During the course of her figure-eight dance, the scout waggles her abdomen left and right while she performs the straight run. The angle of the straight run relative to true vertical ($\alpha°$) corresponds to the angle ($\alpha°$) of the resource relative to the sun as viewed from the colony. Thus, the top of the comb is a symbol for the sun and the angle of the straight run a symbol for the resource's location relative to the sun. Distance to the resource is symbolically communicated by the tempo of the dance. The nearer the resource, the faster the dance. Recruits, if they visit the site and agree with its quality assessment, will return to the nest and reinforce the message with their own waggle dances.

piping noise the dancer makes while she performs her straight run. As the distance to the resource increases, the tempo slows down, the number of waggles increases, and the duration of piping increases. Conversely, the nearer the resource the faster the dance and briefer the piping.

It was once thought that a second recruitment dance is performed by scout/dancers, the round dance, that communicates resources within 50 meters of the nest (thus relegating the waggle dance to resources >50 meters away) (von Frisch and Chadwick, 1967). The round dance was presumed to communicate no directional information but simply indicate "something good nearby," leaving it to recruits to find the resource by odor alone. But thanks to modern audio recording technology, later researchers have discovered that extremely brief bursts of piping do occur when the round-dancer's body is aligned with the resource. Directional information is therefore available. It is better to think of the round dance as a waggle dance in which tempo is so fast (because the resource is so near) that the figure-eight pattern cannot be sustained. The round and waggle dances thus represent a continuum of the same recruitment behavior (Gardner et al., 2006).

Having concentrated so far on the scout/dancer, let's turn our attention to the recruit/dance-follower. Compared with the followers, the dancers are easy to see on a comb and that means it's also easy to measure what they do, whether number of dance circuits, waggles, or piping sounds. But just because a human observer can measure something doesn't mean that that behavior is the cause of a subsequent action. It is in fact a bit difficult to assign cause and effect to components of the dance language. Dyer (2002) said it well: "Dances provide a rich variety of potential communicative stimuli, but it is unknown which stimuli the bees use. In considering the possibilities, note that the features of the dance that help followers find and stay with dancers need not be the features that carry the signal of spatial location." In other words, it can be difficult to distinguish between a signal that actually contains navigational information from a signal that simply advertises for the dance.

To begin, let's consider the sensory pathways by which a recruit could receive information from a dancer. Dyer (2002) considered three: airborne sounds, vibrations picked up from the comb substrate, or direct touch.

The evidence for airborne sound is supported by the fact that bees are in fact capable of detecting airborne signals within the sound frequency of piping. Second, experiments have shown that recruitment rates are reduced in dances where sounds are missing. Third, when researchers from Denmark and Germany made a computer-driven mechanical dancer bee, the presence of airborne sound was necessary before the model could correctly direct recruits to a resource in the environment (Michelsen et al., 1992).

The evidence for vibration signals is the weakest. First, critics have argued that the relative inefficiency of the mechanical bee described above can be explained because the model was not in direct contact with the comb on which the recruits were standing and therefore could not transmit any vibration signal; second, dancers seem to preferentially dance on open cells rather than sealed cells, and open cells are presumed to transmit vibrations more easily. Finally, it is possible to detect and measure very small vibrations in the vicinity of a dancing bee.

In my opinion, the evidence for direct touch, or a tactile pathway is strongest. First of all, the recruits following a dancer frequently make direct contact with her. Second, it cannot be overlooked that however impressive, symbolic, and information-rich the straight-run of the waggle dance may be, it remains that the recruit must be in a position to appraise the posture of that run, apprehend for herself a true vertical, and transform all that information relative to her own body's angle to the dancer—a cognitive performance at least as impressive as the scout's. A part of that transformation could be simplified if the recruit aligns her body with the dancer, thus perceiving directly for herself the angle of the straight run, and indeed experiments have shown that those recruits that align their bodies behind the dancer are more successful at finding the resource than recruits who viewed the dance from different angles. Direct touch is the most parsimonious explanation for how this information flow happens.

The extent to which any or all of these sensory pathways are conduits through which navigational information is communicated remains ambiguous. What does seem certain is that at least some of them are conduits for advertisement. One cannot overlook how difficult it must be to capture an audience in a teeming population of candidate recruits, especially if their colony exists in total darkness like a hollow tree with limited opportunity for visual signals.

Some open-comb nesting *Apis* species in Southeast Asia express dance behaviors unknown in cavity-nesting *Apis*. The open-comb nesters *A. florea* and *A. dorsata* not only waggle side to side, but also up and down, giving their abdomens a flailing appearance when they perform a straight run. This can be understood as a strong visual signal. A comparative analysis of dances performed by open-comb-nesting *Apis* versus cavity-nesting *Apis* supports the generalization that visual signals predominate in open-comb nesters, whereas sounds are more important in cavity nesters where visual signals are less effective.

The difficulty for a human investigator trying to discriminate an advertising signal from an informational signal lies in the fact that the measured outcome for either could be the same—fewer or greater recruits finding the resource. For example, a strong correlation between behavior X and recruitment to a flower patch could mean either (1) X contains the navigational information, or (2) X was simply an effective advertisement for unknown behavior Y that was the real conduit of information.

Even though ambiguity surrounds the mechanisms of the waggle dance, there is no ambiguity to the fact that the waggle dance is a real symbolic language that communicates objective and specific information. No better proof can be offered than the fact that naïve human observers "literate" in the language can use it to find objects in blind field tests. The waggle dance and the group decisions that it engenders are a complex interaction between the cognitive powers of the individual bee and the order that emerges when individuals in large groups are free to access and respond to local information. Just as the evolutionary costs for such a machine must be high, so must the payoffs.

Evolution of Dance Language

Every species of *Apis* is known to perform recruitment dances of a kind similar to the waggle dance. Dance recruitment language is implicitly a *social* behavior; it could not, and would not, have evolved in pre-social species that do not live together in groups. It must therefore be understood as evolution acting upon the group, not the individual.

Evolution is unwaveringly practical and operates on a cost/benefit basis. It is not cost-neutral to engage in symbolic recruitment language. It takes time and energy for a scout to perform a dance, and it takes time for recruits to follow a dance and absorb its information. If the collective calories spent performing and interpreting dances do not exceed the calories gained from subsequent recruitment events, then evolution will not reward this behavior and its genetic codes will not accumulate and become fixed in the population. Sherman and Visscher (2002) first articulated this premise and experimentally showed that *Apis mellifera* colonies whose recruits were deprived of distance and direction information (by experimentally obscuring sun orientation cues) collected less food from feeding stations than those colonies whose recruits were not handicapped. But the benefits were not consistent. Even though distance and direction knowledge increased recruit visitation to feeder stations, about one-third as many recruits could find the stations without this knowledge. Presumably these "naïve finders" were responding to odor signals they had picked up from scouts at the hive and succeeded in finding the source in the field on their own.

Dornhaus and Chittka (2004) noted this ambiguity and reasoned that the benefits of dance language may depend on the kinds of environment in which the bees evolved. Pointing out that floral resources tend to be more patchy in the tropics and more evenly distributed in temperate zones, those authors proposed that precise dances would be more important in tropical zones. In other words, sophisticated recruitment regimes would not evolve if floral resources were ubiquitous. So they proceeded to do an experiment comparing recruitment performance of *Apis mellifera* colonies deprived, or not deprived, of sun orientation cues[1]—and they replicated this experiment in three different habitats: a Mediterranean shrubland habitat in Spain, a site of mixed agriculture and meadows in Germany, and a dry deciduous forest in India.

True to their prediction, foraging success (weight change of hives) was only improved when precise orientation cues were made available at the tropical Indian site. Weight changes in the two temperate sites, Spain and Germany, did not depend on whether scouts or recruits could accurately communicate sun-oriented navigation cues. The authors concluded that their results "are consistent with the hypothesis that the honey bee dance language is an adaptation to the tropical conditions under which the genus *Apis* diversified, and may no longer be essential for efficient foraging in some temperate habitats."

Results such as these strengthen the traditional interpretation that *Apis* has been subject to tropical adaptations for much of its biogeographic his-

1. The experimenters used modified hives in which windows could be opened or closed, allowing or preventing direct view of the sun, and combs that could be turned horizontal or vertical, allowing or preventing the vertical posture that lets bees symbolically represent the direction of the sun as the top of the comb. Closed windows and horizontal combs prevented any means of communicating distance or directionality.

tory. Indeed, the ongoing discussion about ancient *Apis* range expansions can benefit from behavioral data considered alongside other factors such as the hypothesized European origin of the genus, ancient climatic patterns, and whether the lineage leading to *A. mellifera* evolved directly in Europe or first passed through Southeast Asia (chapter 22). In any case, a basal state for small single-comb open nesting has been sustained in the latest phylogenies with the giant single-comb open nesters and cavity nesters diverging later and sister to one another (see figure 22.1).

This behavioral evidence brings us back to our subject at hand. All modern *Apis* dance, but there is a range of adaptions from ancestral to derived that shed light on the evolutionary forces that shaped the behavior. The dwarf honey bee subgenus *Micrapis*, basal with respect to all other *Apis*, performs dances on a so-called "dance platform"—the section of horizontal comb that wraps around the branch from which the comb hangs (figure 14.2). This horizontal platform makes it easy for a scout to communicate direction: when she does the straight-run of her waggle dance, she simply points in the direction of the source (Koeniger et al., 1982). When the ancestor of the giant rock bees committed to single vertical-hanging combs with no horizontal dance platforms, scouts could no longer directly point to the source; this necessitated the evolution of symbolic representation of the sun's direction as true vertical, in other words, orientation to gravity (Dyer, 2002).

In the giant bees, modern subgenus *Megapis*, we first see auditory signals, "piping" used by scouts to reinforce information content in their dance. It is believed that piping was an adaption to dances performed in the dark, and in modern *A. dorsata*, we have both nocturnal foraging as well as dances performed under curtains of bees on the comb—each a situation where auditory reinforcement could be beneficial (Kirchner and Dreller, 1993). In the cavity nesters, subgenus *Apis* including *A. mellifera*, we likewise see the full suite of behaviors—gravity orientation, symbolic representation of the sun, and piping, all of which are understandable adaptations for dances performed in a dark cavity. Interestingly, *A. mellifera* retains the ability to orient on a horizontal surface (Michelsen et al., 1986; von Frisch and Chadwick, 1967), a behavior only expressed during swarming when scouts are informing followers of the location of a suitable nest cavity.

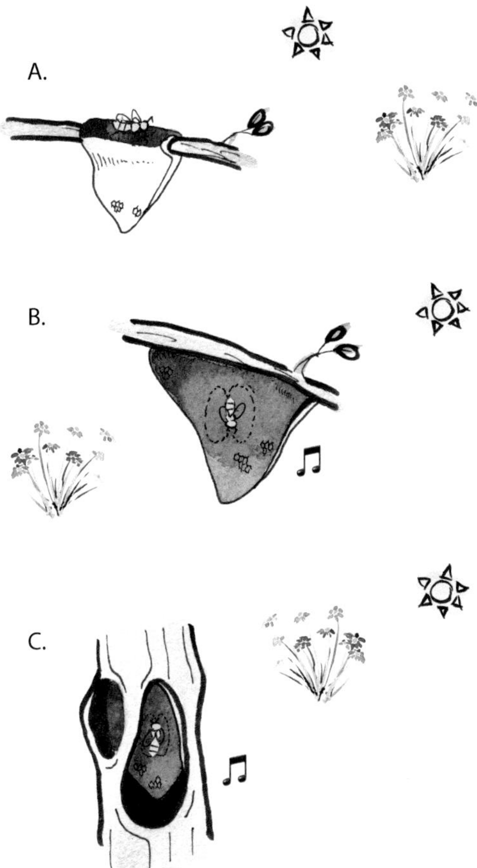

FIG. 14.2. There is a relationship between the phylogenetic history of honey bees and their corresponding dance languages. The first group **(A)** is subgenus *Micrapis*, basal with respect to the rest of the genus, the dwarf honey bees that live on single open combs that wrap around the supporting branch, creating a horizontal "dance floor" (dance floors shaded red here and in **B** and **C**). The horizontal surface lets scouts aim their dance directly at the resource, and dances proceed without regard to position of the sun. The next two are a sister group, beginning with **(B)**, subgenus *Megapis*, the giant honey bees that build a single open comb on the underside of tree limbs or cliff faces lacking the wrap-around horizontal dance surface. The waggle dancer in **B** is performing the straight waggle portion of her dance facing directly opposite the top of the comb, communicating to her sisters that the floral resource is in the opposite direction of the sun relative to the nest. The other pair of this sister group is subgenus *Apis* **(C)**, the cavity-nesting bees. The waggle dancer in **C** is performing her straight waggle run directly facing the top of the comb, communicating that the floral resource is in direct line with the orientation of the sun relative to the nest. Innovations in nest architecture in groups **B** and **C** meant that scouts could no longer aim their dances directly at the resource. The vertical dance posture selected for these groups to evolve a symbolic representation of the sun's position as true vertical, employing gravity cues. Groups **B** and **C** also encounter darkness during their dances, group **B** because scouts sometimes dance under curtains of living bees and forage nocturnally; group **C** because they live in cavities. Dancing in the dark selected for auditory signals, "piping" by scouts as a means to supplement their dance's navigational information. Figure redrawn from figure 1 of l'Anson Price and Grüter (2015).

Speaking of nest cavities raises a legitimate question about the order of events: did the ability to orient dances in darkness enable cavity nesting? Or did cavity nesting induce the need to orient in darkness? The best resolution to this and similar questions comes from a phylogenetic analysis performed by Raffiudin and Crozier (2007) using 10 species of *Apis* sampled from Indonesia, Malaysia, Thailand, Nepal, and Australia and including two "outgroup" bee species for comparison—a bumble bee from Switzerland and a species of *Trigona* from Thailand. Those authors constructed a phylogeny inferring ancestral states for nest site and dance language by comparing gene sequences across groups.

The results were a mix of the expected and unexpected. On one hand, the traditional phylogeny was upheld, showing the dwarf bees sister to the giants + cavity nesters (figure 22.1). But the analysis also inferred that the most recent common ancestor of *Apis* was an open nester, with a silent dance, on a single comb wrapped around the branch, but dancing in a vertical, not horizontal orientation. Slightly more predictably, the common ancestor of the giants and cavity nesters was inferred to have a vertical dance, piping, open nesting with, curiously, several combs; a throwback to this latter behavior is sometimes seen in modern *A. mellifera* even in temperate zones (figure 11.3).

Raffiudin and Crozier (2007) recognized the unexpected signal indicating vertical dancing as a primitive state, and said, "the support for this is sufficiently weak that the question remains essentially unsolved." They suggest that if in fact vertical dancing is ancestral, then the horizontal posture of today's "primitive" *A. florea* may be a derived adaptation to conserve the energy and cognitive innovations necessary for transposing the sun's direction symbolically. If horizontal dancing is ancestral, then this sustains the traditional view that vertical dancing became necessary after innovations in nest architecture—the loss of comb-wrapping around the top of the branch and the move toward cavities—eliminated horizontal dance surfaces.

Finally, we should wonder how dancing became a vehicle for communicating navigational information in the first place. The best thinking is that early honey bee foragers, loaded with nectar and returning to their single-comb open nests, would engage in excited, disorganized movements on the comb. I've wondered if sugar-induced excitement from a belly full of nectar could explain these early "dances," giving a nod to emergent properties, the order that spontaneously emerges from prior conditions (chapter 10). In any case, we see similar forms of this behavior today in the comparatively primitive bumble bees. A bumble bee forager returns to the nest and makes an excited run around the nest, giving off odor signals from the resource just visited. This behavior stimulates other foragers to go searching, but to our knowledge the dance confers no information on direction or distance. Nevertheless, foragers leaving after watching such a dance preferentially end up at sites matching the odor the scout brought home.

From such an early template, we can imagine the subsequent adaptions. Given the evolutionary rewards for precision, especially in the tropics where floral resources are more patchy, we can imagine the benefits to a colony whose scout dances oriented toward navigational cues and whose recruits could bias their subsequent flights according to those cues.

Considering the modern distribution of *A. mellifera* and the evidence that precision is less important in temperate zones where floral resources are more evenly distributed, we may wonder if natural selection will continue favoring precise waggle dances in the western honey bee. Evolution, ever practical, doesn't linger to self-congratulate: if waggle dances increase colony fitness, they will persist. If they cost more energy than they procure, they will not.

Summary

The waggle dance is a means by which scout foragers communicate to potential recruits the distance, direction, and quality of a potential resource. A recruit may then use the information to fly out and inspect the resource herself, and if she agrees about its desirability, she will return to the nest and reinforce the campaign with a dance of her own. In this manner, the foraging force can rapidly mobilize around the most profitable sites and, with voting by direct participation, steer the group toward a decision. Inversely, non-profitable sites receive fewer follow-up dances and, by extension, interest in them decays.

The scout/dancer performs the waggle dance in a figure-eight pattern, during the straight run of which she waggles her abdomen left and right. In cavity-nesting *Apis*, the dancer communicates direction by orienting the angle of her straight run relative to gravity, the top of the comb, to match the angle of the resource to the sun relative to the nest. Distance to the resource is symbolized by the tempo of the dance, the number of waggles performed in a straight run, and the duration of piping noise the dancer makes when she performs her straight run. As the distance to the resource increases, the tempo slows down, the number of waggles increases, and the duration of piping increases. Conversely, the nearer the resource, the faster the dance and briefer the piping. Recruits, the followers of the dance, consider signals from airborne piping sounds, dance vibrations picked up from the comb, and by directly touching the dancer. Additional signals exist that do not code recruitment information but rather advertise the presence of the dancer.

Dance recruitment language must be understood as evolution acting upon the group. All modern *Apis* dance, but there is a vector of adaptions from ancestral to derived that shed light on the evolutionary forces that shaped the behavior. The dwarf honey bee subgenus *Micrapis*, basal with respect to other *Apis*, performs dances on a horizontal comb enabling the dancer to simply aim her straight run directly toward the resource. The subgenus *Megapis* lacks a horizontal dance platform, which facilitated the evolution of a vertical dance orientation using gravity cues to symbolically associate the top of the comb with the direction of the sun relative to the nest. In the cavity nesters, subgenus *Apis* including *A. mellifera*, we see vertical dances, gravity orientation, symbolic representation of the sun, and piping, all of which are adaptations for dances performed in a dark nest.

The evolution of the dance itself is hypothesized to begin with sugar-induced excitation in foragers. We see clues to this behavior today in the comparatively primitive bumble bees. A bumble bee forager returns to the nest and makes an excited run around the nest, giving off odor signals from the resource just visited. Foragers leaving the nest after watching such a dance preferentially end up at sites matching the odor the scout brought home. Given evolutionary rewards for precision and efficiency, natural selection would favor those colonies whose scout dances oriented toward navigational cues and whose recruits could bias their subsequent flights according to those cues.

CHAPTER 15

Group Decision-Making

WHEN I SPEAK TO NON-SCIENTIST GROUPS, I often use an analogy that helps explain what group decision-making is and is not. It goes like this:

Before we arrived here this morning, someone, maybe some of you, set up these chairs we're sitting in now. I wasn't here, but I can guess what happened. Someone, say Tom, grabbed a chair from the stack and put it *here*. Why here? Because Tom had to put it somewhere, and to Tom this seemed the right place. Next came George, and he put his chair here next to Tom's. Why here? Because Tom had put his chair there first. Next came Ellen, and she put her chair next to George's—and so on until the whole room was filled with orderly rows, complete with aisles that conveniently fit the dimension of the human beings who would be using them. Why did the aisles fit? Because it was humans who moved around them as they set up the chairs. One action led to the other. Pre-existing properties, e.g., the walls constraining room size, the location of the podium, task-oriented humans, raw material to work with (the chairs), the body size of the humans doing the work, dictated the outcome—orderly rows and aisles. And it was the actions of individual actors, each responding to their immediate cues, that "decided" the length of rows, the width of aisles, and their exact locations.

Continuing my illustration I say, What I guess was *not* happening was some busybody directing the whole project. I can't imagine President Bob standing in front of the room, pointing: "Tom, put your chair *here*. OK, now George you put your chair *here*, now Ellen *here*," and so forth.

Group decision-making in the superorganism looks a lot like the former but includes elements of the latter as well. Studying group thinking involves parsing out the effects of individuals—how each forages, how each appraises a resource, how each reacts, how the actions of individuals influence other individuals—as well as identifying the tipping points that commit the whole group to decide in favor of A instead of B. This level of thinking is perfectly adequate for the kinds of decisions faced by a honey bee superorganism, such as choosing nest sites, choosing foraging sites, and optimizing the shape and configuration of its comb nest inside a hollow tree.

There are some overarching features that characterize a group engaged in collective cognition. The process depends upon independent assessors (acting individuals) who are numerous enough that errors are averaged out and environmental and group states accurately appraised. The process requires positive feedback loops so that weak signals are amplified and information rapidly spread. There are non-linear (synergizing) interactions among individuals that drive the group to consensus among competing options (Sasaki and Pratt, 2018). Actors involved in group decisions are each (1) independent, (2) free to react appropriately to her local information, (3) interchangeable with other actors to some degree, (4) free to contribute her information to the group regardless of its quality, and (5) free to respond to other actors (Visscher, 2007).

Finally, animal group decisions fall generally into two kinds: combined or consensus (Tindale and Kameda, 2017). In *combined* decisions, an actor decides individually, yet that individual response is influenced by other actors and, in sum with those of others', affects the outcome of the group. A forager, for example, seeks pollen or nectar individually yet is privileged with information gained from dancing scouts back at the nest, as I described in chapter 14. In *consensus* decisions, actors decide together and all abide by the decision of the group. Nest relocation, as I focus on below, is an example of consensus-making.

All of these are characteristics of lateral *heter*archical kinds of thinking, in contrast to the top-down *hier*archical kind of thinking in metazoan animals like ourselves, whose nervous systems are centralized and whose societies often exhibit decision-making that is centralized and top-down. But we must not forget that the individual bee is a multicellular organism whose decisions proceed from a highly centralized nervous system. Thus, thinking in the superorganism is a combination of hierarchical and heterarchical decision-making.

Among the intriguing aspects of the superorganism concept are insights it offers to the evolution of multicellular organisms like ourselves. We will see below that there are elements of our own brains that operate like the superorganism and hint at analogous evolutionary pathways in our shared deep past.

Nest Site Selection: An Archetype of Consensus Group Decision-Making

One of the best-studied examples of decision-making in the honey bee superorganism is the process a swarm uses to select a nest site. It offers examples of decision-making that are hierarchical (individual bee) as well as heterarchical (the group). This is a critical, once-in-a-lifetime decision, in contrast to less weighty ones like whether to forage at this flower patch or that one. A natural consequence of this is the importance of unanimity: when it comes to food plants, foragers can spread themselves among many choices, but when it comes to the nest cavity, the swarm must choose one. Errors in this process are the likely cause of beekeeping stories about "confused swarms" that refuse to settle or relocate or otherwise act in an incoherent fashion.

Lindauer (1957), working in Germany, established that swarms use recruitment dances to finalize nest site decisions in a manner similar to colonies choosing foraging sites. The immediate actors are the same—the scouts who discover and appraise potential sites and return to the nest or clustered swarm to perform a recruitment dance, and the recruits who respond to the dance and fly to the site to inspect it for themselves (figure 14.1).

As I described in chapter 12's "Comb Design and Construction" section, a complex process like this can include fixed action patterns in individual bees, that is, behaviors that are instinctive and "hardwired" into individuals. In our present case, the individual scouts use a mixture of instinctive knowledge and accidental serendipity to discover and appraise a candidate cavity. We know, for example, that tree cavities and their associated knot hole entrances are more readily occupied if the knot hole is in open sun, the cavity exudes an odor of a previous colony, and the cavity is high in the canopy (figure 18.1). But we cannot assign to scouts any higher-level cognition regarding these choices; it may simply be that scouts prefer flying high rather than low and that odorous cavities in full sun are easier to find.

But once a scout finds a cavity, then what follows can only be understood as instinctive imprints, as reviewed by Visscher (2007). The scout walks around the cavity. If space permits, she flies from wall to wall, repeatedly, as if she were accumulating data to determine an average cavity volume. There is evidence that she uses the visual cue of the illuminated entrance as an orientation point for her measures. Notice that at this point, we've switched our focus to the cognitive ability of the individual bee and her central nervous system, and her empirical spatial analysis (which strikes me as impressive).

Although nest site scouting begins in the days leading up to colony fission, selection is culminated only after a reproductive swarm has left the parent nest and clustered in a temporary bivouac (figure 3.13). At this point, none but the scouts and their recruits have any knowledge or opinions about the candidate destinations. A scout knowledgeable of a suitable site performs a waggle dance on the surface of the bivouacked swarm; if it's an especially good site she makes multiple repeats of her dance, which increases the probability of recruiting other scouts to the site who, in turn, return and perform their own dances. All the while, there are other scouts advertising for other sites. Because of the need for unanimity, there must be a mechanism for eliminating competing dances, what Seeley (2003) calls "the expiration of dissent." There seem to be three ways this happens: (1) a high dropout rate of dancers, (2) competition among nest sites for a limited pool of scouts (Visscher, 2007), and (3) so-called "stop signals" (Seeley et al., 2012).

Compared with scouts for food sites, scouts for nest sites have very high dropout rates, even if their nest sites are good. On average, fewer than 5% of nest scouts keep dancing for more than eight trips, compared with nectar scouts, 93% of whom keep dancing for more than eight trips. The flip side of this is that nest seekers have a higher switch rate from dance performing to dance following; in nest-seeking bees, up to 40% of dancers subsequently become dance followers, whereas in nectar-foraging bees, the switch to dance following is almost zero (Visscher, 2007). The sum effect of this is to limit the pool of candidate sites but increase in relative terms the number of appraisers for each site, each of whom may return to the nest with her own "opinion" and either offer a reinforcing dance or go check out another, yet the sum of whose opinions should accurately point to the one best site available.

A second driver for eliminating dissent is the fact that the number of nest scouts is small to begin with. Available nest cavities are "competing" for a finite pool of scouts.

A third dissent eliminator is the *stop signal*—a direct act of inhibition practiced by dancers on competing dancers (Seeley et al., 2012). A signaling bee will butt against the head or thorax of the other dancer while delivering a high frequency vibration, the effect of which is to inhibit dancing. The dancer does not immediately stop but will do so after a certain (unknown) threshold number of stop signals is received. It becomes a matter of numbers—as recruits begin to agree on a particular site, their stop signals will have an increasing impact on an increasingly diminishing number of competing dances.

Although unanimity is necessary to the successful outcome of the project, at first, the bivouacked swarm makes its decision not based on consensus but rather on attainment of a quorum for at least one site. In *Honeybee Democracy*, Seeley (2010) pointed out that it would take too much time for all the swarm's recruits to reach consensus, which would, as a matter of course, require each of them to visit all dances, appraise all sites, and agree on one. How then does the superorganism know when a quorum has been reached?

There is reason to think that scouts and the recruits that follow them to a candidate site are sensitive to the presence of other nestmates visiting the cavity. This is a positive reinforcement I liken to the warm feeling one gets when you arrive at a party and find friends already there. Compare that to the awkwardness of being the first to show up. For bees, this experience is called *quorum sensing*, and it is understood to reinforce their perceptions that the group is nearing agreement that the site is acceptable (Seeley and Visscher, 2004).

Once a quorum is reached for a site (estimated at 75 scouts), these scouts-in-solidarity force the issue by provoking the swarm to take to the air. This is done by making piping signals—which excite the clustering bees to warm their flight muscles, and by making buzz-runs—which "release" group lift-off [figure 16.2(D)]. Once airborne, most swarm participants are just as ignorant of their destination as before, but this time, there is a quorum of scouts and their recruits who know where they want to go. These individuals begin signaling by "streaking" through the disorganized cloud with fast, straight flights pointing directly toward the new home [figure 16.2(E)]. These "streakers" are visible to anyone who has watched a bivouac embark for its new home, and they tend to be toward the top of the whirling mass of bees. By repeating their streak flights, the scouts provide a visual cue to their nestmates who respond by orienting their own flights in the direction of the signals. The visual cue of bees arriving in or near the entrance of new site, as well as their volatile Nasanov gland excretions, further orient newcomers to their new home [figures 11.4 and 16.2(F)].

It is interesting to note that the mass flight behavior of a swarm mirrors elements of apparent coordinated flocking behavior in birds (Couzin et al., 2005), that is, that each individual in flight seeks to maintain a uniform distance between herself and her neighbors. Evasive action by individuals at the periphery against predators or obstacles are rapidly adjusted for by the whole group as each member shifts its course accordingly to maintain distance between its neighbors. Computer models (Janson et al., 2005) have shown that in this manner, the action of only a very few streaker bees is sufficient to guide a swarm of naïve bees to the new nest site.

Even at this stage there may still be dissenters. In such cases, a real group decision, a consensus, may not be final until the group votes by direct participation—a majority entering one cavity and not the other. In such cases, it may also matter which cavity the queen finds first because her pheromones are known to play an important role in the stability of the bivouac and are likely to help orient the swarm to a new cavity.

A final note worth mentioning refers back to idea of the stop signal. It so happens that a parallel phenomenon happens in complex metazoan brains like our own. In both the superorganism "brain" and ours, there are multitudinous "evidence-collecting units"—call them bees or call them neurons—appraising conflicting information and choosing among alternatives. In both types of brains, there are cross-inhibiting features that shut down one set of units in favor of another (Seeley et al., 2012). Demonstrating how nature has converged upon similar answers for similar problems, even across such phylogenetically distant animals as honey bees and humans, is one of the most exciting ways entomology, and the superorganism concept, is informing all of biology today. It's a beautiful example of the unitive quality of all nature.

Evolution of Group Decision-Making Behavior

It is worth reinforcing at this point that group decision-making in the honey bee is an outcome of emergence—the summed decisions and actions of independent actors—bearing the imprints of social selection. In chapter 14, I proposed a plausible scenario for the evolution of one component in this system—recruitment dancing—from an ancestral forager's incipient state of energy excitation following a sugary meal of nectar. Dances, as well as such associated system components as stop signals and quorum sensing thresholds, can only be understood as adaptations to natural selection acting on the group. To the extent a component improves the colony's survival over other colonies in its breeding population, there will be positive selection optimizing that component's execution.

Alongside positive selection on such particulars as dances, dance interpretation, piping, stop signals, streaking, and quorum thresholds, natural selection is expected to favor cognitive pathways that preempt potential problems and conflicts. Tindale and Kameda (2017), in their review of the evolution of group decision-making, pointed out that social insects are vulnerable to *coordination losses*—inefficiencies stemming from poor coordination among nestmates of their cognitive and behavioral inputs toward group performance. For example, consider the positive feedback described in the previous section on nest site selection, by which recruits responding to a recruitment dance return to the nest and, if they agree with the scout's assessment of the site, reinforce that decision with a dance of their own. The positive feedback is adaptive only if it amplifies correct information, but it could just as easily amplify errors unless there were checks in place to mitigate such hazards.

The solution seems to lie in natural selection favoring an optimum mix of interdependence and independence of appraisers. A computer simulation (List et al., 2009) showed that a recruit is interdependent in the sense that she is more likely to visit a site advertised by a previous scout. But the duration of the *recruit's* (now acting as a scout) dance—increasing duration being positively correlated with increasing site quality—is determined solely by the recruit's own perception of the site. In other words, she was influenced by another to visit the focal site, but her opinion of it is her own. In this manner, errors are tamped down, a cascade of faulty information avoided, and the group arrives at more optimum decisions.

Summary

Group decision-making in the honey bee is an outcome of emergence—the summed decisions and actions of independent actors—and bears the imprints of social selection. Actors involved in group decisions are each (1) independent, (2) free to react appropriately to local information, (3) interchangeable with other actors to some degree, (4) free to contribute information to the group regardless of its quality, and (5) free to respond to other actors. Animal group decisions fall generally into two kinds: combined or consensus. In combined decisions, an actor decides individually yet the response is influenced by other actors and, in sum with those of others, affects the outcome of the group. A foraging honey bee, for example, seeks pollen or nectar individually yet is privileged with information gained from scouts at the nest exhibiting recruitment dances. In consensus decisions, actors decide together and all abide by the decision of the group.

Honey bee nest selection during reproductive swarming is an example of consensus-making. Recruitment dances and other system components, such as piping, stop signals, streaking, and quorum thresholds, can only be understood as adaptations to natural selection acting on the group. To the extent a component improves the colony's survival over other colonies in its breeding population, there will be positive selection optimizing that component's execution. Alongside such examples of positive selection, natural selection is also expected to favor cognitive pathways that preempt errors, conflicts, and inefficiencies stemming from poor coordination among nestmates. The solution seems to lie in natural selection favoring an optimum mix of interdependence and independence of appraisers.

As an example, a recruit is interdependent in the sense that she is more likely to visit a site advertised by a previous scout. But she is motivated to amplify the advertisement only if she "agrees" on its quality. In this manner, errors are tamped down, a cascade of faulty information avoided, and the group arrives at more optimum decisions.

CHAPTER 16

Mechanisms and Evolution of Reproductive Swarming

IN RECENT YEARS, the subject of swarming has grown complicated in beekeeping circles in ways I never could have imagined as a young scientist or even younger beekeeper. Ever since C. L. Farrar's pioneering work in 1937 showed a positive relationship between colony population size and honey crops, beekeepers have rejected the whimsical notion that "a swarm of bees in May is worth a ton of hay" in favor of no swarms at all—the pursuit of unnaturally large colonies that make unnaturally large honey crops. Swarm management figures prominently in every beekeeping how-to book or chapter, a literature to which I have actively contributed, promoting the orthodoxy of swarm control as vigorously as I could (Delaplane, 2007, 2015). So when I feel compelled to pull back on my enthusiasm for swarm control in light of emerging evidence for its health risks to bees, I feel I can say so from a position of earned authority.

In beekeeping contexts, evidence is mounting that swarming has beneficial health effects on colonies; when colonies are left free to swarm, they tend to have lower levels of brood diseases and parasitic mites (Loftus et al., 2016; Royce et al., 1991). This seems to be a combined effect of age-based asymmetries in the bees that join the swarm (Gilley, 1998), the disruption in brood rearing, lowered nest congestion in the old cavity, and the prospects of a clean disease-free new cavity—all of which conspire to reset the parasite clock backward. No doubt similar healthful dynamics were at work in natural history, shaping the evolution of reproductive swarming, arguably the fulcrum around which the annual life cycle of *Apis mellifera* turns. In the following discussions, I have been greatly assisted by the excellent review of Grozinger et al. (2014).

Independent versus Dependent Colony Founding

There are two modes of colony founding available to eusocial insects (Cronin et al., 2013). By far, the majority practice *independent* colony founding, in which a solitary queen (or a royal couple in termites) disperses alone (or together) and single-handedly founds a new nest. Independent colony founding, in turn, happens in two forms. In *non-claustral* independent founding, the female must leave the nest periodically to forage for herself and her larvae. An example of this is the bumble bee queen who emerges alone in spring, finds a secure semi-subterranean hollow (usually in grassy thatch), collects a clump of pollen, lays a small cluster of eggs inside it, and single-handedly incubates the brood clump with her own body heat. The queen must abandon the brood clump for brief spells while she forages for her own sustenance. The queen lives in this solitary state until the emergence of her first clutch of workers (Delaplane, 2021).

In *claustral* independent founding, the queen (or royal pair) seals herself away in a small cavity and feeds her brood from her own metabolized body reserves (or for termites, from cellulose in the nest substrate). The red imported fire ant *Solenopsis wagneri* gives an example of claustral nest founding. In the weeks between her emergence and dispersal, a young queen accumulates storage proteins and fats, tripling her body weight. She then mates with a single foreign male, disperses a few hundred meters, sheds her wings, excavates a small enclosure, and seals herself in. Therein, she produces a few tens of eggs in isolation and metabolizes her energy reserves and now-useless flight muscles to feed her first clutch of larvae. These individuals become the first foragers for the new colony, freeing the queen from ever needing to leave the colony again (Tschinkel, 2006).

In the second mode, *dependent* colony founding, one or more queens disperse from the parent colony, accompanied by and dependent on a cohort of nestmate workers. Honey bees are the archetype of dependent nest founding, also called colony *fission*, but it occurs in other eusocial species under a range of variations. Some species donate not just workers and nectar (in the workers' crops) to the new queen/colony but also brood of all stages, as in the ant *Cataglyphis cursor* (Chéron et al., 2011) and pollen and nest-building materials as in the stingless bee *Trigona* (*Tetragonula*) *laeviceps* (Inoue et al., 1984).

Of the two modes, independent founding is the riskier option, especially non-claustral founding, which requires the incipient queen to forage outside her shelter. As a result, numerous lineages have lost the solitary phases of colony founding in preference to dependent founding.

Dependent founding occurs across a range of eusocial insects but does not easily sort by generalizable criteria such as taxon or complex versus simple eusociality. It has evolved at least three times in wasps, numerous times in ants, and twice in bees (Cronin et al., 2013). In bees, it occurs exclusively in the tribes Meliponini and Apini that express complex eusociality, whereas ancestral independent founding is retained in the simple eusocial Xylocopinae, Bombini, and Halictidae. The exclusively complex eusocial ants express the whole gamut of options—independent founding, dependent, and mixed.

Foremost, it is important to emphasize that dependent colony founding buffers the efforts of a pioneering queen with a stable cohort of highly related nestmate workers who immediately perform the duties of foraging and nest construction and defense. For ants, it is thought that high levels of intraspecific competition and predation select for dependent founding owing to the need for an effective workforce from the colony's beginning (Cronin et al., 2013). These risks and drivers certainly seem to apply as well to *A. mellifera*, especially for African biotypes for whom predation rates are high (Fletcher, 1978; Rinderer, 1988) and for temperate biotypes for whom natural nectar periodicity agitates the well-known phenomenon of "robbing," aggressive honey theft of a weak colony by a more populous neighbor. The relevance of robbing as an actor on social evolution was shown by Johnson and Nieh (2010), who found evidence that stop signals in the honey bee's dance repertoire (chapter 15's "Nest Site Selection: An Archetype of Consensus Group Decision-Making" section) function in part to shut down an unprofitable attack by an aggressor colony if the neighboring colony's population is deemed large enough to defend itself. Dependent colony founding in the honey bee can be understood as a similar, but defensive, response to such forms of predation.

The sorting factor that may come nearest to explaining the distribution of dependent versus independent nest founding across eusocial species is wingedness (Cronin et al., 2013). Adding to the other ecologic benefits of dependent founding, a species in which all castes possess wings can escape habitats that have become unfavorable and more quickly exploit new ones. Dispersal is more limited with species with wingless workers, raising risks from intraspecific resource competition and inbreeding depression.

A major limiter to the broader occurrence of dependent colony founding is its formidable energetic costs compared with the modest costs of producing independently dispersing queens. A colony of temperate *Apis mellifera* can typically issue no more than one viable swarm per year (chapter 3's "Temperate Honey Bee Life History" section) whereas the interval between colony reproductive events in stingless bees can be as much as 20–25 years (Slaa, 2006), an adaptation to nest site scarcity and high rates of intraspecific competition (Roubik et al., 2018).

In chapter 5, I discussed the "point of no return," a milepost in complex eusocial evolution at which workers and queens diverge morphologically to the point where neither is independent of the other and the whole colony can be considered a Darwinian unit of selection (chapter 5's "Simple and Complex Eusociality" section). The persistence of independent colony founding in the complex eusocial ants would seem to violate this category were it not for the many other distinctives ants possess that are exclusively complex (table 5.3). But the milepost stands without ambiguity in the honey bees and stingless bees, whose queens have completely lost their ability to function without workers.

Mechanisms of Dependent Swarming in the Honey Bee: Primer Conditions

Swarming is, foremost, reproduction at the scale of superorganism in distinction to reproduction at the scale of individual: the sequence of mating flights and multiple inseminations practiced by young queens and drones; the subsequent egg depositions by the mated queen; and the developmental stages of egg → larva → pupa → adult experienced by every member of the colony.

It is also true that the two types of reproduction are qualitatively different: individual reproduction, at least for workers, is sexual, involving the union of egg and sperm and resulting in new chromosomal combinations, whereas superorganismal reproduction is asexual, analogous to binary fission of the kind practiced by single-celled organisms such as *Paramecium* spp. and bacteria, literally a division of one body into two, each with identical genetic information. But it's that part about identical genetic information where the analogy with single-celled organisms breaks down. Although reproductive fission in the superorganism does involve splitting one colony into two, as we will see later, it also results in new gene combinations. A new swarm is not a clone of the parent colony from which it divided.

A successful swarm event is a coordinated interaction between the workers and the queen. It is also an extremely costly form of reproduction compared, for example, with those insects who deposit hundreds of eggs on the underside of a leaf, offering them little or no parental care, resulting in a shockingly small percentage of the young surviving to reproductive opportunity. The opposite strategy is for an organism to produce very few offspring but invest heavily in each one. Humans follow this strategy, as do honey bee superorganisms. In the case of the honey bee superorganism reproducing by fission, we can gain an appreciation of this cost by comparing the biomass of the parent colony to the biomass of the swarm it produces. Winston et al. (1981) found that colonies with ~20,000 workers were able to issue swarms with ~16,000 workers in a first, or prime,[1] swarm. This is a staggeringly large

1. A colony may swarm more than once. The largest and most successful swarm is the first, also called "prime," swarm. This usage is not to be confused with the "primer" conditions I am describing that prepare a colony for the swarming event.

investment, large enough to imperil the survival of the parent colony. This is why swarming is biased toward the earliest time of year when floral resources are richest. The steady food supply improves the chances of recovery for the parent colony and successful establishment for the swarm. Each has a better chance of building up a food supply in time for the next winter.

Swarming also operates at two time scales—a preparatory, default, or primer stage that may last weeks or months, and a trigger or releaser stage that can last hours or even minutes. For these two time scales, I borrow language used to describe insect pheromones as *primers* or *releasers*. Primer pheromones operate by affecting an insect's long-term physiology or readiness for a function or task. In honey bees, a good example of primer pheromone is queen mandibular pheromone (QMP) that keeps worker bee ovaries in a perpetual state of underdevelopment (Hoover et al., 2003). In contrast, releaser pheromones operate at a short time-scale and cause immediate expression of a behavior or state. A good example from honey bees is the alarm pheromones that trigger workers to search for a moving target (Wager and Breed, 2000). This section focuses on the colony's long-term conditions that *prime* the colony for swarming.

To begin, we need to understand that reproduction is the default setting for any colony of honey bees coming out of winter. The queen seems to be sensitive to changes in day length, and when days begin to lengthen in the depths of winter, she breaks her hiatus from egg laying, and the workers respond by fixing brood nest temperatures at about 35°C (they were rather lackadaisical about it prior to this point) and secreting brood food and feeding it to the larvae (Winston, 1987). This growth begins modestly but increases as new workers emerge to join the effort. This growth is also fueled entirely by stored food. This is why late winter/early spring is the most dangerous risk period for starvation.

The stage is now set for five primer conditions that seem most responsible for regulating swarm preparations (figure 16.1): (1) increasing colony size, whether measured by combs or bee population (Winston, 1987), (2) increasing nest congestion of workers (Lensky and Slabezki, 1981), (3) increasing population ratio of young to old bees (Winston and Taylor,

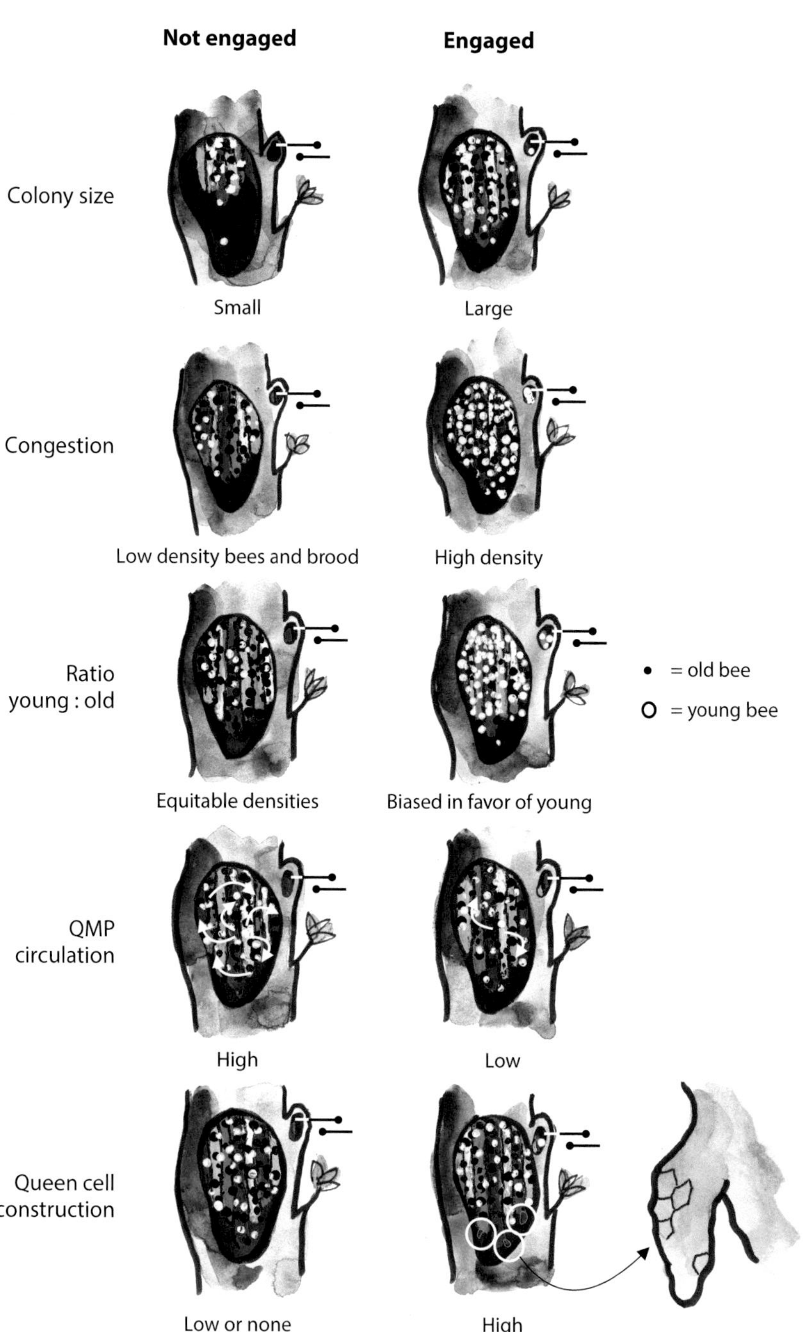

FIG. 16.1. There are at least five primer conditions, or colony states, that regulate swarm preparations. These primers act on a time scale of weeks or months and set the stage for the eventual swarm event which operates on a time scale of hours or minutes. These primer states are, from top to bottom: (1) colony size, whether measured by worker population or combs, (2) brood nest congestion, whether worker population or brood, (3) high ratio of young to old workers, (4) decreasing efficiency of queen pheromone movement throughout the nest, and (5) queen cell construction. Large, congested colonies with a high ratio of young to old bees, with low efficiency of QMP distribution, and high rates of queen cell construction are optimally engaged for swarming.

1980), (4) decreasing efficiency of QMP movement throughout the nest (Naumann et al., 1993), and (5) production of queen cells (Allen, 1956).

The interesting thing is these factors are highly correlated and interdependent. Changes in one affect, as a matter of course, changes in another so that determining a cause and effect stream is difficult. It is easy, for example, to see how increasing colony size can lead to increasing nest congestion, given a finite nest volume. Moreover, individually, no single one of these factors is a *sufficient* trigger for swarming, although of the five, queen rearing is a *necessary* trigger for swarming. A colony may (Winston et al., 1991) or may not (Simpson, 1957) swarm in response to nest congestion; a colony may (Winston and Taylor, 1980) or may not (Simpson, 1957) swarm in association with a young worker population. But a colony will not swarm unless there are queen cells. Queen cell construction is a fitful process, and in the weeks preceding a swarm, the colony may construct or destroy any number of candidate cells, presumably in response to changing foraging conditions.

Although researchers have not yet pieced together a holistic regulatory cascade for swarm preparations, we do know some of the component thresholds and interactions. The initiation of queen rearing is triggered when worker densities crest 2.3 workers per milliliter (Lensky and Slabezki, 1981) or when 90% or more of the colony's brood comb is in use (Grozinger et al., 2014). Thus, queen cell construction is regulated in part by density of adults and density of brood. Moreover, initiation of queen cell construction seems to be affected by the distribution of QMP in the nest, a distribution that is accomplished by tactile worker-to-worker contact. It has been established that QMP and other queen extracts inhibit queen cell construction in queenless colonies (Simpson, 1957) and that queens in congested colonies spend less time on the nest edges where swarm queen cells tend to occur (Lensky and Slabezki, 1981). In addition, workers in congested colonies have lower levels of detectable queen pheromones on their bodies (Naumann et al., 1993). These lines of evidence make it reasonable to presume a higher-than-normal stimulus to rear queen cells in congested colonies, especially at nest edges.

Insight into the primers of swarming came from a modeling study that analyzed the interactions of three of them: population size, congestion, and ratio of young to old bees. The computer model triggered swarming only when the action threshold for each variable was met (Fefferman and Starks, 2006). Thus, there is a tight correlation among them, and it seems to me that at least two of the three are tributary to the overarching importance of QMP distribution and its regulatory effects on queen cell production.

One can hypothesize what a holistic regulatory cascade might look like in its barest form, and following a theme here in Part 2, I predict that it will include emergent properties inherent to the honey bee superorganism: (1) the queen's instinctive resumption of mid-winter egg laying sets the stage for rapid population growth and increasing ratio of young to old workers; (2) caring for this rush of new brood tends to concentrate workers in the center of the nest; (3) an expanding brood nest, growing population, and concentration of workers over the brood serve to extend the social edge of the colony further from the queen; (4) the queen's preference for warm temperatures in the center of the brood nest, along with an increasing worker density, combine to decrease the average amount of QMP received per worker; (5) with QMP's inhibiting effect on queen rearing diminished, workers in the colony attempt to rear queen cells, succeeding or failing in proportion to the quality of nectar flows; and (6) eventually optimum conditions for swarming converge—viable queen cells are present, nectar income has stabilized, and a sufficient biomass of workers is present to execute colony fission.

At this point, we are ready for the next section on swarm releaser conditions. But first, a final thought is worth mentioning, and that is the suggestion by Fefferman and Starks (2006) that "all proposed swarming triggers occur as a function of the *ultimate* cause of a colony reaching replacement stability, the point at which the queen has been laying eggs at her maximal rate." If so, then we can imagine workers using such cues as worker density, brood density, and relative scarcity of QMP as signals that the queen, and by extension her colony, has achieved maximum reproductive capacity. At this point, the only way to increase the superorganism's fitness is to produce a second queen and issue a swarm.

Mechanisms of Dependent Swarming in the Honey Bee: Releaser Conditions

Releaser conditions are those that happen in near time that trigger a swarm event on any particular afternoon. In the last section, we left off with the colony dense with bees and brood, whereupon workers increasingly agitate the queen with "vibration dances" that cause her to increase her rate of movement, lose weight, and regain the capacity for flight. At the same time, the high worker density slows the equitable spread of queen mandibular pheromone among the workers, who respond by rearing queen cells. Cells are begun or torn down according to the quality of the nectar flows, until a day finally arrives when optimum conditions converge: the population is high, ripe queen cells are present, nectar availability has stabilized, and it's a beautiful sunny day.

On the days leading up to that propitious afternoon, a handful of individuals sets in motion the chain of events that releases the swarm event. Already by now, some of the colony's scouts have diverted away from flowers and begun looking instead for nest cavities (Rangel et al., 2010). These individuals begin agitating for greater excitement in the nest by giving other workers and especially the queen "piping" signals—audible vibrations applied to a surface or to another bee. The frequency of piping signals delivered to the queen increases dramatically up to the very hour of the swarm's departure (Rangel and Seeley, 2008). Additionally, scouts begin "buzz runs"—provoking action in nestmates by running through clusters of inactive bees while audibly buzzing their wings. Buzz runs tend to be aimed at the nest entrance, and it appears that buzz-running stimulates other nestmates to do the same. Eventually, the activity reaches such a pitch of excitement that the swarming cohort takes to the air, some of the workers pushing the old queen out to join them [figure 16.2(A)]. All this seems frenzied, dramatic, and decisive—and it needs to be, because it is critical to the success of the project that all swarming bees participate together in one coordinated action. Failure—by which we mean numerous small unviable swarms, numerous aborted attempts, or failure to swarm altogether would at best cheat the colony out of reproducing, or at worst jeopardize the survivability of the queen and the parent colony because of lost production and wasted energy.

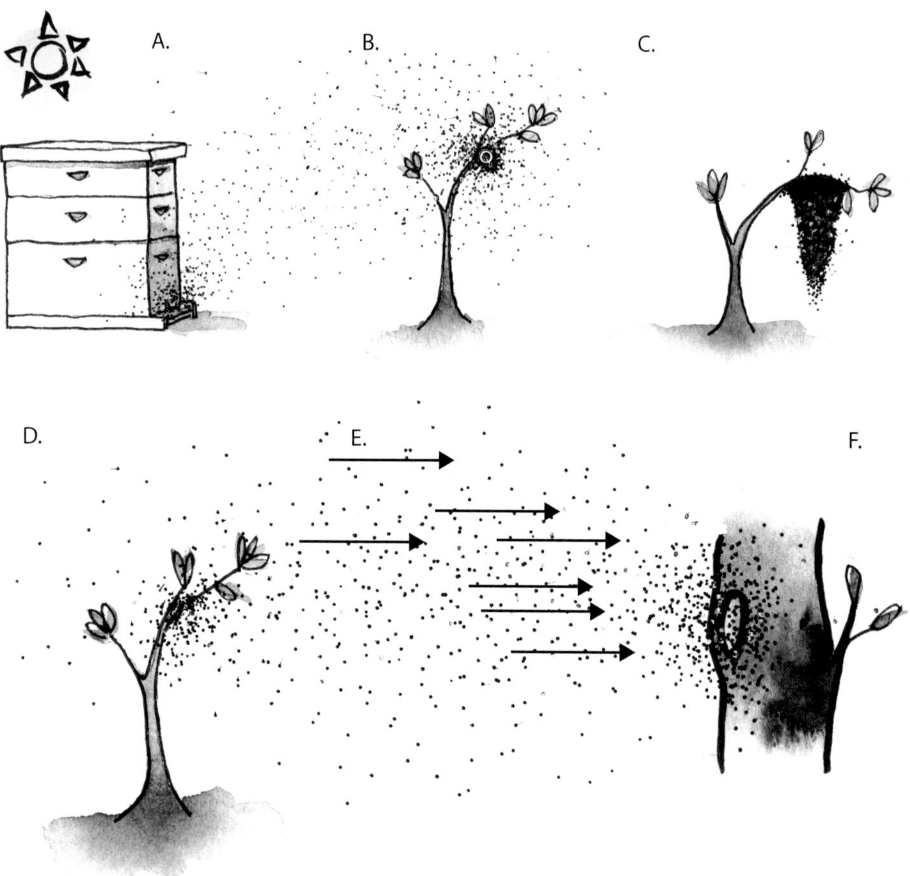

FIG. 16.2. The releaser conditions that trigger a swarm on any particular afternoon are a convergence of high bee densities in the parent colony, the presence of one or more ripe queen cells, stability of nectar availability, and a warm sunny day. On the day of departure, workers begin a series of repeated buzzing, running actions toward the nest entrance; they "pipe" other workers and especially the queen with audible vibrations at increasing frequency, then eventually they leave the hive, pushing the mother queen out along with them (**A**). The queen (**Q**) settles onto some object such as a bush or tree limb (**B**), where her pheromones help stabilize the swarm into a temporary cluster called a bivouac (**C**). At this bivouac, the final decision for a nest site is worked out through a rapid appraisal of competing recruitment dances. A quorum of scouts advocating for a site provokes the bivouacking bees to once again take to the air (**D**). At this point, only the quorum bees know where they are going, and these scouts streak through the cloud (**E**) in rapid directional flights that point toward the new home site (**F**), usually a hollow tree, empty hive, or other safe cavity. The visual cue of bees outside the new cavity and their air-borne Nasonov pheromones further help orient their sisters, and most importantly the queen, to their new home.

The queen quickly lands on some object, apparently at random [figure 16.2(B)]. She is not a very good flyer and, like the vast majority of workers swarming with her, she has no knowledge of the group's destination. Beekeepers can attest to the randomness of these so-called swarm bivouacs: they can settle anywhere, including automobiles, lawn furniture, even major-league baseball stadiums in the middle of televised games. In any case, once settled, the queen's pheromones are critical to stabilizing the cloud of bees which within minutes settles into a tight cluster around her [figure16.2(C)].

To my knowledge, the evolutionary adaptiveness of these temporary clusters has not been systematically studied, but from a methodological point of view, they do solve some important problems—namely, that a bivouac provides a social context for settling who is committed to the swarm and where the swarm is going.

It's likely that many of the bees that participate in the initial rush out the entrance fail to commit to the swarm and end up back at the parent nest. We know that different subfamilies in a colony are more prone than others to join a swarm (Kryger and Moritz, 1997), and the bivouac is the perfect opportunity to sort all this out. This disparity in subfamily participation also means that a honey bee swarm is not a clone of the parent colony from which it divides.

The role of the bivouac as the stage for group decision-making on a new home site has been the focus of considerable research and was winsomely described in Seeley's (2010) book *Honeybee Democracy*. Scouts, some of whom already have knowledge of candidate sites from their personal explorations in the days leading up to the swarm, reinitiate site explorations and appraisals, returning to the bivouac to advertise their locations in a process described in chapter 15's "Nest Site Selection: An Archetype of Consensus Group Decision-Making" section.

Evolution of Dependent Colony Founding in *Apis*

To propose an evolutionary roadmap for honey bee reproductive swarming, we will follow a pattern common to all such analyses: we look for behaviors that are common across the taxon (called *conserved* characters) then focus on variants of that behavior that are particular to the species we're studying. We presume that the common behavior is the more ancient, whereas the

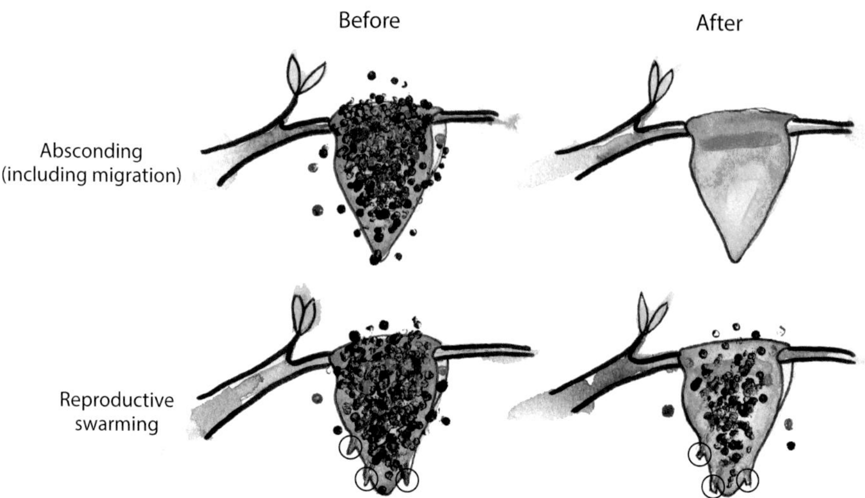

FIG. 16.3. The periodic mass exodus of adults is universal across the genus *Apis*. This diagram represents the open nester, *Apis florea*; its single comb simplifies illustrating the principles. All species practice absconding (top row) and reproductive swarming (bottom). Absconding involves the complete removal of the adult population to a new nest site. There are two types—"simple" absconding which is nest abandonment in the face of acute disaster such as flood, fire, or predation, and "prepared" absconding, functionally the same thing as "migration," which involves regular seasonal colony movement in response to changing floral resources. Reproductive swarming, in contrast, requires the rearing of queen cells (see circles) and does not empty the parental nest of workers.

variant is more modern or derived. In the case of the genus *Apis*, there are two ancient behaviors that appear similar to reproductive swarming—absconding and migratory behavior. And with *A. mellifera*, we are advantaged with the knowledge that it is a relatively derived member of the genus. These are the starting points for our discussion on the evolution of reproductive swarming. Again, I am greatly aided here by the review of Grozinger et al. (2014).

There are three behaviors known across *Apis* that involve all or a part of the adult bees leaving the nest and moving to a new site: (1) migration, (2) absconding, and (3) reproductive swarming (figure 16.3). The first, migration, is a pattern known only in tropical *Apis*, including tropical *Apis mellifera* in Africa, and involves the regular movement of a colony in response to seasonal availability of foraging resources. The second, absconding, happens when the entire colony's adult population uproots from the parent nest, usually in response to some disturbance, and relocates to a new nest site. Migration and absconding are functionally the same thing in that each is a response to changing environmental conditions.

Migration can be called "prepared absconding" because it is closely tied to seasonal floral changes, and bees often prepare for it weeks in advance by slowing foraging and brood production. In contrast to migration, true absconding, sometimes called "simple" absconding, is a response to calamity—the colony relocating because of flood, predation, or some other acute crisis. Reproductive swarming is distinguished from the other two by the fact that it is always accompanied by the production of new queens, and only a portion, not the whole, adult population departs for a new nest site.

Next, let's consider that true migratory behavior—regular colony relocation in response to seasonal fluxes in foraging quality—is predicted to have evolved under tropical conditions in which floral resources are patchy in space but seasonality is low (Hepburn, 2006). In other words, something is always blooming but a colony may need to move to find it. And today, we find migratory behavior exclusively in tropical *Apis*. The primitive open-air, single-comb nesting species *Apis dorsata* and *A. florea* are famous for their regular colony relocations in response to declining local resources. In the case of *A. florea*, the migration may occur laterally for distances of up to 25 kilometers (Soman and Chawda, 1996), and in the African subspecies *A. mellifera scutellata* and *A. m. adansonii*, migrations can occur at elevation gradients up and down mountainsides (Fletcher and Tribe, 1977; Smith, 1961).

Temperate environments, on the other hand, are more spatially uniform in their distribution of forage plants (Dornhaus and Chittka, 2004), but those forage plants are highly seasonal in their nectar output. Nectar availability based on time rather than space, and the overwhelming advantages of a stationary nest heavily invested with food for winter, are strong selective discouragements against a migratory lifestyle. In short, temperate *A. mellifera* lost the ability to migrate as they settled into Europe. Today, no modern temperate-evolved European *A. mellifera* has retained its ancestral ability to migrate.

Migration, absconding, and reproductive swarming all share striking similarities in mechanisms and outcomes across the genus *Apis*. Each involves scout bees who seek new nest sites and recruit nestmates toward those sites. The dances are, not surprisingly, different across species and across functions. We have amply discussed the waggle dance used by scouts to communicate food sources and favorable nest sites (chapter 15). But tropical *Apis* also perform "migration dances" in the weeks leading up to seasonal migration (Lewis and Schneider, 2008). These dances fail to adhere to the figure-eight pattern of the classic waggle dance; instead of circling back to their starting point and repeating a waggle run, migration dancers simply move forward and resume the waggle. In this manner, they roam throughout the nest interior, spreading their message like so many minstrels. The dance does not communicate precise nest location nor distance, but rather the general direction in which lay the scout's focal site. The distances communicated are long, but distances communicated by the same dancer can vary as much as 5–10 kilometers. It may be that migration dances fall under the category of a "primer" function as I discussed earlier in this chapter: they communicate "we'll be flying a long distance before too long in *this* direction."

Curiously, even temperate *A. mellifera* are known to perform migration dances at two weeks prior to reproductive swarming, but the numbers of dancers are far lower than migration dancers in *A. m. scutellata* (which retains the ability to migrate); neither do they reliably communicate direction or distance (Lewis and Schneider, 2008). The authors of that study offered suggestions about the function of this curious dance but also conceded that it may be an obsolete vestige of *A. mellifera*'s ancient tropical heritage.

Migration, absconding, and swarming also share the collective adult flight exodus provoked and guided by scout bees. The status of the temporary bivouac phase is, however, ambiguous. In cavity-nesting *Apis*, the bivouac serves as an important arena for settling "who's committed" to the swarm and for staging the final group decision-making process for an optimum nest site. The single-comb open-nesting migrating *A. florea* of Southeast Asia, however, appears to lack the sophisticated layers of decision-making required of the more derived cavity nesters. *A. florea* build their small single comb on a twig or branch of a bush or tree. With so many twigs and branches to choose from, extreme precision is not so important for *A. florea* as it is for *A. mellifera*, for whom the choice of an adequate cavity can mean life or death. Instead, the dances of *A. florea* communicate general direction, and the cloud of bees may form a bivouac numerous times, "deciding as they go" whether any particular twig is suitable. If they encounter any disturbance such as predatory ants, the bees simply take to the wing again and move a little further (Makinson et al., 2011).

With this background, I offer a plausible scenario for the evolution of group fission involving ancestors of *Apis* transitioning from subsociality to complex eusociality. In the early stages, there was ecologic advantage to free-living sisters who stayed at the nest to help their mother (chapter 5's "Evolutionary Routes toward Eusociality" section). With increasing group size and caste evolution, tropical conditions selected for coordinated resource-dependent nest relocation—migration—elements of which included scout bees who recruit nestmates toward new nest sites, weeks-long preparations involving reduced foraging and brood production, and collective adult flight. To these ancestral elements of seasonal migration were added synchronized production of new queens and the relocation of some, but not all, of the worker population. Migration was lost in European lineages of *Apis mellifera*, but temperate conditions and extreme resource seasonality selected for winter-hardy physiology and cognitive abilities that enabled workers to recognize when the colony has achieved its maximum growth capacity and to coordinate a reproductive fission event during a period of stable nectar availability.

Summary

There are two modes of colony founding available to eusocial insects: (1) independent colony founding, in which a solitary queen (or a royal couple) disperses alone (or together) and single-handedly founds a new nest, or (2) dependent colony founding, in which one or more queens disperse from the parent colony, accompanied by and dependent upon a cohort of nestmate workers. Honey bees practice dependent nest founding, also called colony *fission*, a key advantage to which is the buffering effect of a stable cohort of highly related nestmate workers who immediately perform the duties of foraging, nest construction, and defense for the pioneering queen. Dependent colony founding in the honey bee, also called reproductive swarming, is an extremely costly form of reproduction with colonies of ~20,000 workers able to field swarms with up to 16,000 individuals.

Swarming is directed toward the earliest time of year when floral resources are richest. The steady food supply improves the chances of recovery for the parent colony and successful establishment for the swarm. A reproductive swarm is preceded by five primer conditions that act on a time scale of weeks: (1) increasing colony size, (2) increasing nest congestion of workers, (3) increasing population ratio of young to old bees, (4) decreasing efficiency of QMP movement throughout the nest, and (5) production of queen cells. Releaser conditions are those that happen in near time that trigger a swarm event on any particular afternoon. On a day during a period of consistent nectar flows, scout bees with knowledge of available nest cavities in the vicinity will agitate nestmates to take to the air in a large cohort, accompanied by the old queen, and eventually direct them to the new cavity, after which workers immediately set to work building new comb and the queen resumes laying eggs. A newly emerged daughter queen inherits the original mother colony.

A plausible scenario for the evolution of group fission has been offered, involving ancestors of *Apis* transitioning from subsociality to complex eusociality. In the early stages, there was ecologic advantage to free-living sisters who stayed at their natal nest and helped their mother. With increasing group size and caste evolution, tropical conditions selected for coordinated resource-dependent nest relocation—migration—elements of which included scout bees who recruit nestmates toward new nest sites, weeks-long preparations involving reduced foraging and brood production, and collective adult flight. To these ancestral elements of seasonal migration were added synchronized production of new queens and the relocation of some, but not all, of the worker population. Migration was lost in European lineages of *Apis mellifera*, but temperate conditions and extreme resource seasonality selected for winter-hardy physiology and cognitive abilities that enabled workers to recognize when the colony has achieved its maximum growth capacity and to coordinate a reproductive fission event during a period of stable nectar availability.

CHAPTER 17

Mechanisms and Evolution of Colony Defense Behavior

BEEKEEPERS BECOME SKILLED at reading the signs of a defensive colony and handling bees in a way that minimizes stings. It reduces to little things like brushing away bees with a flick of one's finger before picking up a frame, reaching in from the edges of the hive instead of reaching across the hive (the latter is perceived by bees as a threatening silhouette against the sky), recognizing the difference between a forager landing on one's arm, confused and disoriented, versus a buzzing soldier ready to sting, and knowing when to give a puff of smoke to disorient the bees' perceptions of alarm pheromone and de-escalate a defensive cascade (Visscher et al., 1995).

It's also worth pausing to think about the marvel of being able to go through a bee hive like I just described. There seems no reason for them to tolerate us. A giant mammal, armed with smoke and tools for breaking and entering, constitutes nothing less than the biggest disaster possible in their evolutionary universe—yet tolerate us they do. Many of us reading these pages are acquainted with the joys of watching bees go about their business while we hold a frame in our hands—the queen laying eggs in broad daylight, a court forming about her, a waggle dance in this section of the comb, nurse bees feeding larvae in another. Experiences like this, I believe, are evidence of the evolutionary effects we humans have had on *Apis mellifera*. The shared ancestral ranges between humans *Homo sapiens* and honey bees *Apis mellifera* spans the entire existence of our species. We are the newcomer, first diverging into modern *Homo sapiens* in Africa a mere 140,000 years ago and emigrating into Europe 61,000–44,000 years ago (Shriner et al., 2014). For *Apis mellifera*, on the other hand, dating estimates for its presence throughout Africa and Europe begin as early as 3.5 million years ago (Dogantzis et al., 2021) to 780,000–540,000 years ago (Arias and Sheppard, 1996; Carr, 2023), even though its place of origin has been argued variously as Africa (Tihelka et al., 2020; Whitfield et al., 2006), western Asia (Dogantzis et al., 2021), or Europe (Carr, 2023).

In the beginning, it is certain that the relationship between humans and bees was adversarial, that of predator and prey. But we have cultural evidence that by 7,000 years ago, humans were harvesting honey and wax (Crane, 1999); the relationship was beginning to transition from predation to management. Given that this transition likely pre-dates our material evidence for it, it seems reasonable to me that humans have been "managing" *Apis mellifera* for at least 10,000 years, all the while applying artificial, even unintentional, selection for productivity and gentleness, allowing us in the 21st century the pleasure of handling a quiet frame of bees.

Mechanisms of Colony Defense Behavior

I will talk more about the evolution of social nest defense in the next section, but for now, I want to focus on the behavior itself. In spite of a widespread cultural presumption otherwise, stinging is the last thing a bee wants to do. Nest defense, like an immune response, is a costly undertaking, especially for the individual worker bee who dies after she stings. This is an expression of altruism as I talked about in chapter 6. That chapter laid out the evolutionary forces that push such extremes of self-giving, but for now, let's focus on why it is lethal.

The loss of the sting—so-called sting autonomy (Hermann, 1971)—is an adaptation against vertebrate predators. It's no good against other insects or invertebrates, who have sclerotized cuticles that resist penetration. The sting's barbed tip allows it to lodge in the flesh of a vertebrate assailant like a harpoon (indeed, it looks like a harpoon under magnification), and as the bee pulls away, the stinger remains, along with its closely associated poison sac. If one can endure it, it's instructive to watch an embedded stinger in one's skin for a few moments; the poison sac is pumping like a tiny heart, its independent involuntary muscles contracting to inject venom. Meanwhile, there is a sudden pronounced behavior change in the bee that just delivered the sting. She becomes recklessly aggressive, beating against one's veil, circling one's face, buzzing in your hair—all behaviors adapted to intimidate the intruder. In other words, she has doubled her defensive output; while her sting is autonomously doing its job, she is free to engage in additional defensive behaviors. Magnify this by tens or scores or hundreds of assailants, and onc has a very effective colonial defense response. But the down side is the irreparable injury caused to the individual stinging bee, who dies a few hours later.

However, I am guilty of reversing the story's sequence. What I have just described is the final step in a cascade of earlier behaviors, each a step of increasing engagement as the colony's threat persists or intensifies. It is not adaptive for a bee colony to leap to the ultimate and most costly step in the cascade—stinging—if milder responses will repel the invader.

Figure 17.1 outlines this generalized cascade. One or more guards first perceive a potential threat, either an alien honey bee or vertebrate, and orients her body toward the threat in order to appraise its danger. The appraisal process involves zigzagging flight, or in the case of a nest-invading insect, inspection with the antennae (antennation), either of which behaviors may be sufficient to deter the invader and de-escalate the cascade. If, however, the guard perceives a threat, she switches to alerting behaviors. These include emitting alarm pheromones, which are quickly volatilized into the air and received by "recruits"—similarly aged bees who are then alerted to perform their own surveillance and appraisals. The alerting behaviors can occur inside the nest, with the worker exposing her sting chamber to release pheromone.

Alerted bees have a characteristic posture: the bee rears up, and her front legs may be elevated; her mandibles are open, antennae waving, wings extended, and sting exposed. Alerted workers may be facing different directions, demonstrating that the response at this point is non-directional. If

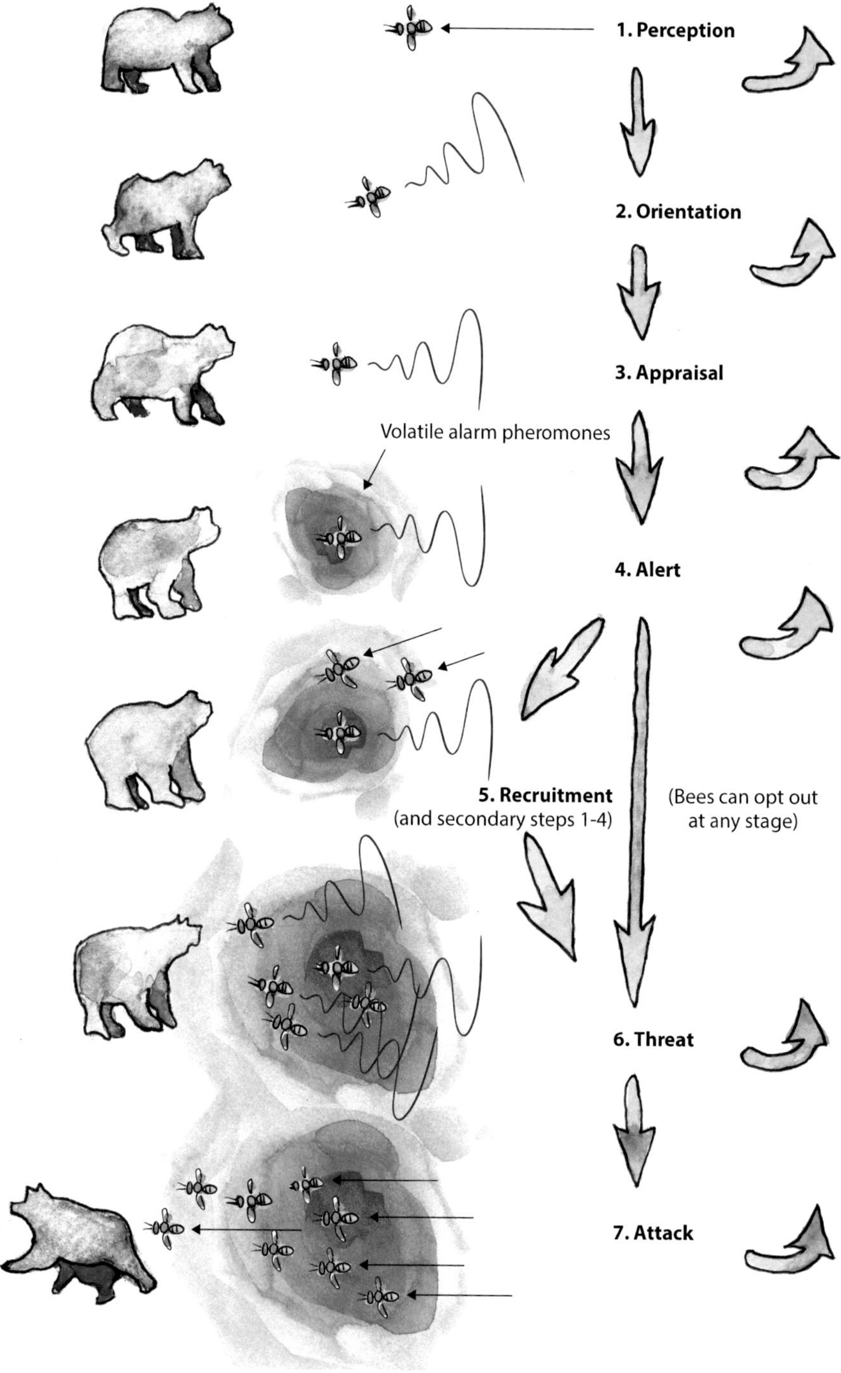
1. Perception
2. Orientation
3. Appraisal
Volatile alarm pheromones
4. Alert
5. Recruitment
(and secondary steps 1-4)
(Bees can opt out
at any stage)
6. Threat
7. Attack

alerted bees also detect the threat and perceive it as a risk, they contribute to the growing alarm and recruit more defenders. Engaged bees at this point perform increasingly threatening behaviors in an attempt to deter the invader. Bees will only attack if these efforts fail. In that case, they will buzz the invader, burrow in its fur, pull its hair, and ultimately sting. The embedded stings themselves release alarm pheromone, but again, these pheromones do not provide orienting information but simply excite surveillance in new recruits. Actual sting targeting seems to depend more on motion and color contrasts (Wager and Breed, 2000).

It's important to place this defensive cascade in the context of division of labor—that well-known feature of honey bees by which individuals perform a predictable sequence of tasks as they age. I talked about this in chapter 8 and established that even though the sequence of labors is predictable in general, many workers show a genetic proclivity toward specialization. This seems to be the case in colony defense.

Much of the foundational work on this was done by Breed et al. (1990). It appears that defensive behavior is made up of at least three subsets of actors: entrance guards, defenders (also called "soldiers," bees that fly out to deter an invader), and defenders who actually sting. These are behaviorally and genetically distinct groups. Guards tend to be younger, whereas defenders

FIG. 17.1. (FACING PAGE) The defensive cascade of behaviors employed by a honey bee colony to deter a vertebrate threat: (1) perception, a suspected threat is noticed by one or more guards; (2) the guards spatially orient toward the threat in order to (3) appraise whether it is a viable risk, or in the case of other honey bees, discriminate it from nestmates. These appraisals are done by reading visual and olfactory cues; large dark objects, high visual contrasts such as eyes, ears, and nostrils, rapid movements, and CO_2 emissions identify vertebrate threats; insect invaders may be antennated to discriminate nest odors. If these sensory appraisals do not intimidate the invader, then guards (4) alert nestmates to danger using volatile alarm pheromones and alert postures. Pheromones do not provide orienting cues; instead they (5) prime recruits to initiate their own surveillance, upon which recruits perform their own secondary steps of perception, orientation, and appraisal. If sufficiently stimulated, these recruits contribute their own alarm actions. Alerted, recruited, and motivated bees then undertake increasingly (6) threatening behaviors such as threatening zigzag flight or aggressive posturing, whereupon if the threat persists they graduate to (7) attack behavior—active biting, burrowing, buzzing, and as a last resort, stinging. There is a genetic component to the propensity of individuals to engage in any of these behaviors, and at any step some individuals may opt out of the cascade. Figure modified from figure 1 of Breed et al. (2004).

are older, the same age as foragers, even though defenders show less proclivity to forage. These genetic differences can be resolved to the level of subfamily; in other words, supersister groups with the same father are biased toward sharing one or more of these traits. Breed et al. (1990) drew an interesting connection to the so-called "lazy bee" syndrome. As far back as the 1960s, Lindauer (1961) noted that a large fraction of the colony's adult bee population appears to be engaged in no work at all. Breed et al. (1990) suggest that these so-called lazy bees may in fact be soldiers—older adults genetically primed for nest defense and therefore disinclined to leave the nest to forage, waiting in reserve if needed.

The main job of guard bees, on the other hand, appears to be excluding non-nestmate honey bees from the entrance (Breed et al., 1992), but there is evidence that they also participate as recruits in the event of attack from a large invader (Moore et al., 1987). In general, however, Breed et al. (1990) believed that entrance guarding is an evolutionary response to non-nestmate invaders, namely robber bees, whereas defenders or "soldiers" are an adaptation to deflect larger vertebrate threats.

A worker's choice to participate in a defense cascade is not only genetically determined, but responsive to environmental cues both in the nest and outside it. Worker responsiveness to a defensive stimulus can be affected by her exposure to juvenile hormone (Harris and Woodring, 1999), and multiple, apparently redundant, alarm pheromone compounds may synergize with one another to enhance recruitment. Bee defensive response is also sensitive to such environmental effects as air temperature, season, and time of day (Breed et al., 2004).

We cannot neglect in our discussion the most famous bees of all when it comes to defensive behavior—the African honey bees (AHBs) of the Americas. For over 50 years, these bees have commanded public attention for their extreme expressions of colony defense. AHBs not only attack at very slight provocations, but they attack in overwhelming numbers and pursue their invaders for great distances. The sum of evidence suggests that AHBs follow the same general defensive cascade as described here; however, at each juncture, they respond to lower action thresholds; they respond in greater proportions; they respond faster, and they recruit more efficiently.

The AHBs have been present in the United States since 1990, and it seems safe to say that they are making their mark on American beekeeping. Colonies along the Gulf Coast and American Southwest may express typical African behavioral extremes, but it's still possible in most cases to moderate the behavior by requeening. It's also safe to say that their northern advance has significantly slowed, and many experts now believe we're seeing a latitude-based limit to their range expansion similar to that seen in southern Argentina (Kerr et al., 1982).

Evolution of Colony Defense Behavior

The defense cascade has been adapted to maximize the probability of preserving the colony at the lowest cost in terms of lost bees. The existence of a cascade also highlights the fact that a bee colony experiences a range of dangers of increasing intensity. An invading insect is not as dangerous as a human predator, and the cascade permits a range of responses scaled to the danger.

Ordinary human experience also makes clear that nest defense differs according to the life history of the insect. It's no accident that the *social* wasps and bees are the ones who elicit the largest and most frightening defense reactions. A solitary mud dauber wasp certainly possesses the tools; females possess stings, which they use to paralyze spiders and feed to their larvae. However, human stings from mud daubers are exceedingly rare. The same is true of virtually all solitary wasps and bees, including cicada killers, miner bees, carpenter bees, potter bees, and mason bees. It's not that solitary bees don't have predators, it's just that it is too risky for a solitary fertile female to engage in direct confrontation with a predator. It's far safer to abandon that nest and attempt another one elsewhere. This is also why solitary bees, especially soil-nesting miner species, tend to nest in large aggregations near one another (figure 17.2). The mass aggregation lowers the risk for any one nest being attacked by a predator or parasite. For a few brief days every year, the intensity of their localized flight can rival that of a strong honey bee nest, and the concentration of their burrows gives the false appearance of a social colony. Yet a human observer can walk through them unscathed like a boat cleaving water. Despite appearances, the bees are solitary—each an autonomous reproductive female unwilling to risk her life attacking a large predator.

FIG. 17.2. Many solitary soil-nesting bees make their simple earthen burrows in close aggregations to one another, as in this specimen of *Colletes hederae* in Ireland. The close proximity of nests decreases the odds of any one being attacked by a parasite or predator. When examined closely, these soil aggregations are each made of simple, single burrows (figure 4.2). Each is the effort of an autonomous reproductive female, and the burrows do not interconnect. Solitary species tend to emerge at the same time each year, and for a few brief days, their intensity of localized flight can give the false impression of a social colony. PHOTO: FRANCIS RATNIEKS.

So we've already begun tracing the earliest steps of an evolutionary roadmap for honey bee nest defense. No nest defense at all, the solitary bees show us, is a reasonable option if the individual is the sole reproductive agent and nest making is relatively cheap: flight over fight. However, to this humble beginning, it's easy to add the first refinement—nesting together in large aggregations. This improves the benefit to risk equation by lowering risk, i.e., the likelihood that any one burrow will be attacked, just like herd behavior lowers the likelihood of any one wildebeest being attacked by the lion.

But things start changing rapidly once the species begins moving up the road toward sociality. E. O. Wilson (1971) spelled out three ascending criteria for *eusociality*, or *true* sociality: (1) cooperative brood care, such that individuals share one nest and help tend the common brood, (2) reproductive division of labor, such that some individuals abandon reproduction in order to help others reproduce, and (3) overlapping generations (Chapter 5, "Degrees and Qualities of Sociality" section). As lineages of species move up this gradient, it's possible for some to express only Steps 1 or 2. The first step, cooperative brood care, happens when two or more reproductive females, perhaps a mother and daughter, begin sharing one burrow. This moment in evolutionary history was recalled in Chapter 7's "Restraining Genetic Conflict in the Honey Bee Superorganism" section, when I pointed out that a daughter of a singly mated (to one male) mother can pass on 50% of her genes in one of two ways, namely, by staying at the nest and helping her mother raise more of her siblings or by striking out on her own. From the discussions in chapter 6, we know that staying and defending a common nest is a powerful benefit not to be rejected lightly.

So it is no overstatement to say that nest defense was there at the beginning; it helped kick-start social evolution. And once a lineage starts up E. O. Wilson's path of ascending criteria, its defense behavior can be analyzed in light of those same criteria.

An example of defense behavior at Wilson's (1971) first level, cooperative brood care, can be seen in a modern solitary bee in which investigators in New York showed a reproductive advantage in those nests that were shared by more than one female compared with nests that had only a solitary occupant. Nest sharing permitted continuous guarding of the entrance, which restricted entry of a parasitic cuckoo bee that would have otherwise usurped the nest (Abrams and Eickwort, 1981).

We see an example of Wilson's (1971) second level, reproductive division of labor, in a modern carpenter bee living in Asia and Africa, in which the division of labor is integrally linked to nest defense. This carpenter bee begins nesting as a solitary female—indeed it can complete its entire life cycle in a solitary state. However, nests can be usurped at any time, at which point the species can adopt a primitive sociality. The usurping female tolerates some of the remaining brood; thus, a female may emerge and live in a nest with a bee who is not her mother. The dominant female lays all the eggs and does all the foraging while the subordinates guard the nest entrance.

These young guards are called "hopeful reproductives" and may eventually challenge the dominant female and take over egg laying.

The formerly dominant female, now dismissed from the brood-rearing areas of the nest, adopts guard duties, efforts of which no doubt help some of her own developing brood cells (Hogendoorn and Velthuis, 1995). But the question remains, why does a dominant female allow a defeated female to stay at the nest as a guard when there's risk that the guard will reassert reproductive dominance? Investigators showed that this risk was offset by the benefits of the guards repelling same-species robbers and allowing the dominant female more foraging time (Hogendoorn and Velthuis, 1993). These benefits were especially acute at times of environmental dearth or intense competition for nest sites.

So we see that at the very earliest stages of the honey bee's social evolution, nest defense was a good investment, so it's no surprise that defense was retained, refined, and amplified in those species that achieved eusociality—all three of Wilson's criteria. And the point about amplification speaks to our earlier observation that we see the most convincing expressions of nest defense in the complex eusocial species. These are the superorganisms for which survival of the nest is synonymous with survival of the entity—Darwin's unit of selection. At this point, any threat to the colony is a threat to the *colony* and not just the individuals that make it up, and so the response is a *colonial* response involving numerous actors advantaged with shared information and pheromone-assisted vigilance and coordination. Not only are the stakes higher for a stationary, complex, and resource-rich superorganismal nest, but the predators tend to be larger and more dangerous—vertebrates such as *Homo sapiens*. In light of such a cost/benefit calculation, nothing will do short of a human-sized defense strategy.

As we survey defensive strategies across Wilson's (1971) categories of sociality (see figure 5.1),[1] we find behaviors shared in common, and these, we

1. Looking back again at figure 5.1, we can now see that the evolution of sociality was intimately tied to nest defense. These diagrams show a stylized nest of a soil-tunneling bee. A central tunnel has one or more lateral branches, each terminating in an earthen cell in which the female deposits a ball of pollen and lays a single egg on it. In time, the larva consumes the pollen ball, pupates, and emerges from the cell as a new adult. A solitary female in a single nest (A) leaves her nest vulnerable to predators,

presume, are the ancient ones. The behaviors of stinging, biting, and buzzing are surely primitive, as is the fundamental ability to distinguish nest mate from non-nest mate. But there are defense characteristics unique to the eusocial species, and these we hold as more modern, or derived—the outcome of the selection pressures mentioned above. These include the barbed sting or "sting autonomy" (Hermann, 1971) and the complex mixtures of alarm pheromones, numbering at least 30 different compounds and remaining a puzzle to scientists who wonder at the need for such redundancies (Breed et al., 2004).

Eusociality has also complexified defense behavior through division of labor and genetics. In chapter 8's treatment of division of labor, I explained how in honey bees, maternal hormone expression has been pushed back in development so that young bees express maternal characters such as elevated levels of the yolk protein vitellogenin, care of larvae, and defense of nest. As young bees feed larvae, their vitellogenin levels are depleted, and this pushes maturing bees toward more non-maternal activities such as nectar foraging (Page et al., 2006). These dynamics set the stage for the well-known fact that honey bees progress through a predictable series of tasks as they age. It seems to me that guarding behavior, occurring mid-life around 15 days of age (Moore et al., 1987), must signal the end of the "maternal" phase of a worker's life.

One group of investigators (Breed et al., 1990) has shown that guard bees are behaviorally and genetically distinct from "soldiers"—those bees who actually fly out and attack nest invaders. Moreover, the soldiers are old bees—the same age as foragers, but genetically disposed to stay at a nest, ready to defend it if necessary. Breed et al., (1990) hypothesized that guarding is an adaptation against same-species robbing, whereas soldiering is an adaptation against large vertebrate predators. If so, then guarding strikes me as the

---

parasites, or usurpers every time she is away on a foraging flight. With cooperative brood care (B), two or more reproductive females share a nest and guard duties. In semi-social species (C), we see the beginnings of reproductive division of labor where only one female forages and lays eggs, and her subordinates guard the nest. Some of these subordinates may be "hopeful reproductives," biding their chance to take over egg laying from the dominant female, who in some cases will be tolerated in the nest as an additional guard. Only in the complex eusocial species, in which generations overlap and daughters remain at the nest to help their mother, do we see larger populations, larger food stores, more intense defense reactions, and more complex defense adaptations including the distinct behavioral castes of guarding and soldiering.

more primitive state—a natural outcome of the age-based task sequence, itself a recapitulation of the ancient maternal tasks of caring for young and defending a tunnel-like nest. Soldiering, on the other hand, is an innovation adapted for the heightened stakes for nest defense represented by the eusocial superorganisms.

When one thinks about it, the two really are distinct behaviors. Nest guarding can be a passive affair, nothing more than physically blocking a burrow entrance with one's head, or at most flaring one's mandibles and biting, whereas soldiering involves threatening zigzagging flight, buzzing, hair pulling, burrowing in hair, stinging, and engaging in the exaggerated post-stinging threat behaviors that sting autonomy affords—the harpoon-like irretractable sting itself a morphological novelty.

It is well-known to any bee breeder that colony defense behavior is strongly under genetic influence. Defensive response is heritable and one of the most common targets in breeding programs (Guzmán-Novoa and Page Jr., 1999). And nowhere is that range of genetic variation more evident than in the extraordinary differences between European honey bees and the well-known African honey bee, whose explosive colony-level defensive responses are the stuff of newspaper headlines. I cannot end a discussion on evolution of defense behavior without addressing the stark differences between these two groups.

The most common explanation for the extreme defensiveness of African honey bees is the intense predation pressure endured by honey bees in sub-Saharan Africa (DeGrandi-Hoffman et al., 1998). However, Tom Rinderer (1988) argued that predation pressures between Europe and Africa are not so fundamentally different, to the extent that both regions share selection pressure from the most dangerous predator in the history of planet Earth, *Homo sapiens* (Rinderer, 1988). The nature of that relationship, however, has been profoundly different between the two continents based on differences in seasonality. In temperate Europe, regular seasonal nectar flows and long winters selected for permanent perennial colonies, whereas in tropical Africa, a succession of smaller, less predictable nectar flows selected for smaller colonies and frequent absconding. When times are bad, African bees simply relocate.

Humans in Europe were inclined to manage permanent colonies for large annual harvests, inadvertently selecting for gentleness by culling the most defensive colonies. Conversely, the absconding nature of the bees in Africa, never staying stationary for 12 months, encouraged predation by humans—indiscriminate killing over management; thus, there was no selection for gentleness.

Rinderer (1988) also posited an explanation for AHB's ferocity based on the comparative randomness and unpredictability of nectar flows in Africa. "Through evolutionary time," he argued, "selective forces should favor bees that conserve collected resources well enough to survive long dearths. Part of this conservation is likely to be increased defense." However, I feel this hypothesis is unsatisfying because it could apply equally to temperate bees in Europe, who arguably have even longer dearth periods between annual nectar flows during which they must defend a precious food hoard.

I think a more plausible explanation ties into the derived soldier caste described above. If it is true that soldiers and foragers draw from the same age cohort of workers, then it seems reasonable that the cohort fraction in each behavioral caste would be responsive to environmental selection. A habitat that yielded small nectar crops and dangerous predators would select for a comparatively higher fraction of soldiers, whereas the reverse would be true if the habitat yielded larger nectar crops and less intense predation. A large and idle "army" may thus be itching for a fight, so to speak, which may be anthropomorphic, but I think it captures the hair-trigger and spectacular defense reactions we see in the African bees.

Summary

A colony defensive reaction happens in a cascade of increasing engagement as the threat escalates. It is not adaptive for a colony to leap to the ultimate and most costly step in the cascade—stinging—if milder responses will repel the invader. At first, one or more guards perceive a potential threat and orients her body toward the threat to appraise its danger. The appraisal process involves zigzagging flight or inspection with the antennae, either of which may be sufficient to deter the invader. If, however, the guard perceives a threat, she switches to alerting behaviors, which include emitting alarm pheromones which are received by "recruits"—similarly aged bees who are then alerted to perform their own surveillance and appraisals. Alerted bees have a characteristic posture: the bee rears up, and her front legs may be elevated; her mandibles open, antennae waving, wings extended, and sting exposed.

If alerted bees also detect the threat and perceive it as a risk, they contribute to the growing alarm and recruit more defenders. Engaged bees at this point perform increasingly threatening behaviors in an attempt to deter the invader. Bees will only attack if none of their earlier steps succeed. Their attacks involve buzzing the invader, burrowing in its fur, pulling its hair, and ultimately stinging. The embedded stingers themselves release alarm pheromone, exciting new recruits.

For a solitary female, risking her life to defend a simple nest may be too costly; far better to abandon the nest and begin another. But in early social evolution, accumulating genetic and ecologic benefits for daughters who stayed at a nest to help their mothers selected for colony-level defense responses. Not only are the stakes higher for a stationary, complex, and resource-rich superorganismal nest, but the predators tend to be larger and more dangerous. As we survey defensive strategies across taxa of increasing sociality, we find behaviors shared in common, such as stinging, biting, and buzzing, as well as the ability to distinguish nest mate from non-nest mate, which we presume are ancient. Defense characteristics concentrated in the eusocial species such as the barbed stinger, complex mixtures of alarm pheromones, and a "soldiering" caste distinct from guard bees we hold as more derived—outcomes of social selection.

CHAPTER 18

Mechanisms and Evolution of Social Immunity

Mechanisms of Social Immunity

Bearing in mind that superorganisms do the sorts of things that organisms do, it should be no surprise to learn that there are integrated behaviors and phenomena in the honey bee colony that deter the entry and proliferation of pathogens and parasites in the nest. There are parallels to the immune systems of organismal creatures such as ourselves (Cremer and Sixt, 2009), but when the focus is on the superorganism, the term most commonly used is "social immunity."

Before we get started, I need to point out that immunity against pathogens exists in the honey bee at two levels of organization: the level of individual bee and the level of colony. Individual bees express innate immunity with a variety of means shared with other insects. These include initial defenses such as antimicrobial secretions on the integument, microbe-hostile gut chemistry, and the physical barrier of the gut lining. If pathogens breach these outer defenses, they next encounter cellular and humoral (body fluid) resistances such as enzymes that degrade pathogens, cells that engulf pathogens, antimicrobial peptides, and cells that promote melanization—a process analogous to scar tissue formation in mammals that walls off invading cells (Hoffmann, 2003).

Even though honey bees boast a full repertoire of these immune classes, they are relatively impoverished in the gene richness of those classes. Compared with two widely studied solitary insect groups—fruit flies in the genus *Drosophila* and mosquitoes in the genus *Anopheles*—honey bees possess about one-third fewer immune genes (Weinstock et al., 2006), and the genes they retain appear to be very ancient. In other words, honey bees have not kept pace with solitary insects when it comes to richness and diversity of immune responses. The two best explanations for this are that (1) bees tend to be attacked by a fairly small number of highly coevolved pathogens, thus narrowing the range of demands on immunity, or (2) the innovation of social life—and simultaneous evolution of social immunity—have downgraded selection pressures for individual innate immunity (Evans et al., 2006). For the individual, it appears, social life decreases costs for disease resistance.

For our purposes, we want to focus on social immunity. But it is important to understand that individual innate immunity is still alive and well, acting as another layer of immunity in the colony. I organize my following synopsis of social immunity with the "sequential lines of defense" approach used by Cremer et al. (2007).

Let's look first at the possible modes of parasite transmission. Epidemiologists recognize two. The first is parasite transmission that happens between parent and offspring. This is called *vertical* transmission, and in the case of honey bees, it is best understood as transmission between a parent colony and its swarm. The second mode of transmission happens between two members of the same generation. This is called *horizontal* transmission, and in honey bees, this happens when parasites are transmitted between colonies. Of the two, horizontal transmission is expected to select for more virulent pathogens (Bull et al., 1991).

The first line of social defense is to prevent the "uptake" of parasites by individual nest members. Because nest predators and parasites come from outside the nest, the members most prone to parasite uptake are the foragers. In honey bees, one way to limit parasite uptake is to narrow the range of individuals engaged in this risky behavior, and in a normal colony, foraging is indeed restricted to the oldest and most expendable individuals. If an old forager becomes infected, its short remaining lifetime limits its opportunity for spreading the parasite.

A second line of defense is to prevent or reduce parasite "intake"—the entry of parasites into the nest. At a basic level, this is exercised in the choice of bees to occupy cavities. Entrances to these cavities tend to be small, ranging from 10 to 40 square centimeters in area (Seeley and Morse, 1976), which restricts access points for nest invaders and limits the surveillance demands on guard bees (figure 18.1). However, as we mentioned, many parasites gain entry to colonies not by direct assault but by catching a ride on a forager. This is the exclusive mode of entry for the parasitic *Varroa destructor* mite, which is otherwise incapable of independent movement between colonies. Guard bees inspect returning foragers and repel those infected with pathogens including viruses, but I am unaware of any direct evidence that guard bees restrict entry of *Varroa*-laden nestmates.

There is, however, evidence that entrance guard bees are sensitive to *Varroa*-induced chemical changes in the cuticle of forager bees and react more aggressively toward freeze-killed bees that had carried one or more *Varroa* mites for seven days compared with control bees that never carried mites (Cappa et al., 2016). In general, the choice of nest site and guard policing behaviors constitute an important line of social defense.

FIG. 18.1. Entrance to a natural honey bee nest in a tree hollow. The bees have smoothed the wood around the entrance and infilled cracks with propolis. PHOTO: GEORGIA P. ZUMWALT.

If these first lines of defense fail, or if a parasite gains a foothold through vertical transmission, then the third line of defense focuses on preventing the parasite from getting established in the nest. This is where a battery of hygienic behaviors come into play, and for beekeeping audiences, I must clarify that I'm talking about general hygiene—not the specific form that has become a familiar management tool against *Varroa* (Spivak and Danka, 2020).

For now, we're including one of the most well-known examples of insect-applied antimicrobials—the use of plant resins. Honey bees collect tree resins (figure 18.2), return them to the nest, mix them with beeswax, and apply them inside cells and onto nest cavity walls, at which point we call the substance propolis. The substance has antimicrobial properties, which serve to reduce pathogen load in the nest environment. However, it has also been recently shown that propolis reduces the expression of immune response genes in seven-day-old bees. The significance of a *reduction* in immune response was due to an overall reduction in bacterial loads in the experimentally propolis-treated colonies, but more generally, a highly charged immune

FIG. 18.2. A forager bee collecting tree resin.

system is not only indicative of a pathogen problem, it is also exhausting on the bees to sustain (Simone-Finstrom and Spivak, 2010). Propolis de-escalates the cost of pathogen intake, first by direct antimicrobial action and second by reducing the need for the bees to ramp up a costly immune reaction. Colonies whose bees are in a sustained state of immune response produce less brood (Evans and Pettis, 2005).

There are other behaviors in the colony that constitute hygienic resistance to parasite establishment in the nest. These include the well-known "undertaker" bees who remove corpses of dead nestmates from the nest. And venom, it turns out, has more uses than its well-known function in defense; there is evidence that bees apply it to their beeswax combs and onto their own integuments, apparently benefiting from the venom's antimicrobial properties (Baracchi and Turillazzi, 2010). And lastly, there is evidence that the Cape honey bees of southern Africa, *Apis mellifera capensis*, "socially encapsulate" invading small hive beetles with propolis prisons in an action analogous to scarring and abscess formation in mammals (Neumann et al., 2001).

In the event a parasite becomes established in the nest, then the colony attempts a fourth line of defense: limiting the parasite's spread between colony groups. The probability of a healthy colony member becoming infected is a product of its susceptibility, its contact rate with an infected individual, and the infectivity of that individual (number of infectious propagules it carries). The most direct way to reduce infectious propagules is to pick them off and kill them—and the best example of this for bees is the well-known grooming behavior against *Varroa* mites. Bees expressing this heritable trait can detect and remove mites off their own bodies or bodies of nestmates and sometimes lethally bite them (Guzman-Novoa et al., 2012). A similar strategy is employed with so-called hygienic lines of bees that are capable of detecting compromised cells of brood, opening them up, and removing the infected pupa and its associated pathogens or parasites (Spivak and Gilliam, 1998).

A higher-order expression of this fourth line of defense happens with the fact that members of a colony do not randomly distribute themselves throughout a nest but instead compartmentalize themselves into recognizable zones based on age and reproductive status. Young bees, the brood, and the queen are always central in the nest, whereas older hive bees and foragers predominate at the periphery. Because social interactions are more common within, rather than across, these compartments, this has the effect of localizing parasites and limiting their spread (figure 18.3). This has been called "organizational immunity" (Naug and Smith, 2007), and it is an example of an emergent property (chapter 10), the kind of order that emerges spontaneously given enabling pre-existing conditions.

Another higher-order example of limiting a parasite's spread invokes genetic diversity, and here we harken back to polyandry, the subject of chapter 9, the queen's habit of mating with many males, which causes her workers to be genetically diverse. Genetic homogeneity, i.e., sameness, would be a dangerous situation in a dense aggregation of individuals like a social insect colony. One virulent pathogen could sweep through the nest with devastating results. But genetic diversity not only increases the likelihood that individuals will possess innate resistance mechanisms to a variety of pathogens, it also increases behavioral repertoires that add up to social immunity.

A good example is hygienic behavior. It is not just one behavior, but rather a suite of behaviors—the ability to detect abnormal brood, the ability to uncap it, the ability to remove the contents, and a low tolerance threshold for abnormal brood that stimulates the possessor to engage in the process. There are at least six genetic regions responsible for these behaviors (Oxley et al., 2010), and a multiply mated queen has a better chance of delivering all necessary genes to her colony. It is no surprise that high rates of queen polyandry have been associated with lower disease incidence in colonies (Tarpy and Seeley, 2006).

As a fifth and final line of defense, we can hypothesize on colony strategies that limit vertical transmission of parasites to a colony's swarm offspring. One line of evidence for this is the fact that workers infected with *Vairimorpha* sp. (formerly *Nosema*) microsporidians (Tokarev et al., 2020) remove themselves from tending the queen (Wang and Mofller, 1970). Because it is the old queen that moves with a swarm, this can be interpreted as a strategy for reducing *Nosema* infection risk to the swarm offspring.

The existence of natural selection against horizontal transmission seems less likely, at least from the point of view of an infected colony. There is no obvious evolutionary advantage to protecting a neighboring colony from your

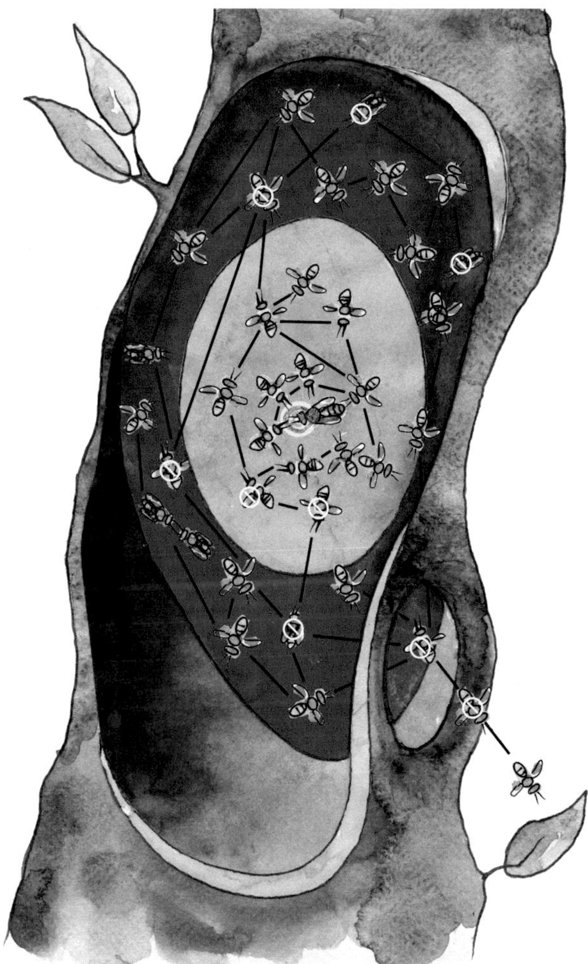

FIG. 18.3. Bees in a colony compartmentalize themselves into zones based on age and reproductive status. Young bees, the brood, and the queen (with halo) are centered in the nest (light zone) whereas older maintenance workers and foragers predominate at the periphery (darker zone). Social interactions (shown by connecting lines) are more common within, rather than across, these compartments; this has the effect of localizing parasites and limiting their spread, an outcome called "organizational immunity." The zones are not equal in value. The brood, queen, and young bees are more important than the older bees on the edge. It is from the older edge cohorts that foragers are drawn who leave the colony (lower right), and as the most expendable cohort in the colony, it is foragers that are most likely to experience parasite uptake. By extension, foragers' tendency to linger in the periphery tends to restrict infected individuals (⊘) to the periphery. Image redrawn from Cremer et al. (2007).

infection unless that colony is closely related or its proximity increases the chance for your re-infection. If natural selection responds to horizontal transmission at all, it is probably active as a defensive measure in step one above—reducing uptake of parasites. This may be one explanation why colonies in nature separate themselves from one another at rather large distances, ranging from 304 to 4,848 meters (from data cited by Nolan and Delaplane, 2017).

Evolution of Social Immunity

The term "social immunity" first showed up in 2007 in a landmark paper by Cremer et al. (2007), who defined social immunity as the "collective action or altruistic behaviours of infected individuals that benefit the colony." Three years later, it was pointed out that such a term should be reserved for immune behaviors that are explicitly *social*—that is, the result of natural selection acting on groups (Cotter and Kilner, 2010). Self-grooming, small nest entrances, and removing waste, for example, could just as easily be expressed by a solitary mother in her simple earthen tunnel. Group responses of any kind, including holdovers from ancient solitary living, are better called "collective" responses, a safe term that makes no claims whether the trait is socially selected or solitary.

In a more recent and refined definition of social immunity, Meunier (2015) took into account solitary versus social origins and clarified that the term must be limited to a *group*: "any collective and personal mechanism that has emerged and/or is maintained at least partly due to the anti-parasite defence it provides to other group members." With this wording, an immune trait may be solitary in origin, but if it has been modified by group selection, we may now include it under "social immunity." The point of these clarifications is to pinpoint what exactly is selected for; who benefits; who pays, and did it evolve in response to social or solitary living?

Now that we have the terms settled, we can begin considering the various immune mechanisms (figure 18.4) and their evolutionary histories. First, we must determine if a trait is a product of group selection or a carry-over from solitary life. This is complicated by the fact that we cannot be certain that an immune response we witness at the colony level began as an immune response at the solitary level. Social evolution is replete with examples

1. Preventing uptake
- Big distances between colonies
- Restrict foraging to oldest bees (A)

2. Preventing intake
- Guards reject infected (⊘) foragers (B)
- Small nest entrances (C)
- Nesting in protective cavities (D)

3. Preventing establishment
- Undertaker behavior (E)
- Antimicrobial propolis
- Antimicrobial venom
- Social encapsulation of nest invaders (F)

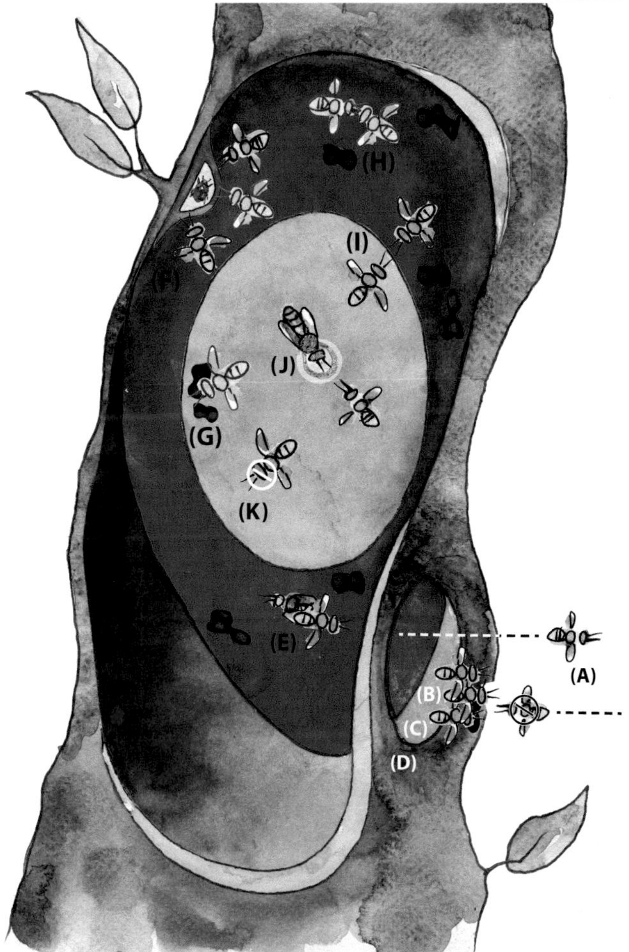

4. Preventing spread between member groups
- Hygienic cell-cleaning behavior (G)
- Grooming behavior (H)
- Age compartmentalization (I)
- Polyandrous queen (J)

5. Preventing spread to offspring swarm
- Infected bees (⊘) don't feed queen (K)

FIG. 18.4. Mechanisms of social immunity in a representation of a natural bee colony in a hollow tree. The dark ring represents the nest periphery surrounding the pale center of the nest. Nest members segregate themselves according to age and reproductive status, with older workers in the nest periphery and younger workers, nurses, and the queen in the center. With organisms and superorganisms alike the immune reactions follow a sequence of increasing engagement as the threat moves from outward to inward.

of solitary genes being co-opted into rather different uses in a social context. And finally, understanding a mechanism's evolutionary pathway implies understanding how it is regulated in modern colonies. Immune responses are powerful things, and overreactions can be more damaging than the initial parasite threat. The vertebrate immune system contains complex feedback loops that contain overreactions, but our knowledge of such feedback loops in bee colonies is rudimentary.

One of the most interesting questions has been the focus of Meunier (2015), who asked whether social immunity is a cause or effect of social life. One can imagine either scenario: (1) a limited cavity congested with individuals is a recipe for disease outbreak, constituting strong selection pressure for evolving social immunity, or (2) forms of social immunity are emergent in early pre-social groups, providing stability and removing obstacles to the evolution of sociality.

Meunier (2015) analyzed the literature on 30 mechanisms of immune response and their occurrence across insect groups that are eusocial, non-eusocial but group living, or solitary. By overlaying the behaviors across this range of social types, he was able to support an hypothesis that social immunity was a prerequisite for "increased complexity in the nature and frequency of social interactions." He goes on to say:

> In particular, eusocial insects express the individual mechanisms used by solitary species to limit parasite uptake from the environment . . . to which they add the collective ones used by non-eusocial species to limit both parasite establishment in the nest . . . and parasite transmission to the brood . . . as well as specific ones that limit parasite spread between adult group members.

What's taking shape here is a picture more similar to the second scenario above—that individual forms of immunity were active in the earliest evolution of insect societies, providing stability that encouraged the increasingly layered forms of social interaction that were culminating in complex eusocial life. As those primitively individual forms of immunity began conferring fitness at the level of colony, they were further acted upon by those same social forces, rendering them into their modern forms we see today and, as products of group selection, now qualify as examples of "social immunity." And finally, we cannot exclude the possibility that some of the social immunity mechanisms we see today are wholly modern—the derived products of social, not solitary, selection. Undertaker behavior and age compartmentalization may fit this description.

But at this point in time, we cannot unambiguously say that *any* immune mechanism has been socially selected for its benefit to the group. Part of the problem is that some immune mechanisms are known to express other functions in insects, and it is experimentally difficult to separate out the immune benefits from the non-immune benefits—or costs—to get a clear picture. As one example, consider adult-to-adult grooming: a groomer not only helps rid her sister of parasites, but the groomer also helps spread the chemical odor signature of her nest, which collectively helps the colony recognize "self" from "intruder" (Zhukovskaya et al., 2013). Another example is the benefit of queens that are polyandrous, i.e., mated to many males. The genetic diversity caused by this queen state is known to improve colony-level disease resistance (Tarpy, 2003) but also other fitness measures such as brood production (Delaplane et al., 2021) and foraging recruitment (Girard et al., 2011).

One can appreciate the practical importance of clearing up these ambiguities when one considers the beekeeping industry's longstanding interest in breeding bees for disease and parasite resistance. If beneficial non-immune mechanisms are confused for immune mechanisms, then progress in selective breeding will be seriously hampered. On the other hand, what if a supposed immune mechanism has an unknown and unidentified cost that sabotages its benefits against the parasite? This is why understanding "all the moving parts" is such a good research investment—it gives the best possible answers up and down the stakeholder chain.

Summary

A colony expresses at least five stages of increasing engagement as a parasite or pathogen threat escalates: (1) limiting parasite uptake by relegating out-of-nest jobs to the oldest and most expendable members and spacing natural colonies far apart, (2) preventing parasite intake by choosing defensible nest cavities and guarding nest entrances, (3) preventing parasite establishment by coating nest interiors with natural antimicrobials and removing dead nestmates, (4) limiting the parasite's spread by self- or allo-grooming, removing infected brood, compartmentalizing member age groups in the nest, and acquiring a genetically diverse workforce, and (5) limiting vertical transmission of parasites to a colony's swarm offspring. However, knowing mechanisms such as these is not the same as assigning them to social selection.

The study of social immunity is a relative new science with much foundational work remaining. It has been pointed out that the term should be reserved for immune characters that are explicitly the result of natural selection acting on groups. Self-grooming, small nest entrances, and removing waste, for example, could just as easily be expressed by a solitary mother in her simple earthen tunnel. The question has been asked whether social immunity is a cause or effect of social life. One can imagine either (1) a limited cavity congested with individuals being a recipe for disease outbreak, constituting strong selection for evolving social immunity, or (2) forms of social immunity emerging in early pre-social groups, providing stability and removing obstacles to the evolution of sociality. Available meta-analyses suggest a pathway nearer to the second scenario, but at this point we cannot unambiguously say of any immune mechanism that it has been socially selected for its benefit to the group. Part of the problem is that some immune mechanisms are known to express unrelated functions. As one example, genetic diversity caused by a queen mating with multiple males is known to improve colony-level disease resistance but also other fitness measures such as brood production and foraging recruitment.

CHAPTER 19

Nest Symbionts

THE NEST OF A COMPLEX social insect can be so environmentally stable, both as a structural shelter and an oasis of optimum temperature and humidity, that it has been called "a factory in a fortress" (Oster and Wilson, 1979), the "factory" referring to the production of a worker force that in turn procures the energetic resources to sustain the colony's reproduction, and the "fortress" in the case of the honey bee referring to the hollow cavities that scout bees seek out and appraise for their optimal volumes, insulative properties, and defensibility. These hollows are usually in trees, range from 30–60 liters in volume, and have small entrances. Once bees occupy them, the workers scrape away the dead soft wood, coat the sound wood and interior cavity walls with propolis, then proceed to build up to three square meters of parallel combs (Seeley and Morse, 1976). This is the "skeleton," if you will, of the honey bee superorganism, the substrate on which the bees through behavioral means regulate temperature and humidity conditions for optimum living conditions. These cavities are then, by direct behavioral extension, rendered into ideal environments for insect adult life and immature development.

Other species think so too. In honey bees, as with many other highly social insects, there are tenants in the house—whole bestiaries of organisms

(Kistner, 1982) who live in the same nest, benefiting from the safe harborage, and making a living off the bees or nest detritus. Such close relationships between different species are called *symbioses*, and within that label, the relationships can range from *parasitic* (one species benefits at the expense of the other), to *commensal* (one benefits, the other is not harmed), to *mutualistic* (each benefits). Additionally, these categories can be further categorized as *obligate* (the relationship is exclusive and necessary to at least one of the parties) to *facultative* (one or both can enter or leave the relationship as opportunity permits). Commensals in the context of social insect nests are commonly called *inquilines*.

It is helpful to never forget that genes are inherently selfish and that natural selection rewards, without bias or partiality, on the basis of one criterion only—whatever works. In this case, "working" means whatever heritable characters promote the survival and transmittal of the genes that code for them. It's easy to read cold-hearted selfishness in a parasite that takes the life of its host, but even the most benign of mutualists is playing by the same rules—only in this case, cooperation with its symbiotic partner has proven to be the best strategy for proliferating its genes. Once the calculus of gene transmittal tips in a different direction, the mutualist will be quick to adapt—on purely selfish grounds. So, when we talk about symbionts, it's good to remember that these categories are, in the words of Hughes et al. (2008a), "a continuum of costs and benefits with parasitism at one end and mutualism at the other."

Many authors have observed that large social insect colonies are like miniature ecosystems with layers of species and multi-trophic interactions similar to that observed in human-scale terrestrial ecosystems. The same laws of ecology apply so that, for example, whether it's a social insect colony or a hedgerow in England, the number of species living in either ecosystem tends to increase with increasing size and age of the ecosystem. Large, long-lived, and stable systems tend to "collect" species over geologic time, many of which species may enter into symbiotic relationships with others, thus further enlarging and complicating the relationship webs.

In the ants and termites, we find the largest number, diversity, and complexity of nest symbionts—and for these very reasons: large size, long life, and stability. With virtually infinite space, the subterranean nests of ants and termites become diffuse and vast, a feature that encourages the evolution of secondary reproductives and large populations. Their nests include examples of sophisticated animal architecture, such as heat chimneys, that maintain favorable microclimates for insect nurseries and fungus gardens grown by the insects for food—among the most remarkable examples of symbiotic coevolution known in biology. Their highly diversified and numerous soldier castes provide waves of defensive protection, and in some species, the reproductive pair live their post-dispersal lives protected in hardened mud bunkers deep in the nest.

In the end, these contrived habitats generate positive feedback loops that further perpetuate large populations, long colony life, and stability. These same conditions render complex social nests largely predator-free—yet another positive reinforcement, in that life history theory predicts that low predation pressures encourage selection for long life (Keller and Genoud, 1997). Is it any wonder that opportunistic species would evolve mechanisms to escape detection, to invade, to colonize, and to integrate into these attractive "fortress factories?"

And, reflecting on a theme stated elsewhere herein, we see how features of the honey bee superorganism recapitulate the evolution of free-living organisms such as ourselves, so that by studying them we learn about us. For just as the honey bee superorganism has picked up symbionts in its long natural history, so too has our own lineage of *Homo sapiens*. There are "nest invaders" in our bodies, tenant species who run the gamut of parasitic (hookworms) to commensal (hair follicle mites) to mutualist [an array of microbiota essential to normal human physiology (Dethlefsen et al., 2007)]. And, whether in the honey bee superorganism or human organism, these symbionts have themselves shaped the evolution of their hosts. We could not be what we are without our symbionts.

Neither does our metaphor depend on microscopic invaders; we can do away with the body size gap if we consider dogs, cats, farm animals, and other "commensals," the so-called *peridomestic* species who live in our homes and communities and with whom we show signals of coevolution (Chambers et al., 2020; Herbeck et al., 2017). Dog–human symbioses alone have ranged from mutualistic to commensal to competitive over the course of our long shared history (Pillay et al., 2022).

The analogies between organisms and superorganisms are instructive for understanding general principles of evolution, and sometimes those analogies are remarkable. But the two levels are not equivalents, and when it comes to symbionts, their differences mean that host/symbiont evolution follows different paths. To begin, most free-living organisms do not have a "factory in a fortress" lifestyle that so effectively buffers stressors and eradicates predatory pressure.

Hughes et al. (2008a) hypothesized that in complex social species, parasites tend to become less virulent and mutualists less cooperative compared with the same players in non-social or simple social species.

Parasites tame down for several reasons, not least of which is the favorable selection for long life afforded by the factory in a fortress lifestyle. From the parasite's point of view, a short-lived host that could die at any time from a number of causes should be exploited rapidly, but a long-lived host represents an opportunity for slower, more sustained exploitation—hence, selection for low parasite virulence (Anderson and May, 1981). Second, in spite of the fact that early epidemiologists predicted that large colonies of densely populated, highly related individuals should be a recipe for epidemics, the reverse has proven true: colony-level selection for group immunity (chapter 18) has effectively dampened an escalating arms race between host and parasite. The sum of these effects is the prediction that large social colonies accumulate large numbers of diverse low-virulence parasites.

Mutualists are predicted to become less cooperative in large social colonies owing to the ecological principle that species accumulate over time in stable habitats. Some of these newcomers may enter into symbiotic relationships, either parasitic, commensal, or mutualistic, with the earlier mutualists. Following their own selfish interests, players in the new arrangement may revise life history strategies to be less beneficial to the host.

Hughes et al. (2008a) gave an interesting example of a kind of symbiotic "arms race" that can occur in the fungus-growing ants and termites. These large and successful social insect groups are widely separated in their phylogenetic history, yet each has adopted a sophisticated and literal form of animal agriculture—the culturing of elaborate fungal gardens deep inside their nest for food—an example of *convergent* evolution, when unrelated species arrive at a similar solution to a similar problem. The habit of fungus farming is tens of millions of years old, rendered stable by the hosts' evolved abilities to protect their fungal mutualists from disease pathogens. But these safety nets have nevertheless not prevented the accumulation of opportunistic fungal invaders that attack the mutualist fungi. In response, certain ant species have evolved new mutualisms with bacteria that attack the fungal pathogen, growing the bacteria on their bodies in fact. Not to be outdone, the bacteria themselves are attacked by yet another nest invader, a yeast, rendering the bacteria less effective. I can't help but think of the poem by Augustus De Morgan (1915):

> Big fleas have little fleas upon their backs to bite 'em,
> And little fleas have lesser fleas, and so, *ad infinitum*.
> And the great fleas, themselves, in turn, have greater fleas to go on;
> While these again have greater still, and greater still, and so on.

The picture painted by Hughes et al. (2008a) is a time vector of increasing species number and diversity inside the complex social insect nest. They become "biodiversity hotspots" within their larger landscapes—little enclaves of environmental stability, independent to some degree of ambient conditions. They are predicted to "produce highly diverse communities of relatively avirulent pathogens and moderately benign mutualists." Hughes et al. (2008a) essentially unpacked an intuitive principle that parasite virulence is expected to decrease when hosts are free to coevolve resistance or tolerance (Gandon et al., 2002), leading me to state frankly that we can expect increasing host/parasite stability as the age of the relationship lengthens.

Descriptions such as "biodiversity hotspots" certainly apply to the overachieving ants and termites, but things are a little more understated with the bees which, as a group, are conspicuous for their relative scarcity of nest symbionts. Kistner (1982) offered some explanations for this: the comparatively small populations of bee colonies (tens of thousands) compared with ants and termites (tens of millions); a higher spatial density of termite colonies (mounds "as far as the eye can see") compared with honey bee colonies spaced at a scale of square kilometers (Seeley, 1985; Visick and Ratnieks, 2022); the small volume of bee nests, equipped with propolis envelopes, and small highly defensible entrances; and the habit of bees for removing refuse and dead bees from the nest, whereas ants and termites often incorporate debris into the structure of the nest.

Nevertheless, I think that even the few honey bee nest symbionts that exist give evidence of the kinds of evolutionary dynamics noted by Hughes et al. (2008a) for nest symbionts in general. In figure 19.1, I have listed the macrobiotic nest symbionts of the western honey bee, most of whom are familiar to beekeepers everywhere, and ranked them by their virulence.

Starting at the bottom, we have the facultative nest commensal/mutualist, the greater wax moth, *Galleria mellonella*. Even though wax moths are listed as pests in every beginning beekeeping book, their role is more nuanced than that, to the point that I suggest them as the only honey bee nest symbiont that performs mutualistic benefit for its host. Granted, they are first and foremost opportunistic scavengers. They are almost exclusively associated with honey bees, but they are occasionally found in bumble bee nests. The adults lay their eggs in cryptic cracks and crevices in the nest, whereupon the larvae hatch and tunnel through combs, eating pollen and other protein refuse. It is only when a colony is weakened and its adult population plunges from an unrelated stressor—usually queen loss—that the moth larvae, now unrestrained, explode and cause complete destruction to the combs and their contents. It is in this context that most beekeepers encounter moths and, justifiably associating them with dead colonies and destroyed combs, relegate them to pest status. But at a population scale, this same behavior constitutes an ecosystem service for the next swarm occupant of a hollow cavity following the death of the previous colony. My University of Georgia colleague, Lewis Bartlett calls *G. mellonella* "the vultures of the honey bee world," disposers of dead carcasses and pathogens they may transmit (L. Bartlett, personal communication). Feral colonies in general have lower levels of pathogens compared with managed colonies (Bailey, 1958; Gilliam and Taber, 1991; but see Thompson et al., 2014), and the frass of wax moths recovered from dead natural colonies has extremely low levels of disease bacteria or spores (Bai-

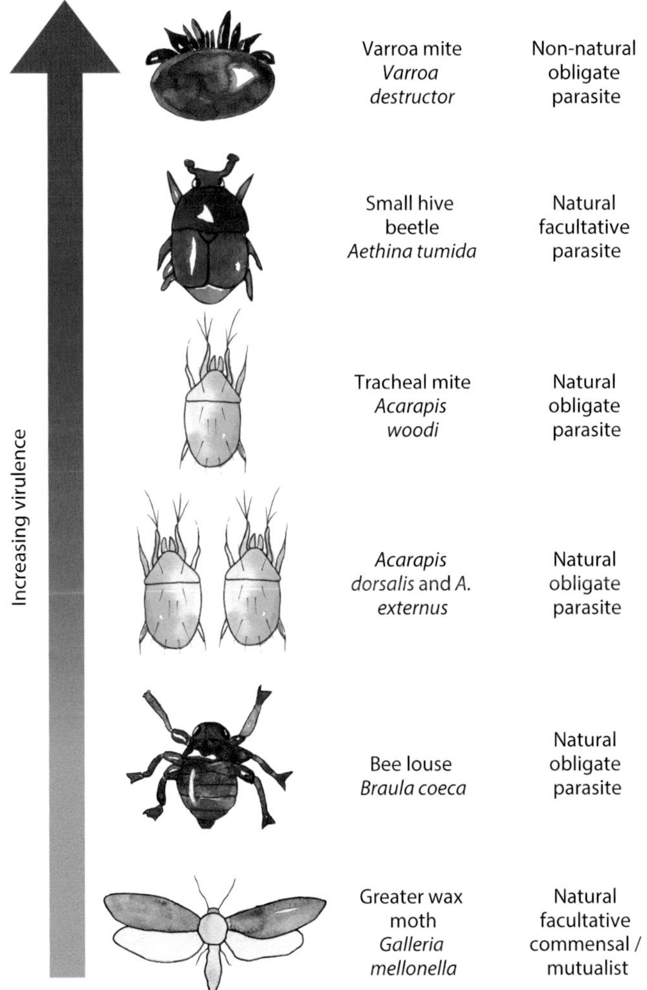

FIG. 19.1. A summary of the macrobiotic nest symbionts of the western honey bee, *Apis mellifera*, ranked in order of virulence expressed toward their host. The arrow indicates increasing virulence as one moves toward the top of the figure.

ley, 1958; Gilliam and Taber, 1991). A USDA scientist, E. F. Phillips [cited by Gilliam and Taber (1991)], pointed out in 1907 that wax moth larvae do not eat scales of American foulbrood, but even if they eat the surrounding comb so that scales drop to the floor where they are subsequently covered up by debris, that would have a suppressing effect on the proliferation of the disease. This is a weak stab at mutualist status, but it may be as good as it gets with honey bee macrosymbionts.

Moving up the virulence scale in figure 19.1, we next encounter the obligate parasite, the bee louse, *Braula* spp. These are flightless flies found exclusively in honey bee nests (Grimaldi and Underwood, 1986). They practice chemical mimicry, taking on the cuticular odor of the bees to avoid detection (Martin and Bayfield, 2014). They are particularly attracted to queens, with infestations of over 100 per queen reported in the literature (Grimaldi and Underwood, 1986). Affinity for the queen is probably adaptive, given that she is the longest-lived individual in the colony and receives frequent attention and feeding from her attendants. The bee louse is associated with little (Strauss et al., 2014) but non-zero fitness costs to the host colony. High numbers on the queen are reported to reduce egg production (Crane, 1990), and beekeepers in Jordan direct control measures at them, citing improved honey production (Al Ghzawi et al., 2009).

Braula schmitzi in Ecuador has been found to carry particles of acute bee paralysis, but there is no evidence that the virus replicates and amplifies in the louse or transfers to the host bees (Avalos et al., 2019). From a purely economic viewpoint, the bee louse has been considered a beekeeping pest because the tunneling activity of its larvae can spoil the pristine appearance of marketable comb honey. Because it cannot be considered a mutualist, I default to considering it mildly parasitic.

Next up the virulence scale are the microscopic mites *Acarapis dorsalis* and *A. externus*. These mites are natural obligate parasites of honey bee adults. In a supreme expression of niche specialization, *A. dorsalis* spends its time on the pronounced dorsal groove of the adult bee's thorax, whereas *A. externus* occupies the neck area. Both feed on bee blood, presumably by piercing soft intersegmental tissues. The literature appears nearly silent on any measurable fitness costs to bee colonies from these parasites.

This is not the case for their sister species, the tracheal mite, *Acarapis woodi*, next up on our virulence scale. Not content to occupy the exterior folds of its host's body, *A. woodi* made the leap to entering the internal tracheal (breathing tube) network of its host where, in that protected environment, it pierces the tracheal wall, drinks its host's blood, and lays eggs and produces a brood of young. Infested tracheae are heavily scarred, and susceptible infested individuals rapidly perish.

Authorities have debated for decades whether *A. woodi* was responsible for a bee epidemic in early twentieth-century England called Isle of Wight disease. Between the years 1984 and 1990, there was a verified outbreak of *A. woodi* in North America. Colonies would degenerate into weak disorganized clusters with extremely high rates of overwintering loss. State regulators responded in some cases with draconian apiary depopulating programs that some argued, during and since, were more damaging than the parasite itself.

Morse and Eickwort (1990) put forward the tantalizing hypothesis that *A. woodi* is in fact a very new species, having recently diverged from one of its external cousins, *A. dorsalis* or *A. externus*, going so far as to pin down a time and place: "near the end of the last century [the nineteenth] in the United Kingdom." Evolutionary biologists do not typically traffic in such recent time scales, but those authors' premise is sound in that it is a simple innovation to move from living externally on the thorax of one's host to entering the tracheae, whose largest branch exits via a spiracle on the side of the thorax. It is easy to imagine how the safe harborage and easy access to food afforded by such a move would be highly adaptive, at least in the short term.

Next, we encounter a nest symbiont with a more recent and dramatic apicultural history—*Aethina tumida*, the small hive beetle (SHB). First detected in the United States in Charleston, South Carolina, in 1996 (Hood, 2004), the SHB is a sap beetle, member of the beetle family Nitidulidae known for their strong affinity for fermenting plant products. Subsequent genetic analysis confirmed its origin from southern Africa (Evans et al., 2000), where it presumably speciated as a dietary specialist on the fermenting products and nest detritus of honey bee nests. Adult beetles enter bee nests and deposit their eggs cryptically, in a fashion similar to wax moths. Beetle larvae then spread across combs, eating pollen, honey, and even brood. In uncontrolled infestations, the action of beetles causes unchecked honey fermentation and formation of a slime on the surface of combs.

Unlike wax moths that limit their depredations to fatally weak colonies, SHBs are able to contribute directly to colony dwindle and decline. There is evidence that their speciation event is recent, perhaps even on a scale similar to the divergence hypothesized above for *A. woodi*. This is because *A. tumida* is not (yet) an obligate dietary specialist on *A. mellifera* nest contents. The beetles clearly prefer bee nest products, but they will opportunistically deposit eggs and complete a life cycle on exposed fruit (Buchholz et al., 2008; Ellis et al., 2002).

At the apex of figure 19.1, we land on the most persistent and damaging nest symbiont of all—the non-natural obligate parasite *Varroa destructor*. Having been introduced onto the western honey bee, *A. mellifera*, from its natural host the Asian honey bee *A. cerana* in the 1960s, *Varroa* found in modern commercial beekeeping the ideal situation—a populous new host lacking virtually any natural resistance to its depredations. It is an un-evolved for, non-natural relationship, which means that the geologic age of the host/parasite relationship is virtually zero. *V. destructor* and *A. mellifera* have not had enough time to coevolve a more stable symbiosis (Gandon et al., 2002).

The case histories of *A. woodi* and *A. tumida* are instructive for understanding how geologic time can explain modern day observations on parasite virulence. In both cases, the North American honey bee populations evolved resistance rapidly to these new pests. It is generally agreed that depredation pressure from SHB is less now than it was even 10 years ago, and in the case of tracheal mites, it is simply quite difficult to find a positive sample for teaching purposes. These outcomes are a direct result of the natural relationship between parasite and its host. I can suggest that *A. woodi* and *A. tumida* do, in fact, represent recent speciation events, in scenarios that look like this:

From a common ancestor in western Asia around 6.3 million years ago, future African and European lineages of *A. mellifera* diverged and began pursuing different histories, the Y, L, and A lineages moving west into Africa and the M and C lineages moving northwest into Europe [figures 22.7(F) and 22.8(E)]. It's a good bet that by 3 million years ago, *A. mellifera* occupied most of Africa and Europe [figure 22.7(F)]. The European lineages carried with them the ancestral lineage of the parasitic mites *Acarapis*. During a prolonged succession of glaciation events lasting from around 2.58 million to 11,700 years ago, the European subspecies of M and C evolved into their modern forms, including two representatives well-known to modern beekeeping, the northern European honey bee *A. m. mellifera* and Italian bee *A. m. ligustica*.

Long thereafter, perhaps even in historic times; perhaps—dare I be so precise?—between the seventeenth-century human-assisted introduction of M lineage *A. mellifera* into North America and the early twentieth century, European *Acarapis* speciated a virulent form, *A. woodi*, capable of lethally exploiting the host's tracheal system. But owing to its inherited obligation to life on *A. mellifera*, it was profoundly maladaptive for the incipient *A. woodi* to continue such unchecked reproductive exploitation of its host. In both the United Kingdom and North America, the rapid decline of their respective tracheal mite epidemics can be explained as the rapid adaptations of the host and parasite to moderate a virulent mutation that disrupted what was otherwise a relatively ancient, stable, albeit mildly parasitic symbiosis.

A similar situation is playing out with SHBs. Sometime after the arrival of *A. mellifera* in Africa, an ancestral *Aethina* species began exploiting the fermenting plant sugars it found in nests of *A. mellifera*. In what was no doubt a kind of arms race, the two species adapted to one another until a form of commensalism evolved between them, explaining why to this day African honey bee subspecies resist SHB more easily than the M- and C-derived subspecies in North America (Ellis et al., 2003).

One might say that in these cases, North American *A. mellifera* had between a few decades to up to 3 million years of coevolution to catch up on, a project that seems to be working, especially in the case of *Acarapis woodi*. But in the case of the *Varroa/mellifera* host-shift, worldwide *A. mellifera* populations have between 7 and 20 million years of coevolution to make up for [figures 22.1 and 22.7(F)]—the time since the divergence of the *A. mellifera* lineage from the rest of its Asian cousins (Asia being the home of *Varroa*). *Apis mellifera* has not had enough time to coevolve stable host resistance.

Summary

Many organisms have evolved the capacity to live alongside a social insect host in its nest. Such close relationships between different species are called *symbioses*, and these relationships can range from *parasitic* (one species benefits at the expense of the other), to *commensal* (one benefits, the other is not harmed), to *mutualistic* (each benefits). These categories can be further categorized as *obligate* (the relationship is exclusive and necessary to at least one of the parties) to *facultative* (one or both can enter or leave the relationship). Commensals in the context of social insect nests are often called *inquilines*. Parasite virulence decreases if hosts are free to coevolve resistance, and an expression of this principle is seen in the ecologic stability that accumulates over time in complex social insect nests. Therefore, we can expect increasing host/parasite stability as the age of the relationship increases.

In the honey bee colony, there are a handful of symbionts who together range across the scale of increasing virulence in patterns consistent with this expectation. Although the greater wax moth *Galleria mellonella* is considered a pest in a beekeeping context, I offer it as the nearest example of a macroscopic facultative mutualist owing to its capacity to consume beeswax comb and detritus, rendering a cavity cleaner for the next occupying swarm. The obligate inquiline bee louse, *Braula* spp., is largely harmless, but because it steals food from the mouths of queens, it must be considered mildly parasitic. The external obligate parasitic mites *Acarapis dorsalis* and *A. externus* feed on adult bee hemolymph; their near relative *Acarapis woodi* evolved the virulent capacity to enter, feed, and reproduce in the host's breathing tubes.

The facultatively parasitic African small hive beetle *Aethina tumida* was first detected in the United States in Charleston, South Carolina, in 1996, and in European honey bee populations where it has been introduced, it stresses bee colonies with its adults and larvae who scavenge on stored honey, pollen, and bee brood.

The most damaging relationship occurs with the introduced obligately parasitic mite *Varroa destructor*. Having been moved away from its natural host, the Asian honey bee *A. cerana* in the 1960s, *Varroa* found in European *Apis mellifera* a new host lacking any natural resistance to its depredations. This outcome is consistent with the extreme distance in time (7–20 million years ago) between the divergence of the *A. mellifera* lineage from its Asian nearest relatives to the time humans transferred *Varroa destructor* onto *Apis mellifera*. There has not been enough time for host resistance to coevolve.

CHAPTER 20

Senescence and Mortality

IT IS PART OF THE MYTHOLOGY OF BEEKEEPING, a compelling story that we want very much to be true. It comes up in hallway discussions and in the question and answer (Q&A) time after lectures on sustainable beekeeping. It evokes an image of a simpler time when beekeeping wasn't much more than controlling swarms and supering up for the honey flow. I'm talking about reports of long-lived wild bee colonies. It usually refers to a bee tree with a long record of activity: "This bee tree's been here since Grandpa's time." Of course the standard answer is "Yes that's interesting," but we have no way of knowing if it's an *uninterrupted* colony—one expressing a continuous line of genetic descent. Such an interruption would be easy to miss. All it takes is for a colony to die out in winter and a new swarm take its place, reoccupying the cavity. To a casual observer it would appear unchanged from the previous season. If there was time in between for wax moths to clean out the old combs, all the better because now the new swarm is the beneficiary of a clean cavity.

But the myth persists, and one reason, I believe, is because it hints at the holy grail of immortality, and who of us is immune to that siren song? And if there were ever an organism for which such a promise seems likely, it's the honey bee colony with its habit of reproductive fission—colony splitting via the annual swarm—placing it in the same category of organisms that reproduce by budding, like the so-called immortal hydra (Bosch, 2009). Even the luminous sociobiologist E. O. Wilson lumped honey bees with other social insects, army ants and stingless bees, into those he deemed "immortal" on the basis of this reproductive strategy (Wilson, 1971). And after all, what's to keep a colony from dying once it has settled into a snug cavity surrounded by reliable nectar and pollen plants that bloom year after year and one's evolutionary environment happens to not include such modern threats as *Varroa* mites and pesticides?

We will come back to that question in a moment, but for now let's step back and ask why should the honey bee superorganism, or any organism for that matter, die at all? If evolution is all about maximizing the delivery of one's genes to the next generation, how can senescence and death possibly contribute to that plan? The hydra excepted, but why isn't Earth positively teeming with immortals?

Let's eliminate the easy answers first. To begin, organic life on this planet is predicated on the capture and transfer of energy up and down food chains or webs. Primary producers are plants that capture the sun's energy and convert it to carbohydrates. Plants are in turn eaten by herbivores, who are eaten by carnivores, and so on, up to the apex predators who in turn are eaten by decomposers who disintegrate their bodies, returning their elemental constituents to the nutrient web. Not to put too fine a point on it, but organisms on this planet eat one another. One might say that death is engineered in. But that also means that death is the seedbed for ongoing life.

Mortality also enters the picture when genomes compete for limited resources. Resource limitation is a huge selection factor—whether for food, nesting sites, or mates—and the jostling of natural selection means that some individuals, and whole lineages, are out-competed and die. And finally, we must point out the obvious fact that life is full of dangers that have nothing to do with food webs or competition. Organisms die from ordinary day-to-day hazards and from unpredictable catastrophes like world-colliding meteors, volcanoes, and climate change (Glen, 1990; Urban, 2015).

That things die is a point hardly worth bringing up. But what about senescence? Getting old? Why has evolution produced organisms that fall apart even if they manage to avoid predators, life-threatening competition, or meteors? At age 62, I'm starting to take more than academic interest in the question.

If evolution is a general explanation for what we see in living things, then it should be able to explain the existence of senescence—a stage of life in which reproduction declines, the individual is enfeebled, and life expectancy falls.

Biologists have come up with two theories for the evolution of senescence, which are not mutually exclusive. There are subtle differences in emphasis, but each places a premium on reproduction as the currency of value in natural selection (Kirkwood and Rose, 1991).

The first theory builds upon two premises. The first premise is that natural selection does not act with equal strength throughout the life of an individual. It is strongest when an individual is young because at any given moment, a young individual has a higher *reproductive value*—defined as predicted number of future offspring (Williams, 1966)—owing to the simple fact that a young individual has more life left ahead. The second premise is the fact that one gene may code for two or more effects—a phenomenon called *pleiotropy*. If a pleiotropic gene codes for a beneficial effect early in life but exerts a negative cost later in life, natural selection will respond more to the "good" effect when young than to the "bad" effect when old, owing to the greater strength of natural selection on the young. In this way, negative pleiotropic effects accumulate in the aging individual, unresisted by natural selection. These bad accumulations include loss of fertility along with the host of other handicaps that define senescence and place it chronologically after the fertile period. This is called the theory of *antagonistic pleiotropy*.

A second theory emphasizes that there are fundamentally two types of cell in metazoan animals like humans and honey bees—the reproductive or *germ* cells and all the rest, called *somatic* cells. Every cell contains all of the organism's genes, but only the germ cells divide into sperm and eggs and are directly responsible for producing the next generation. From an evolutionary point of view, therefore, the germ cells are the most valuable. All the rest—the muscles and bone and cuticle and nerves and organs—are simply the vehicle, called collectively the *soma*. Evolution has worked heavily on optimizing the allocation of energy and resources between reproduction versus maintaining the vehicle for reproduction. And evolution, ever ruthlessly practical, has concluded that all that's needed is a soma that stays in good enough condition to reproduce. Any maintenance beyond that point is wasted energy, resulting in no additional offspring. Senescence is the result of accumulated defects to the soma. This theory is called, rather prosaically, the *disposable soma* theory.

With this background, we can now return our attention to the honey bee. First, let's go back to that idea of immortality—a colony that never dies, just divides. To get straight to it, this doesn't happen. But it does constitute an example of how thinking about the honey bee superorganismally can add clarity to our thinking. If the queen is analogous to the germ cells, then her daughter workers are analogous to the soma. As long as the queen is producing eggs, the entire colony, germ cells and soma, is genetically mutually reflective. But once a daughter queen is produced in the annual swarm cycle, her subsequent progeny no longer reflect the genotype of the previous queen, at least to the same degree. The successor is a daughter, of course, but not a clone; and moreover the daughter is mated to a new cohort of males. The genetic continuity has been interrupted; we can no longer view the new entity as the same superorganism (Moritz and Southwick, 1992). From equation 6 in part I, we can surmise that the nestmates of the succeeding colony share only $r = 0.135$ genes in common with nestmates of the original colony. If we struggle to name the entity that persists in the same hollow tree year after year, then perhaps "successor" is a better choice, or at best "descendant" if we want to capture the modicum of relatedness that transmits.

We also quickly see that the theories for senescence discussed above strain to accommodate the biology of the honey bee superorganism. Honey bee colonies in nature follow a pattern of annual and predictable reproduction. In fact, the likelihood of a temperate colony swarming (reproducing) once a year is quite high, 92% (Seeley, 1978). It is hard to see anything that looks like senescence in this pattern of regular yearly reproduction. A hint of it may be happening if we expand our question to include the survival of all those swarms, and in fact we do know that parental colony size is positively associated with the size of offspring swarm and the number of workers the swarm subsequently produces; not surprisingly, the number of workers subsequently produced is positively associated with a swarm's survival (Lee and Winston, 1987).

With this bigger view, if one assumes that "low parental colony size" is a symptom of senescence, we may be able to make a case. But in reality, modern beekeeping is all too full of reasons for low population size that have everything to do with *Varroa* mites, viruses, pesticides, and starvation—and nothing to do with evolutionarily driven senescence, i.e., "growing old." It's possible that enough things kill honey bees nowadays that we are rarely given the chance to observe senescence.

But it is also possible that the theories of senescence do apply; they just need to be adjusted to accommodate the powerful adaptation for annual swarming. This gets us back to that entity, the "successor" I called it, that persists for a while in the same tree hollow. It strikes me that this entity has a better bid for immortality than the colony/superorganism/organism, and it's this entity that storytellers in beekeeping circles and biologists like E. O. Wilson have in mind when they talk about the immortality of "the colony." Yet, we know that even these succeeding natural colonies do not live forever, with an average life span for temperate *A. mellifera* of only 5.6 years (Seeley, 1978). Why should these successors, evolutionarily wired for annual fission and annually reinvigorated with new genetics, not be immortal?

To answer this, we should first consider the possibility that the patterns of life and death in the honey bee superorganism can be explained by events at an ecological timescale, not evolutionary—things like winter starvation and extreme temperatures. The life history of temperate *Apis mellifera* does indeed show a strong adaption for surviving winter (Seeley and Visscher, 1985), yet we know that the majority of first-year swarms do in fact perish (Seeley, 1978). Living as a colonial insect year-round in cold climates is risky business, and perhaps mortality is so high that *A. mellifera* is a poor candidate for the evolution of senescence.

But I think there is still room for the possibility of senescence, or at least a clue of where to go looking for it—and that's the constraints imposed on the colony by the physical limits of an average tree cavity. If the successor changes genetically every year, at least the combs and propolis infrastructure of the nest remain. We know that as a colony grows and adds new comb over time, bees will disproportionately shift brood production to newer comb, relegating previous years' brood combs to honey storage (Free and Williams, 1974). Aging combs have higher pathogen loads (Koenig et al., 1986), and colonies reared on new first-year comb have more brood than do colonies forced to rear brood on old combs (Berry and Delaplane, 2001). It's plausible, therefore, that an aging comb infrastructure limits brood production and the large populations required to produce large, fit swarms (Lee and Winston, 1987). It may reach a point where the energy balance between maintaining the soma and producing a swarm becomes sub-optimal.

If natural selection is sensitive to these dynamics, then this would be consistent with the disposable soma theory. If controlled experiments were to confirm all this, then it would elevate our understanding of the physical infrastructure of the nest as a participant in the species' evolution.

But until this narrative is experimentally verified, the evidence for true senescence is weak. For the time being, we seem to be faced with the curious proposition that the honey bee colony dies, but never grows old.

Summary

Two theories are in currency for the evolution of senescence. The theory of antagonistic pleitropy draws upon two premises. The first is that natural selection does not act with equal strength throughout the life of an individual but is strongest when an individual is young. The second is the fact that one gene may code for two or more effects—a phenomenon called pleiotropy. If a pleiotropic gene codes for a beneficial effect early in life but exerts a negative cost later in life, natural selection will respond more to the "good" effect when young than to the "bad" effect when old, owing to the greater strength of natural selection on the young. In this way, negative pleiotropic effects accumulate in the aging individual.

The theory of the disposable soma emphasizes that there are two types of cell in metazoan animals—the reproductive or germ cells and all the rest, called somatic cells. From an evolutionary point of view, the germ cells are the most valuable. Evolution has prioritized somatic fitness only up to and through an organism's reproductive period. Senescence is the result of accumulated defects to the soma. Each of these theories strain to accommodate the biology of the honey bee superorganism in part because its high rates of climate-induced colony mortality make it a poor candidate for senescence.

But if we extend the idea of the superorganism to what I call "successor" colonies, the colony headed by an annually succeeding daughter queen in the same cavity, each successor related to its predecessor by no more than $r = 0.135$ on average, then we might reasonably predict declining somatic fitness with the aging beeswax combs and accumulating pathogen load in the cavity. But until this idea is challenged experimentally, the evidence is weak for true senescence in the honey bee superorganism.

PART III

Honey Bee Phylogeography

CHAPTER 21

The Corbiculate Bees

APIS. THE GENUS TO WHICH THE WESTERN honey bee *Apis mellifera* belongs is so distinctive in the human imagination that its name is recycled up and down the chain of bee taxonomy. First, moving down the taxon, the genus *Apis* is divided into the small open-nesting honey bees comprising subgenus *Micrapis*, the large open-nesting honey bees *Megapis*, and the medium-sized cavity-nesting honey bees who are granted privilege of retaining the name *Apis* unappended.[1] Moving up the chain, we have the tribe Apini, the subfamily Apinae, the family Apidae, and last of all, the whole superfamily Apoidea to which all bees belong—along with the Spheciformes wasps, who are lumped with the bees owing to shared morphology that marks them as bees' nearest living kin (Michener, 2000). The "true" bees can still be pulled out of this awkward grouping and called the Anthophila. Then there are the extinct genera *Electrapis* and *Hauffapis*—but at this point let's stop, content to leave *Electrapis* and *Hauffapis* comfortably extinct and move on with our

1. Subgeneric names are rarely used except when the context makes it useful. When they are, convention calls for the name to be italicized, capitalized, and in parentheses between the generic and specific names; hence, for example, *Apis* (*Micrapis*) *florea* or *Apis* (*Apis*) *mellifera*.

point that modern *Apis* must be important for a reason, namely, that the origin of *Apis* makes for a fascinating story about how and where sociality emerges.

Many streams of evidence are at play, and the evidence is not tidy nor consistent. The story teaches us that in the end, there is still an element of human subjectivity in judging the various streams of evidence—morphologic, geographic, genetic, behavioral, and fossil—to arrive at the most parsimonious and reasonable explanations for the origin and modern distribution of the genus *Apis*.

We start our story before the emergence of *Apis*, estimated at between 38–28 million years ago (Kotthoff et al., 2013; Smith, 2021), by considering the genus's most recent common ancestor (see figure 1.3). This long-lost species was a member of the corbiculate bees, so named for their possession of a rear-leg pollen basket, more properly called a corbiculum (see figure 4.4). The corbiculate bees are a relatively ancient group of bees, comprised of around 890 species (Martins et al., 2014) and made up of four tribes: the bumble bees (Bombini), the orchid bees (Euglossini), the stingless bees (Meliponini), and the honey bees (Apini). Lacking any proper taxonomic rank, these four tribes are nevertheless considered a natural grouping with a shared common ancestor from whom the corbiculum was inherited. More to our present purposes, simple eusociality was also inherited from that same common ancestor, a state eventually inherited, in its complex form, by the common ancestor of *Apis* (Cardinal and Danforth, 2011).

The corbiculate bees have been the subject of intense interest for years owing to the fact that within these four tribes, we see the full range of social behavior from solitary to communal (Euglossini, E) to simple sociality (Bombini, B) to complex sociality (Meliponini, M, and Apini, A). This, added to the fact that this socially diverse assemblage is nevertheless held together by one strong derived character—the corbiculum—means that the corbiculate bees constitute a tight little taxon within which we can expect to find clues to the origin of social behavior.

Corbiculate Phylogenies

With only four tribes, one would think it is easy to figure out their relationships and the order in which divergences happened. Charles Darwin ruminated on the matter as early as 1859, but it was Charles Michener (1944) who offered the first, and admittedly satisfying, phylogeny based on shared behavior and morphology. This phylogeny was strengthened in subsequent years by additional morphological data from adult stages, larval stages, and fossils, and it has the added virtue of parsimony—that philosophical bias in favor of the simplest of explanations. Parsimony, applied to phylogeny, means favoring a family tree that requires the fewest evolutionary changes, and statistical analyses are available for generating the most parsimonious relationship tree among a group of taxa based on shared derived characters.

In Michener's tree, simple eusociality arose once—with the common ancestor of the corbiculates (Cardinal and Danforth, 2011); complex eusociality arose once—with the common ancestor of MA, and simple eusociality was lost in E[2] for a total of three evolutionary changes (figure 21.1). The B retained their inherited simple eusocial state and were placed in an intermediate position sister to the socially complex MA.

This phylogeny is intuitive, parsimonious, and biologically reasonable, but this tidy state of affairs began to unravel in the early 1990s with the advent of DNA sequencing technology that added enormously to the number of characters—shared genes in this case—available to phylogenists for reconstructing family trees. It quickly became apparent that trees constructed from DNA differed significantly from traditional trees built from morphology. Moreover, the new reconstructions were not parsimonious when it came to the origins of sociality, and neither did they necessarily agree with what had been the assumed direction of change in social evolution. This directionality, called *polarity*, is the vector of change over time in a character state. In the character state "sociality," for instance, we generally presume that small populations, fluid behavioral castes, and overwintering as a solitary mated female are ancient states, whereas large populations, fixed castes, and overwintering as a colony are derived. This is the presumed polarity, or direction of change in the character state "sociality." Moreover, it is attractive to assume that eusociality, once gained, is difficult if not impossible to reverse

2. Preceding the work of Cardinal and Danforth (2011), Michener (1944) would likely have posited a most parsimonious arrangement with simple eusociality appearing with the common ancestor of the B(MA) clade and complex eusociality at the common ancestor of MA for a total of two evolutionary changes. There would have been no need to invoke loss of simple eusociality in E.

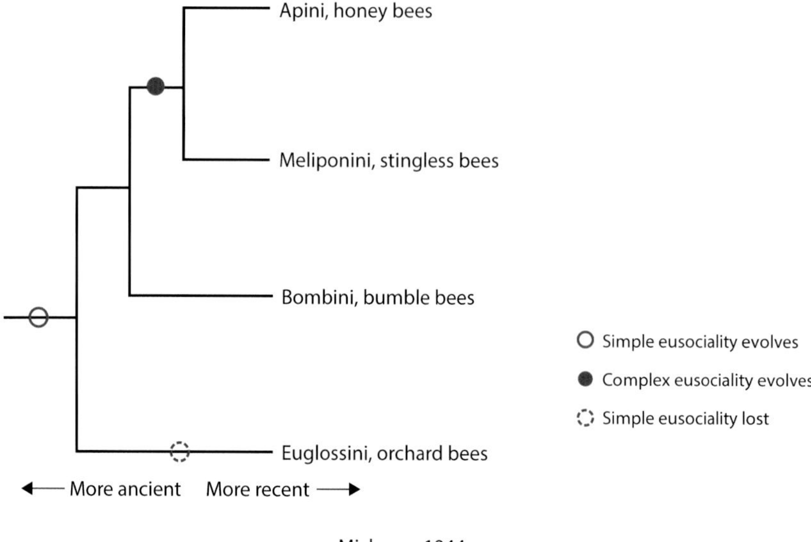

FIG. 21.1. The morphological phylogeny for corbiculate bees of Michener (1944) and associated state changes in eusociality.

owing to life history interdependence of queens and workers. But the new DNA trees for corbiculate bees were blowing these assumptions to pieces.

Figures 21.2 and 21.3 show a phylogeny published by Cameron (1993) based on shared mitochondrial DNA sequences. These figures highlight some of the problems this phylogeny, and others subsequently like it, caused Michener's traditional narrative of social evolution. Cameron (1993)'s phylogeny places the complex eusocial A sister to the rest and lumps the complex eusocial Meliponini with the simple eusocial Bombini as downstream sister group MB, with the solitary/communal Euglossini E in between. Figure 21.2 shows the sequence of events that must have happened if one assumes that complex eusociality derives from simple eusociality: (1) simple eusociality appears in the common ancestor of the corbiculates, (2) complex eusociality appears in A, (3) simple eusociality is lost in E, and (4) complex eusociality appears independently in M. We now have four evolutionary changes in one character state.

Cameron (1993) pointed out that a more parsimonious interpretation is available if one removes the assumption that eusociality necessarily evolves

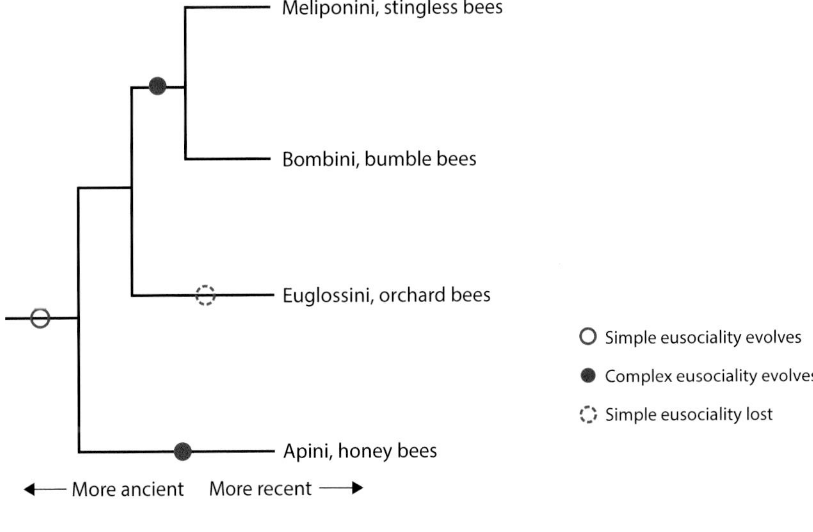

FIG. 21.2. An early molecular corbiculate phylogeny that challenged the morphological phylogeny of Michener (1944). The legend demonstrates points of evolutionary change in social behavior. By assuming that complex eusociality derives from simple eusociality, this phylogeny requires four evolutionary changes.

from more simple forms (figure 21.3). This time there are three necessary events: (1) eusociality appears in its complex form in the common ancestor of the corbiculates, (2) complex eusociality is lost in E, and (3) B convert to simple eusociality. At three instead of four evolutionary changes, this interpretation is certainly more parsimonious, but it raises serious questions about our understanding of social evolution. Can complex eusociality evolve without simpler antecedents? Can complex eusociality, once gained, revert to simpler forms? While reversions from simple eusociality to solitary life are known in bees (Danforth, 2002; Wcislo and Danforth, 1997), no such reversions have been found in complex eusocial species (Wilson and Hölldobler, 2005).

The Cameron (1993) paper is an example of one of the earliest molecular phylogenies and serves to show the kind of evidence streams that help decipher the origins of sociality and relationships. Subsequent molecular studies suggested even more arrangements. Overall, among the 15 relational combinations possible with these four tribes, at least nine have been published

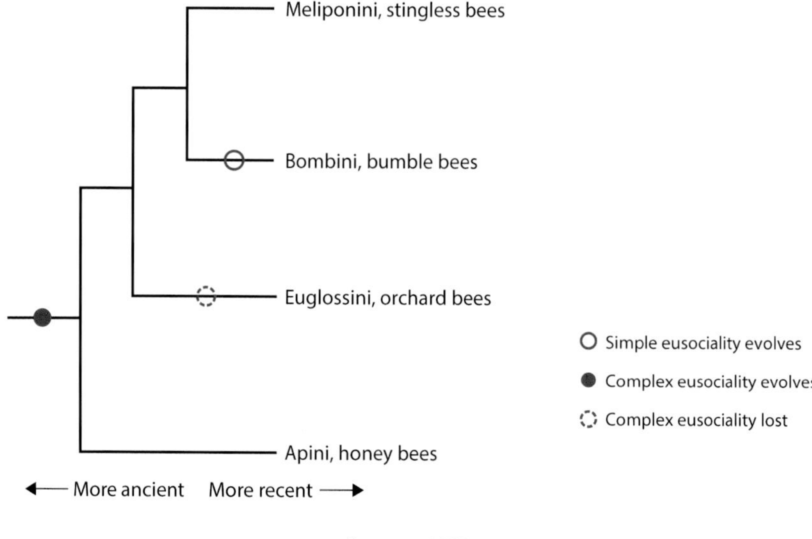

FIG. 21.3. The same phylogeny as in figure 21.2, but without the assumption that complex eusociality requires simple antecedents. This model requires only three evolutionary changes but raises serious questions about whether complex eusociality can regress to simpler forms or even solitary behavior.

as proposed phylogenies using varying combinations of morphology, behavior, or molecular evidence (Cardinal and Packer, 2007). Among molecular studies, most of the controversy has swirled around the relationship between Apini (A) and Euglossini (E), with the sister grouping of Meliponini + Bombini (MB) staying relatively consistent.

Beginning in 2008, the use of significantly enlarged molecular and behavioral data sets improved resolution (Cardinal et al., 2010; Kawakita et al., 2008), and by 2023, three consecutive molecular studies (Almeida et al., 2023; Bossert et al., 2019; Romiguier et al., 2016) had affirmed the MB sister grouping and returned the Euglossini to the basal position recognized by Michener (1944).[3] This arrangement forces a most parsimonious polarity for

3. But it seems unlikely that Michener would have accepted the MB grouping. In his book *The Bees of the World*, Michener (2000, p. 650) wrote, "Molecular approaches do not agree with morphological approaches, especially because the former indicate a

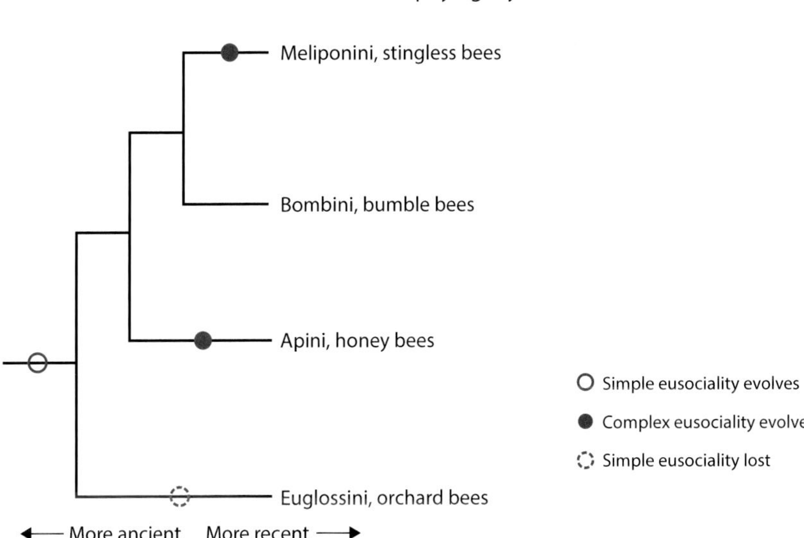

FIG. 21.4. A strongly resolved molecular phylogeny that preserves the durable Meliponini + Bombini clade and returns Euglossini to their basal position in the group. This model requires four evolutionary changes in eusocial behavior.

social change to that shown in figure 21.4: (1) simple eusociality is inherited from the corbiculate common ancestor; (2) complex eusociality evolves independently in the A tribe and (3) in the M tribe; and (4) simple eusociality is lost in the Euglossini. An equally parsimonious arrangement could place the origin of complex eusociality once at the common ancestor of the A(MB) clade, but this would require the Bombini to lose complex eusociality which, as I mention above, is unlikely.

These latest phylogenies, along with ancestral state reconstructions by Cardinal and Danforth (2011), have affirmed the overall antiquity of the corbiculate bees at 78–95 million years ago (figure 1.3), which means that the hallmarks of honey bee social complexity we have been studying in this book are 95 million years in the making. This location in antiquity is intermediate to the social wasps (63 million years ago) (Wenzel, 1990), ants (115–140

sister-group relationship between Bombini and Meliponini. . . . No known morphological character supports this conclusion." Michener died in 2015.

million years ago) (Borowiec et al., 2021), and termites (180–230 million years ago) (Ware et al., 2010).

These analyses also underscore the general and intuitive trend that social complexity is positively associated with greater age in a lineage, so that, for example, social complexity is greater in the corbiculate bees than in the more recent lineages of Allodapini (53 million years ago), *Lasioglossum* (22 million years ago), *Halictus* (sweat bees) (21 million years ago), and Augochlorini (20 million years ago) (Cardinal and Danforth, 2011). Put a different way, deeper time allows the accumulation of more complex modes of social life. This principle may explain the general observation that the highest expressions of social complexity are seen not in the bees, but in the older ants and termites (Wilson, 1971), although I expect that the ecological constraints of a comparably small nest space for the social bees (a hollow tree or single comb on a limb), compared with the spatial infinity afforded soil-nesting ants and termites, may have constrained social complexity in the bees.

The latest phylogenies (figure 21.4) also help explain some of the differences we observe in the highly eusocial honey bees (Apini) and stingless bees (Meliponini). Figure 21.4 demonstrates how each group inherited essential features of eusociality—cooperative brood care, reproductive division of labor, and overlapping generations—from their shared ancestry as primitively eusocial corbiculates, but that when it comes to advanced eusociality, each shows marks of a distinct evolutionary history. For instance, in colony reproduction in the honey bees the old queen leaves with the swarm, whereas it is the other way around with the stingless bees: a young queen leaves with the swarm. When it comes to recruitment to food sources, the Meliponini use odor trails to lead nestmates to food sites, whereas the Apini use their famous recruitment dances. Two independent origins of advanced eusociality can explain differences such as these.

The phylogeny in figure 21.4 also helps explain similarities long observed between the socially dissimilar stingless bees (Meliponini) and bumble bees (Bombini), similarities that were difficult to reconcile with the traditional phylogeny of Michener (1944) (figure 21.1). Each group shares striking similarities in nest architecture (Delaplane, 2021) and the habit of recycling wax from brood cells (Michener, 1974) (figure 4.5). A near phylogenetic kinship between the two helps explain these formerly inexplicable similarities.

Significance of the Corbiculum

The possession of a corbiculum is highly associated with cavity nesting (Michener, 2000), but it remains to wonder what, if anything, is particular to the possession of a corbiculum that inclines a lineage toward eusocial living. A corbiculum is not essential to bee eusociality, as proven by the three eusocial tribes—the Allodapini, Augochlorini, and Halictini—who lack it. But the association of a corbiculum with eusociality seems inescapable. Not only was the evolution of eusociality much earlier in the corbiculate tribes (Cardinal and Danforth, 2011), it is the corbiculate bees who today possess the most socially complex bee species. It was the opinion of Charles Michener (1999) that "it is probably easier to remove pollen from a corbicula than to remove an equal amount of pollen carried within a brush of hairs." If so, then one can imagine a certain ergonomic economy in pollen handling that might favor social life.

There are a few structures technically corresponding to corbicula in other bee groups; however it is only in the "true" corbiculate bees that pollen is first moistened with nectar to hold it in place during transit. The pollen is carried dry in other groups (Michener, 1999). This habit of dampening pollen with nectar strikes me as another preadaptation favorable to social life. It has long been known that bees ferment pollen to acidify it and prolong its storage life in the nest (Pain and Maugenet, 1966), and it has been shown that the bacteria responsible for this fermentation reside in the honey stomach from which the forager regurgitates nectar (Vásquez and Olofsson, 2009). Long-term pollen storage is important for a colonial species that lives together as a group year-round.

Another feature of the corbiculate bees seems important, not so much for the evolution of eusociality, but rather the style of eusociality we have in the honey bees—and that's cavity nesting. It's true that the four tribes comprising the corbiculate bees overwhelmingly use cavities as the mode of nesting. Moreover, the cells are built up, not excavated out of a substrate. The ability to excavate burrows, common across the Apidae, has apparently been lost in the corbiculate bees, even though bumble bees and stingless bees can enlarge existing cavities (Michener, 2000; Sakagami, 1966). This loss was no impediment to the evolution of eusociality—witness the teeming colonies of the ants and termites. But it certainly causes one to wonder how honey bee evolution would have proceeded with soil nesting, to speak nothing about bee*keeping*—or whether beekeeping could even exist!

The hypothesis has been put forward that the ancestor of the corbiculates was an oil-collecting specialist and that the corbiculum was derived from a bristly oil-carrying hind tibia of the kind we see in the modern *Centris* and *Epicharis* comprising the tribe Centridini. The phylogeny upon which this hypothesis hangs requires the corbiculates to be nested inside a paraphyletic Centridini [figure 21.5(A)] (Martins and Melo, 2016; Martins et al., 2014). By this scenario, oil collecting was lost several times in the history of this clade, in the case of the future corbiculates in the context of the hind scopa undergoing modifications for carrying resins for nest making. A bristled scopa is not suitable for carrying sticky resins, and genetic evidence suggests that genes were available for fixing this problem. In modern *Apis* and *Bombus*, shutting off the gene Ultrabithorax (*Ubx*) returns the hind tibia from a state smooth, concave, and bristle-free to one that is bristly (Medved et al., 2014).

The "oil foraging ancestor" hypothesis is satisfying to the extent that our story demands an origin to the modern corbiculum, and its case is certainly buttressed by the *Ubx* gene and its flattening effect on a bristly tibia. But it is not clear to me that modifications for resin collecting were significant drivers of eusociality given the persistence of non-social states in the corbiculate Euglossini. More damaging to the oil-collecting thesis, rather, are two more recent phylogenies restoring *Centris* and *Epicharis* to a monophyletic Centridini [figure 21.5(B)] and removing their shared oil-collecting ancestor from the line of descent to the corbiculates (Almeida et al., 2023; Bossert et al., 2019).

It remains difficult to assign to the corbiculum any compelling role in the evolution of eusociality in the corbiculate tribes. The propensity of the group toward cavity nesting is certainly a credible precondition favoring sociality, but I am inclined to think the corbiculum is incidental to the project. And even if a flattened, bristleless tibia confers social benefits associated with

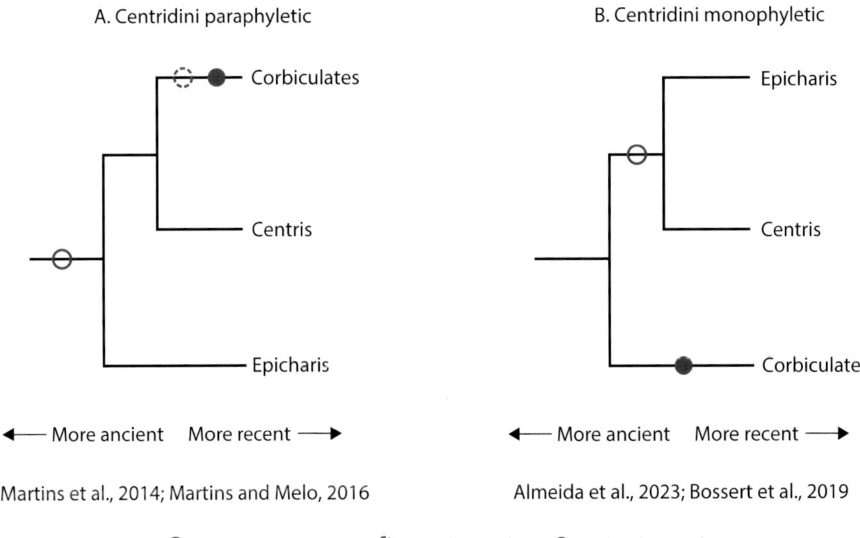

FIG. 21.5. Two phylogenetic hypotheses for the New World oil-collecting Centridini. This tribe is comprised of two genera, *Centris* and *Epicharis*, and constitutes the closest relatives to the corbiculate bees. Phylogeny (**A**) presents the corbiculates as sister to *Centris* with *Epicharis* basal to the group and all three with a common ancestor possessing a bristly hind scopa adapted for oil collecting. Although this hypothesis offers a credible scenario for the origin of the corbiculum, it requires three evolutionary changes in the pollen-carrying apparatus; moreover, it suggests that Centridini is paraphyletic as it nests the corbiculates inside the tribe. Phylogeny (**B**) returns the corbiculates outside of a monophyletic Centridini and outside the line of descent from an oil-collecting ancestor. This scenario is more parsimonious, requiring only two evolutionary changes in the pollen-carrying apparatus.

resin collecting, we need not invoke an ancestral oil-carrying scopa. Given the general action of the *Ubx* gene on leg morphology across a wide range of insect groups (Mahfooz et al., 2007), it seems plausible that *Ubx* could have worked its effect on any apid ancestor possessing a hairy tibia.

Summary

The corbiculate bees are united by their possession of a rear-leg pollen-carrying structure called the corbiculum and are distinctive for representing the full range of sociality within their four member tribes—the orchid bees, Euglossini (solitary to communal), the bumble bees, Bombini (simple eusociality), and the honey bees, Apini, and stingless bees, Meliponini (complex eusociality). The breadth of their social expression and shared possession of a derived corbiculum have made them the target of intense study for clues to the origin of social behavior. Despite containing only four tribes, it has taken decades to arrive at consensus on the group's interior relationships.

Some of the latest phylogenies imply that simple eusociality was inherited from the group's common ancestor, lost in the Euglossini, and evolved into complex eusociality independently in each of the Apini and Meliponini. The possession of a corbiculum is highly associated with cavity nesting and thought to confer ergonomic efficiencies to pollen-handling, but otherwise it remains difficult to assign any compelling role to the corbiculum in the evolution of eusociality.

CHAPTER 22

The Genus *Apis* and Phylogeography of *Apis mellifera*

THE ANCESTRAL HOME of the genus *Apis* is traditionally thought to be Southeast Asia. This view is supported by the fact that this region includes the natural ranges of all extant species but one—the western honey bee *Apis mellifera*, the focus of this book. In recent years, Europe has been put forward as a credible alternative to Southeast Asia. Specialists sort through competing sets of evidence by drawing upon principles of phylogeography and speciation as I discussed in chapter 1 ("Evolution" section). We will see in this chapter that the debate on honey bee origins and phylogeography is still current and imperfectly resolved.

It is worth noting that a proposed European emergence of *Apis* at 38–28 million years ago falls within an Oligocene world with warm higher latitudes sustained from an earlier warm Eocene (O'Brien et al., 2020). The climate of the subsequent Miocene was still generally warm but more variable, with continent-sized ice sheets absent from the Northern Hemisphere (Steinthorsdottir et al., 2021). The ancient warm-weather adaptations frequently implied in *Apis* evolutionary biology and invoked by me in this book are not necessarily challenged by an "out of Europe" hypothesis.

Phylogeography of the Genus *Apis*

Member species of *Apis* are partitioned into three subgenera: the dwarf open nesters *Apis* (*Micrapis*), the giant open nesters *Apis* (*Megapis*), and the medium-sized cavity nesters *Apis* (*Apis*). Figure 22.1 gives a dated topology of the relationships among these groups (Smith, 2021). The most recent common ancestor of the genus *Apis* is thought to have evolved 38–28 million years ago (Kotthoff et al., 2013, Smith, 2021); this ancestral population possessed all of the derived characters that define *Apis* and is itself a member of the genus. The first branching point (node) in figure 22.1 represents the divergence of the *Micrapis* from other *Apis* around 27 million years ago. As such, the *Micrapis* are in a sister position to the rest of the clade. In turn, the *Apis* and *Megapis* diverged from their most recent common ancestor at 25 million years ago and are sister to one another. *Apis* (*Apis*) *mellifera* diverged

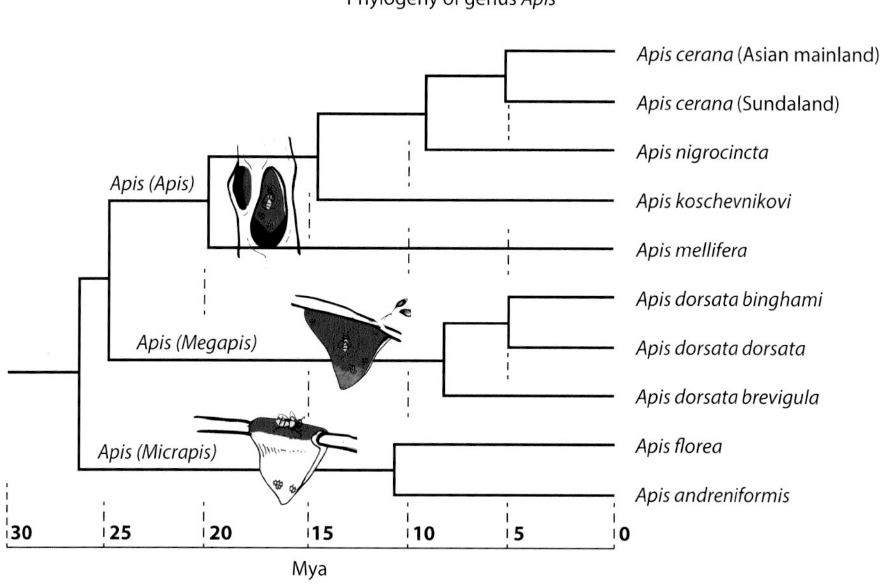

FIG. 22.1. *Apis* phylogeny and divergence times after Smith (2021) showing the three subgenera, the small-comb open-nesting *Apis* (*Micrapis*), the large-comb open-nesting *Apis* (*Megapis*), and the cavity-nesting *Apis* (*Apis*), and major representative species. Divergence times among *Apis* subgenera differ from values in figure 1.3, but the topology shown here has been consistent since Koeniger et al. (2011). Phylogenies for *A. mellifera* subspecies given in figure 22.7.

from the other cavity nesters around 20 million years ago, and as the oldest lineage of the subgenus *Apis* is said to occupy the basal position among cavity nesters. At the terminating branches of this tree, we find representatives of our modern species.

In spite of historical tradition favoring a southeastern Asiatic origin for the genus *Apis* (Han et al., 2012; Ruttner, 1988), the evidence for a European origin is, in my opinion, persuasive. The earliest known fossils of a definitive *Apis* are found in Oligocene (33.9–23.03 million years ago) deposits in Germany and France (Engel, 1999, 2001). Kotthoff et al. (2013) analyzed forewing venation and wing cells from modern species alongside fossil specimens going back 56 million years collected from present day Austria, Germany, Spain, France, Japan, and the United States. They found that the greatest morphological diversity was centered in Europe and concluded on the basis of the fossil record, modern species distributions, and continental drift history that *Apis* originated in Europe and spread from there to Asia and Africa, diversifying in both regions and, in the case of Africa, sending lineages re-migrating back into western Europe. At least one lineage crossed Beringia into North America but eventually went extinct there. The Asiatic center of diversity we see in modern species is an artifact of ancient migrations and subsequent weather-induced extinctions in Europe during global cooling periods.

Kotthoff et al. (2013) argued that from its European progenitors arose a lineage of honey bees persisting into the Miocene (23.03–5.3 million years ago) they called a hypervariable morphotype "encompassing and exceeding the range of variability known from modern species." Within this hypervariable type, they identified lineages that they labeled a CM type having *cerana*/*mellifera*-like wing venation and a D type with *dorsata*-like venation. Thus, the hypervariable bees (called CMD type) were poised to differentiate into the three main lineages known today: the dwarf honey bees *Micrapis*, the giant honey bees *Megapis*, and the medium-sized *Apis*. It is from within the CMD type that the Asiatic common ancestor of extant *Apis* postulated by Smith (2021) most likely arose in the late Oligocene or early Miocene.

Below, I harmonize the models of Kotthoff et al. (2013) and Smith (2021) to offer a scenario for the historic panoply of honey bee biogeography. It's important that we remember that Europe was configured quite differently during the geologic epochs here discussed. Much of the modern-day continent was inundated by an inland sea that figured prominently in the migration routes and distribution of species. This historical biogeography is consistent with the molecular phylogeny in figure 22.1, the topology of which has proven robust for more than a decade (Almeida et al., 2023; Bossert et al., 2019; Koeniger et al., 2011).

The proposed scenario is as follows:

1. Honey bees of the CMD type expand from Europe into Southeast Asia during the Oligocene (*circa* 30 million years ago) using available land routes (figure 22.2). By the late Oligocene or early Miocene (24 million years ago), early branches of the CMD group evolve reduced body size and eventually give rise to *Micrapis*. The lineage leading to *Megapis* diverges from the cavity-nesting lineage.

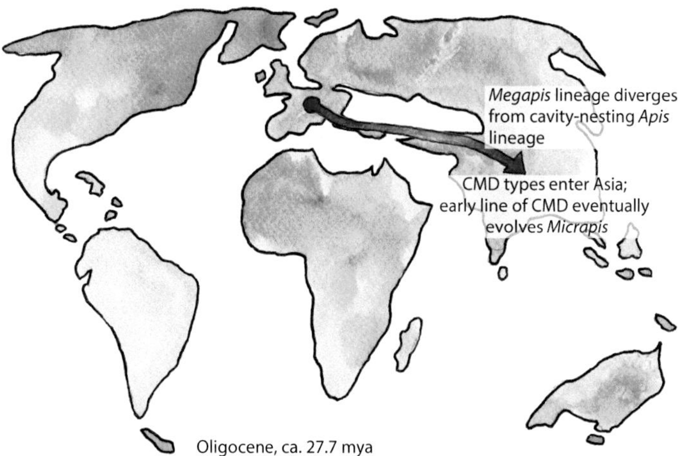

FIG. 22.2. Oligocene range expansion of early *Apis*, modified from figure 2 of Kotthoff et al. (2013). Map redrawn from Scotese and Golonka (1997). In this model, Europe (modern day France and Germany) is the center of diversity and home of the earliest known verifiable fossil specimens of *Apis*. From here, ancient *Apis* colonized Asia in the form of a "hypervariable morphotype," within which Kotthoff et al. (2013) identified members possessing wing venation patterns consistent with modern medium-sized subspecies *cerana*/*mellifera* (CM) and others possessing patterns more consistent with the giant subspecies *dorsata* (D). The "hypervariable morphotype" is labeled CMD.

2. CMD types continue invading Asia from Europe. *Megapis* evolves from D type lineages with enlarged body size (figure 22.3). Members of CM type cross the Bering land bridge and eventually give rise to North American *Apis nearctica*.

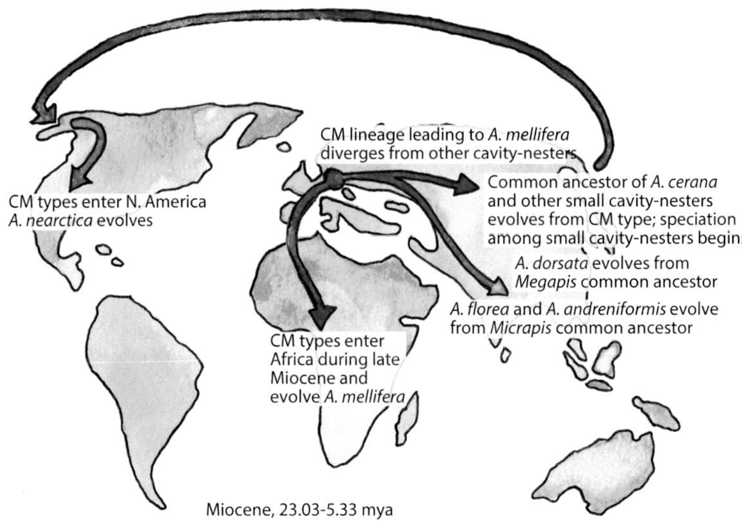

FIG. 22.3. Miocene movement of early *Apis*. See figure 22.2 for credits and labels.

3. The hypervariable CMD morphotype remains widespread across Europe and Asia. By the end of the Miocene (5.33 million years ago), a lineage of CM *Apis* diverges and expands into Africa, most likely crossing the strait separating southwestern Europe from Africa (figure 22.3). *Apis mellifera* evolves in Africa from this medium-sized progenitor, while the common ancestor of the small cavity nesters evolves and begins diversifying in Asia. The common ancestor of modern *Micrapis* expands into India and gives rise to *A. florea* and *A. andreniformis* around 10.5 million years ago (Smith, 2021).
4. During a cooling Pliocene (5.33–3.6 million years ago) and cold Pleistocene (2.58–0.129 million years ago), there are no Apini in western and central Europe, while at the same time, Africa serves as a refugium for early *Apis mellifera*. During warm interglacials and by an early warming Holocene (*circa* 0.012 million

years ago), *A. mellifera* re-colonizes western Europe by way of the Iberian Strait and the Middle East, Anatolia, and eastern Europe by way of extreme western Asia. Climate-driven changes during the Pliocene/Pleistocene cause population fragmentation among the European and southeast Asian cavity nesters, continuing the diversification that began in the Miocene. The same dynamics increase diversification in *A. dorsata* (figure 22.4). North American populations of CM descendants go extinct.

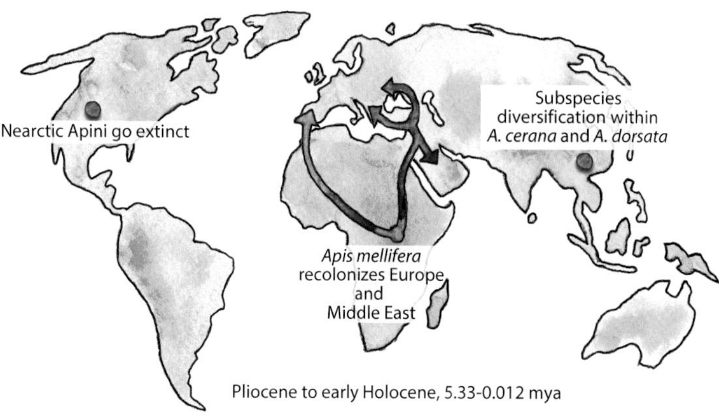

FIG. 22.4. Movement of early *Apis* in the Pliocene through early Holocene. See figure 22.2 for credits and labels.

Aside from the strength of its fossil evidence, I find an "out of Europe" model for the origin of *Apis* attractive for at least two other reasons. First, it may explain what has been a longstanding puzzle with the traditional Southeast Asia origin model, and that is the modern spatial detachment of *A. mellifera* from the rest of the genus. Not only is the natural range of *A. mellifera* widely removed from the Asiatic center of diversity, but there is no evidence of the natural range overlap one might expect if the lineage leading to *A. mellifera* had moved out of Asia. Figure 22.5 illustrates plainly that the center of distribution for *A. mellifera* is west, not east. A European origin for the genus that places the ancestor of *A. mellifera* in the west resolves this problem.

Second, I have long puzzled over why, of all species of *Apis*, only the western honey bee *A. mellifera* is unburdened by a natural Asiatic mite parasite. All Asiatic *Apis* are natural hosts to one or more parasitic mites in the

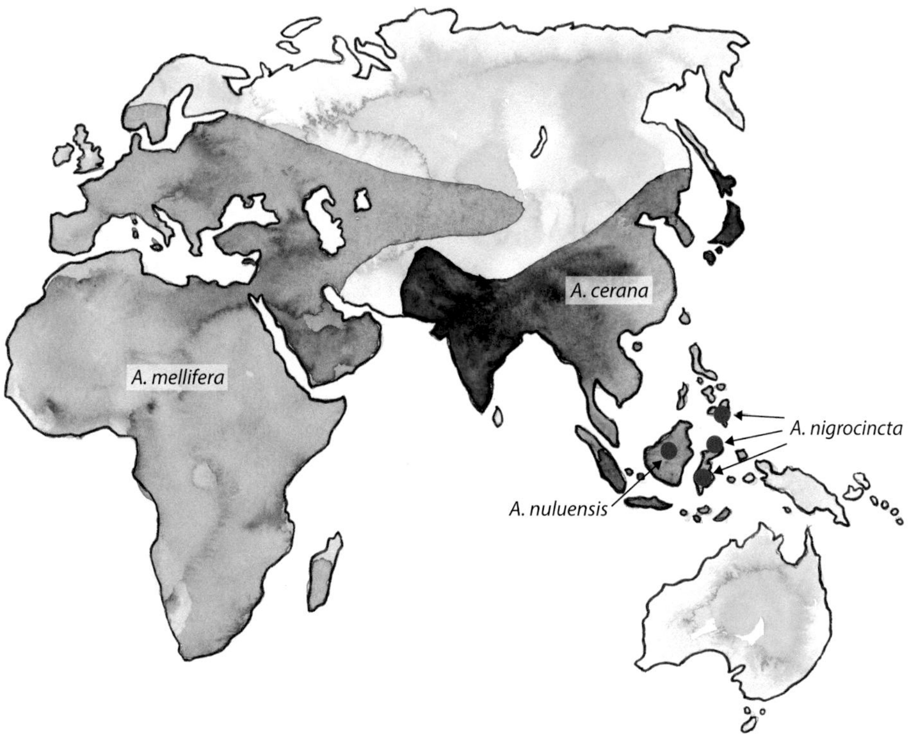

FIG. 22.5. Natural ranges of the eastern honey bee *Apis cerana* and western honey bee *Apis mellifera*. *Apis nigrocincta* diverged from *A. cerana* and today is restricted to the islands of Sulawesi, Mindanao, and Sangihe; a similar event occurred for *A. nuluensis* which today is restricted to the mountain regions of Borneo (Koeniger et al., 2011). An unresolved question is the discontinuity of the range of *A. mellifera* from the rest of its congeners. Map adapted from Ruttner (1988), Sheppard and Meixner (2003), and Chen et al. (2016).

genera *Varroa*, *Euvarroa*, or *Tropilaelaps* [reviewed by Chantawannakul et al. (2016)]. *Apis mellifera*, in unique contrast, was free of Asiatic mites until the late twentieth century, when human-assisted movement facilitated a host expansion in the mite *Varroa destructor* from its natural host *A. cerana* onto *A. mellifera*, where it has since become the most important health problem in managed honey bees worldwide (Rosenkranz et al., 2010). Bee scientists have long pointed out that one reason *Varroa* is so damaging to *A. mellifera* is that this bee is totally non-adapted to this parasite. It seems to me that if the lineage spawning *A. mellifera* had migrated out of Asia, then their own highly adapted Asiatic bee mites would have accompanied them.

But it remains that puzzles like the spatial discontinuity of *A. mellifera* and its lack of an Asiatic parasite could be the result of unknown stochastic events in the long history of speciations and migrations. Arguments for an Asiatic origin of the cavity-nesting *Apis* remain persuasive and persistent across a number of evidence categories, including a recent analysis employing both mitochondrial genomes and niche modeling (Ji, 2021).

Phylogeography of *Apis mellifera*

The natural range of the western honey bee *Apis mellifera* spreads across all of Africa, the Middle East, Europe north to the Arctic Circle, and almost a quarter of Asia. Across that vast stretch of Earth's terrestrial surface, this one species has differentiated, by my count, to no fewer than 32 recognized subspecies or races, 22 of which are mapped in figure 22.6. Moreover, these subspecies cluster naturally, based on morphology and molecular data, into five to eight lineages, which represent shared evolutionary history and migration patterns: the O group representing subspecies from the Middle East and west- to central Asia; the A group constituting subspecies in Africa; the M group including subspecies from northern and western Europe and eastward into central Asia; the C group representing subspecies from southern and eastern Europe, and the Y group from eastern Africa and the Middle East (Dogantzis and Zayed, 2019; Ruttner, 1988). To this original cluster have been more recently added an S lineage from Syria (Alburaki et al., 2013), an L lineage from Egypt, and U lineage from Madagascar (Dogantzis et al., 2021).

In a pattern consistent with our earlier discussions on the corbiculate ancestors of honey bees (chapter 21), our evolving understanding of honey bee genealogy is partly technology-dependent. Early studies were limited to focusing on geographic distributions and shared morphology of current and fossil species. Later studies have taken advantage of improving genomic technologies to include shared genes in phylogenetic reconstructions. And finally, the enormous quantity of molecular data generated by the genomics revolution has engendered the practice of depositing data in publicly accessible banks from which other scientists can draw for their own reconstructions.

In this manner, newer phylogenetic studies are privileged over older ones for the simple reason that they draw from more data and offer better

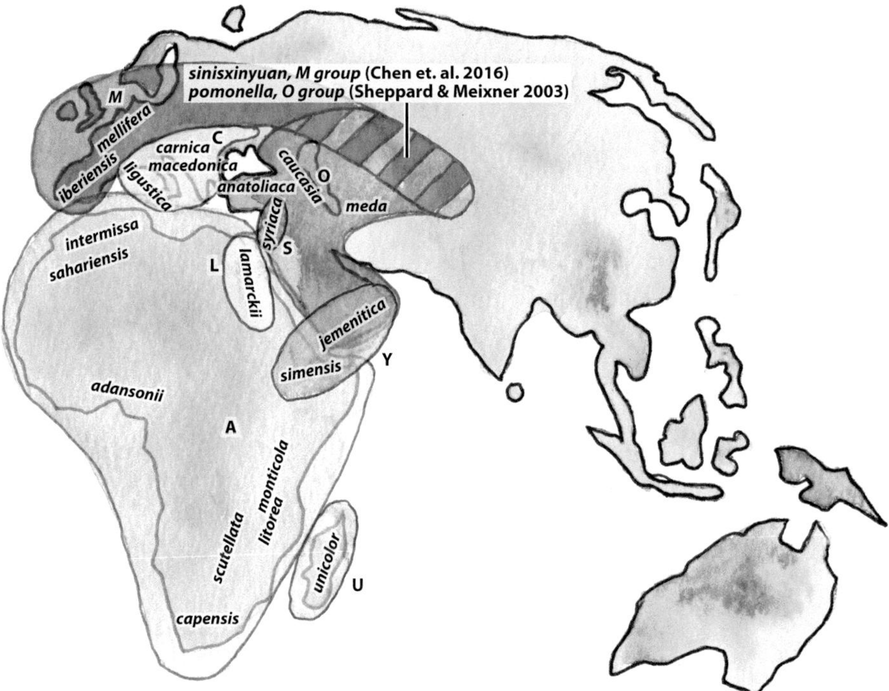

FIG. 22.6. Subspecies of *Apis mellifera* and their natural ranges. The subspecies are further grouped by morphological and molecular characters into five to eight lineages with shared evolutionary and migration histories: the M lineage of northern Europe and Eurasia, C lineage of southern Europe, O lineage of western Asia, S lineage of Syria, Y lineage of the southern Arabian peninsula and east central Africa, A lineage of Africa, L lineage of Nilotic Egypt, and U lineage of Madagascar. Only 22 of 32 recognized subspecies are shown here. All subspecific names are those recognized by Engel (1999) with the exception of post-1999 discoveries indicated directly on the map.

resolution of thorny relationships. Increasingly sophisticated statistical tools are available for discovering networks of shared genes, nearest genetic neighbors, and most parsimonious genetic reconstructions, all of which means we are living in an age of unprecedented power to understand and reconstruct natural history. Another pattern that emerges is a messier history. It's as if the more orderly early models were biased with human expectations of "what should be," whereas later models are better at surprising us with "what actually happened."

In our present case, the five to eight major lineages M, C, A, O, Y (and now S, L, and U) have proved a stable backbone to our understanding of the origin of *Apis mellifera* and evolution of its many subspecies. Originally constructed in the 1970s based on shared morphology and phylogeography (Ruttner, 1988), these lineages and their member subspecies have, for the most part, remained intact with the addition of molecular data. Each lineage is understood as a branch of the *A. mellifera* family tree with its own unique geographic migration history, a branch within which member populations diversified into genetically distinct subspecies. Genetic differences between subspecies within a lineage are smaller than differences from across lineages. Only very recently has this model been challenged with alternative geographic clusters and migration scenarios (Carr, 2023), but for this book, I continue using the traditional paradigm.

Estimates for the timing of divergences of *A. mellifera* into its constituent lineages have ranged between a scale of millions of years to hundreds of thousands. One early estimate recognized only three lineages (M, C, and A) and, using mitochondrial DNA, put the period of their diversification at 1.35–0.33 million years ago (Garnery et al., 1992). Another set of authors, also using mitochondrial DNA, shifted this range (this time for M, C, A, and O) to 850,000–470,000 years ago (Arias and Sheppard, 1996). Wallberg et al. (2014), using whole genome sequencing, posited a divergence among A, C, and M at 300,000 years ago, followed by the divergence of C from O at 165,000 years ago [figure 22.8(C)], and one of the latest, Dogantzis et al. (2021), analyzing single-nucleotide polymorphisms (SNPs) across 18 putative subspecies, found the YLUA clade diverging from the rest at around 6.3 million years ago and M diverging from CO around 5.7 million years ago [figure 22.7(F)].

Data such as these alongside modern subspecies distributions and dates of ancient glaciation are important pieces of evidence for inferring geographic focal points of the divergence events that birthed *Apis mellifera* lineages and routes of their subsequent migrations. One useful tool is the "neighbor joining method" that creates a phylogenetic tree based on allele sharing distances between groups under consideration. Allele sharing distance is a measure of the genetic divergence among groups. Groups with more alleles in common have shorter allele sharing distances, are closely related, and share a recent common ancestor. Groups with fewer alleles in common have longer

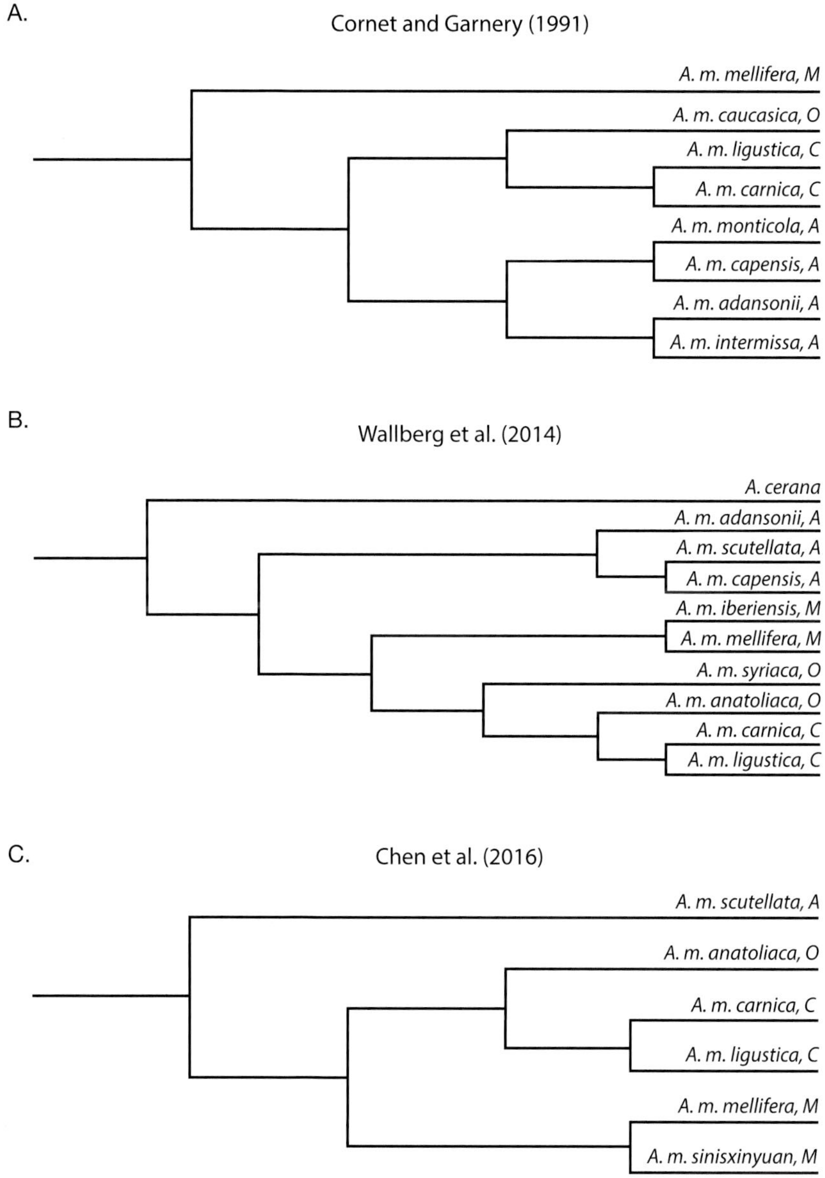

A.
Cornet and Garnery (1991)
A. m. mellifera, M
A. m. caucasica, O
A. m. ligustica, C
A. m. carnica, C
A. m. monticola, A
A. m. capensis, A
A. m. adansonii, A
A. m. intermissa, A
B.
Wallberg et al. (2014)
A. cerana
A. m. adansonii, A
A. m. scutellata, A
A. m. capensis, A
A. m. iberiensis, M
A. m. mellifera, M
A. m. syriaca, O
A. m. anatoliaca, O
A. m. carnica, C
A. m. ligustica, C
C.
Chen et al. (2016)
A. m. scutellata, A
A. m. anatoliaca, O
A. m. carnica, C
A. m. ligustica, C
A. m. mellifera, M
A. m. sinisxinyuan, M

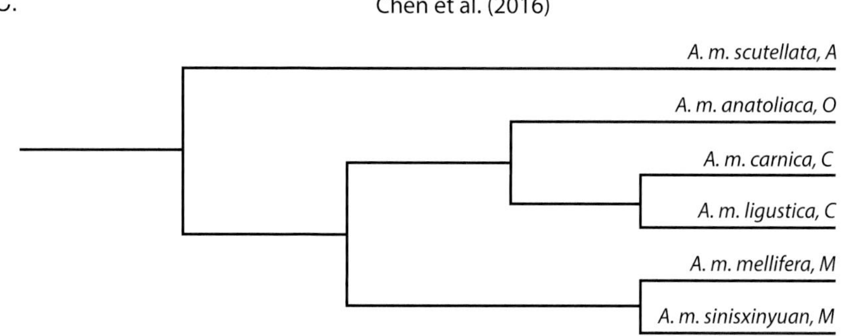

FIG. 22.7. A sampling of phylogenies put forward for subspecies of *Apis mellifera*. The five traditional lineages M, C, A, O, and Y have proven robust across analyses using morphology and nuclear DNA methods, but in Carr (2023)'s (**E**) meta-analysis of complete mitochondrial DNA coding regions, he could not consistently assign subspecies to their traditional groupings. Instead, he inferred four new geographic clusters: Eurasian (including Asia Minor and southeast European), Levantine/Nilotic/Arabian, Mediterranean, and sub-Saharan. Tihelka et al. (2020) (**D**) placed *A. m. syriaca* in a recently-inferred lineage S (Alburaki et al., 2013) and *A. m. simensis* in an African lineage Y. Dogantzis et al. (2021) (**F**) elevated *A. m. lamarckii* of Egypt into a distinct new lineage L and *A. m. unicolor* of Madagascar to a new lineage U. All cladograms are undated except for that of Dogantzis et al. (2021) (**F**), which harmonizes with the timeline given in figure 1.3.

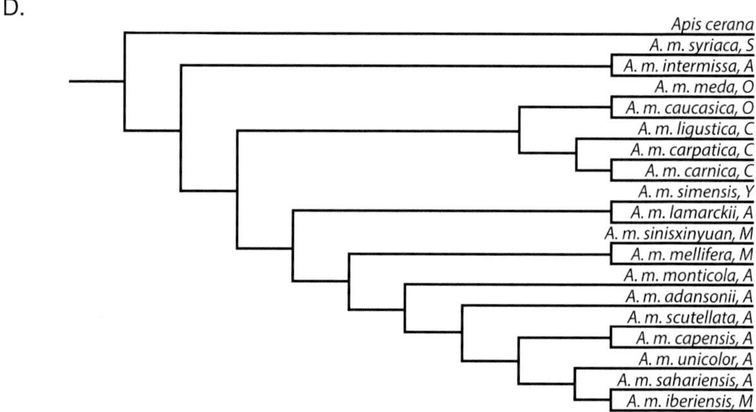

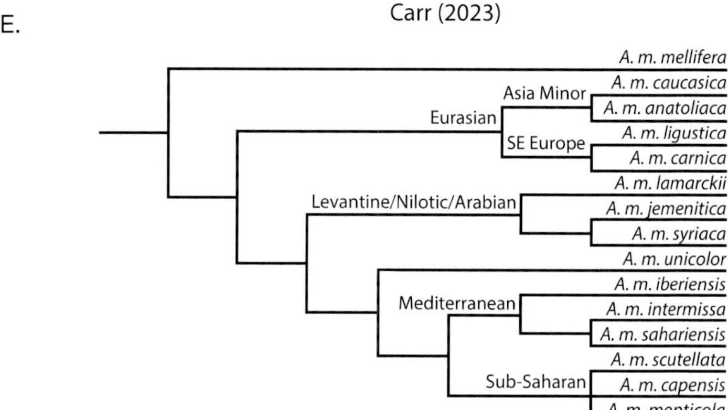

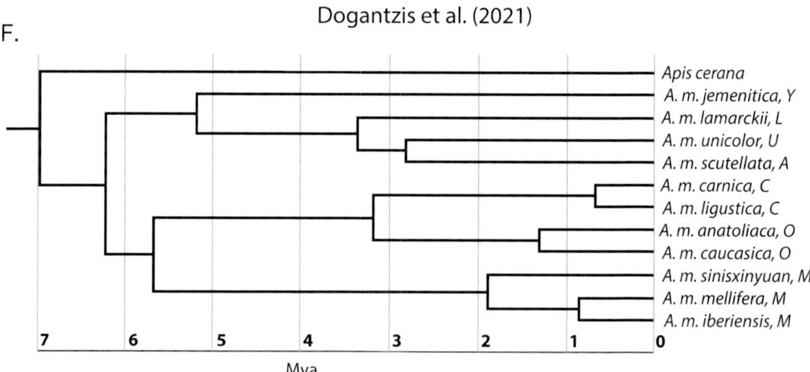

FIG. 22.7. (Continued)

allele sharing distances, are more distantly related, and share a more remote common ancestor.

The neighbor joining method also requires genetic data from an outgroup to serve as a root to the tree that results. Choice of an appropriate outgroup is important; it must not itself be a member of the group under consideration, i.e., the ingroup, but it must be related closely enough that comparisons are meaningful. Another way to think of it is that the most recent common ancestor shared by the ingroup and outgroup must be older than the most recent common ancestor of the ingroup alone. Where the root appears in the final tree gives important clues to where the ingroup originated.

In one well-known study (Whitfield et al., 2006) [figure 22.8(B)], the authors analyzed genotypes of 14 *A. mellifera* subspecies representing lineages M, C, A, and O from their native ranges in Europe, Africa, and Asia. The outgroup was represented by data from two sister Asian species, *Apis cerana* and *A. dorsata*. In the tree that resulted, the outgroup, that is the "root" of the tree, appeared squarely in the data for the African *A. mellifera* subspecies. This was interpreted as evidence that *Apis mellifera* first evolved in Africa. Moreover, modern members of lineages M and C were the most distantly related of the *A. mellifera* subspecies in spite of their modern proximity in Europe, with M "more similar to A than to either C or O." This was taken as evidence for an ancient divergence of M from A, and M's migration into Europe across the Strait of Gibraltar with a later migration of C and O via northeast Africa.

The enthusiasm for an African origin for *A. mellifera* has been dampened by a longstanding counterclaim for a point of origin in the Middle East or Asia [figure 22.8(A)], an argument that seems to be gaining momentum [figures 22.8(C) and 22.8(D)].

Cridland et al. (2017) combined data from publicly archived data sets with new original observations to produce the most comprehensive whole genome data set to that time. Among their many conclusions, those authors endorsed a lineage Y centered in northeastern Africa and the Middle East and first proposed by Frank et al. (2001). This region also corresponds to Cridland et al. (2017)'s best estimate for an origin for *A. mellifera*, with lineages

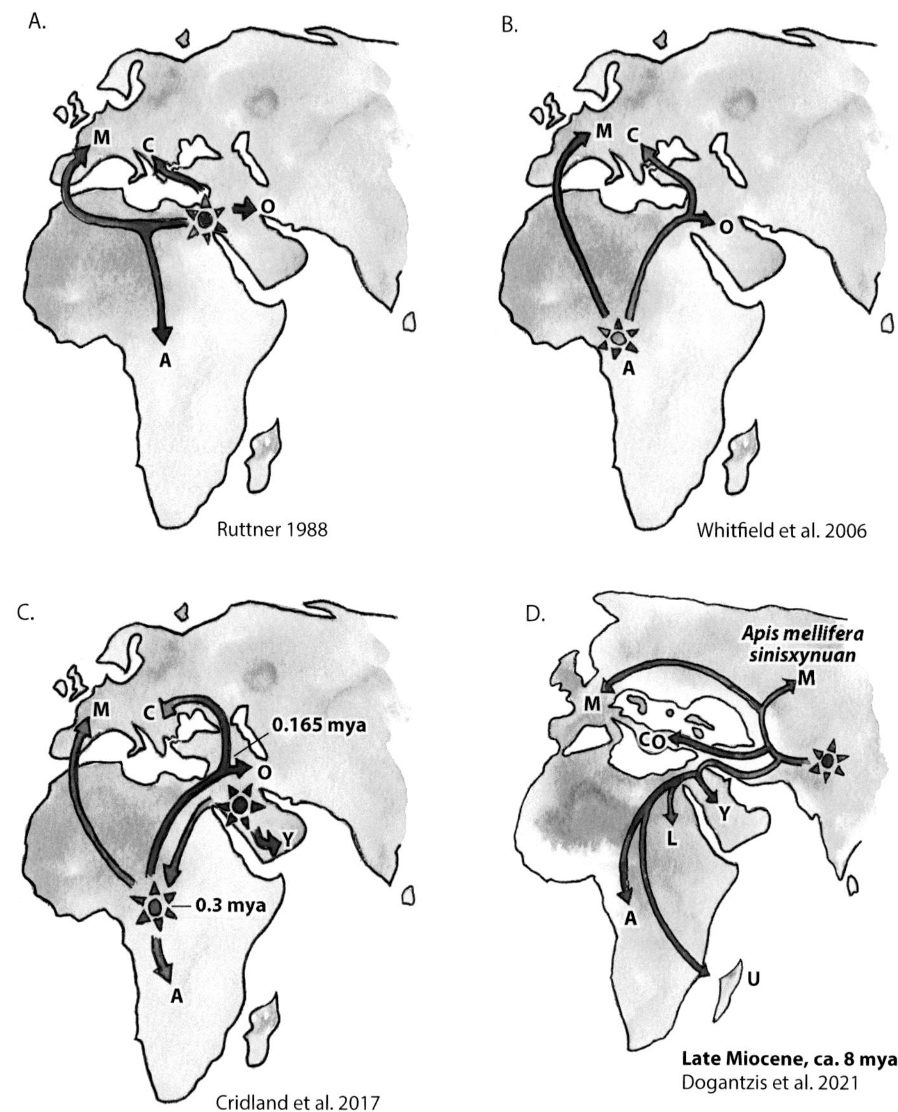

FIG. 22.8. Different models of *Apis mellifera* lineage origins and migrations. The star symbol is the proposed locus of speciation event. Divergence times for **C** are from Wallberg et al. (2014). Map of late Miocene used in **D** redrawn from figure S1 in Marzocchi et al. (2015) and online maps at Markwick (2007).

A and Y constituting its most ancient branches. Consistent with the model of Whitfield et al. (2006) [figure 22.8(B)], Cridland et al. (2017) noted a close relationship between C and O, but when resurrecting lineage Y, they noted its close relationship to A and remote relationship to O—this in spite of the modern proximity of Y and O. Consistent with other models, M retained its distant relationship with C and O.

In what is admittedly a messy phylogeny, Cridland et al. (2017) harmonized their results with a kind of dual-origin hypothesis: *A. mellifera* speciated into existence in northeastern Africa or the Middle East, and A and Y are its earliest diverging lineages. From an ancestral African (A) population, the earliest common ancestors of M and C radiate off and migrate northward, M in a westerly route across the Strait of Gibraltar and C east across the Isthmus of Suez. O subsequently radiates off C. This model is not as contrarian to the "Out of Africa" hypothesis of Whitfield et al. (2006) as may appear at first read, given that Cridland et al. (2017) recognized the significant African origin of much of the variation particular to the M, C, and O lineages. It is the inclusion of the Y lineage and its demonstrable shared antiquity with A that significantly sets the two models apart.

Dogantzis et al. (2021) analyzed patterns of clustering and admixture among SNPs across their sampled populations to create their scenario for descent across the *A. mellifera* lineages [figure 22.7(F)]. Applying a biogeographic reconstruction to the topology shown in figure 22.7(F), the authors calculated the probable ancestral geographic range of the most recent common ancestor at each node. I have used figure 22.7(F) and Dogantzis et al. (2021)'s nodal probabilities to create a plausible emigration route corresponding to their divergence dates [figure 22.8(D)]. The species evolved from an Asiatic cavity-nesting common ancestor, although a precise location is unknown. The YLUA clade and MCO clades separate, YLUA moving west toward Africa, and MCO northwest into Europe. Y diverges from LUA and moves into the Sinai Peninsula, which very likely connected at its southern terminus with eastern Africa in the late Miocene, explaining the modern distribution of Y genotypes across the Red Sea (figure 22.6). LUA move into their respective ranges in Africa and MCO into Europe. The invasion of Europe via western Asia instead of the Afro-Iberian land bridge is supported from high computed probabilities for an Asiatic ancestral range for the common ancestor of MCO, and the fact that the far eastern *A. m. sinisxinyuan* is basal to the rest of M.

Against divergence dates for the lineages ranging from 6.3 million to 165,000 years ago, divergence dates of the subspecies within them range as recently as 13,000 to 38,000 years ago (Wallberg et al., 2014). For context, consider that anatomically modern human beings are known from Africa by 230,000 to 300,000 years ago (de Castro and Martinón-Torres, 2022), and behaviorally modern humans by 77,000 years ago (Henshilwood et al., 2002). Given the range overlap in our two species's evolutionary past, I find it interesting to muse on the natural selection factors to which the other may have responded. I expand on these ideas in Chapter 1's "Contemporaneous relationships between *Apis mellifera* and *Homo sapiens*" section and in the introduction to chapter 17.

Evolution of European *Apis mellifera* Subspecies

In this section, I concentrate on the European subspecies of *Apis mellifera* in a nod to their celebrity as managed pollinators and honey producers. In species like the honey bee in which genetic variation is strongly structured around geography, there are two classical explanations for the ways variants come about from an ancestral species. I described these in chapter 1 as sympatric or allopatric speciation. Sympatric speciation happens when a region is colonized by a highly variable founder population, followed by subsequent loss of local variants. What remains are discontinuous populations that are genetically diverse. Allopatric speciation happens when a subset of a species's population is cut off from others by any kind of climatic or geographic barrier. Rendered unable to interbreed, the two populations embark on separate histories with different selection pressures. Left unchecked, these processes lead to new species. When either of these dynamics is observed in early stages short of full speciation, we are witnessing population variants, biotypes, subspecies, or races of the same species.

In the case of European *A. mellifera*, we have ample and powerful drivers of geographic isolation and examples of its effects on honey bee subspeciation—the repeated glaciers beginning in the Pleistocene epoch 2.58 million years ago up to the interglacials of the Holocene 11,700 years ago.

The Pleistocene, or "Ice Age," encompassed no fewer than 20 major glaciation periods, with the most extensive five or six occurring in the last 900,000 years (Ehlers and Gibbard, 2007). It is in this context that the M and C lineages advanced and settled into Europe, with expansions during the relatively warm interglacials and retreats into milder enclaves, or "refugia," at the next cooling spell.

It is tempting to imagine ice age refugia as redoubts against physical obliteration from sprawling ice sheets. They were certainly that, but as figure 22.9 shows, the last European glacial maximum was limited to only sections of northern Europe. There were more hazards than ice sheets, namely, unfavorable weather and floral conditions. Pleistocene weather was generally drier and colder than present. There was a lot of dust. Depending on the particular interglacial period, the dominant flora could range from forest ecosystems to open vegetation (Zagwijn, 1989), with strong cyclic changes in availability of insect-pollinated plants (Müller et al., 2005). These repeated pulses of severe habitat change offered the ancestors of the modern M and C subspecies plenty of opportunity for divergent selective adaptation.

Moreover, glaciation could be a relatively local phenomenon. There are two examples of this, two events widely separated in place and time—the Riss glaciation over the Alps separating the Italian peninsula from mainland Europe, and the Würmian glaciation over the Pyrenees separating modern day France from Spain (figure 22.9). Each of these glaciations is a locus of attention on plausible evolutionary scenarios for modern day European honey bee distributions. Plausible scenarios are at best testable hypotheses awaiting comparison against other scenarios using every possible line of converging evidence. The biogeographer tries to align dates of species's divergence with dates of geologic or climatic phenomena that could explain the divergence. Not all honey bee models align with the chronology of the Riss or Würmian glaciations! With these caveats, let's consider scenarios for the separation of *A. m. mellifera* from *A. m. iberica* (lineage M) and *A. m. ligustica* from *A. m. carnica* (lineage C).

Most models agree that the histories of lineages M and C are surprisingly dissimilar in time and place: the two diverged from a most recent common ancestor around 5.7 million years ago according to the model of Dogantzis et al. (2021) [figure 22.7(F)], and several models agree that M entered

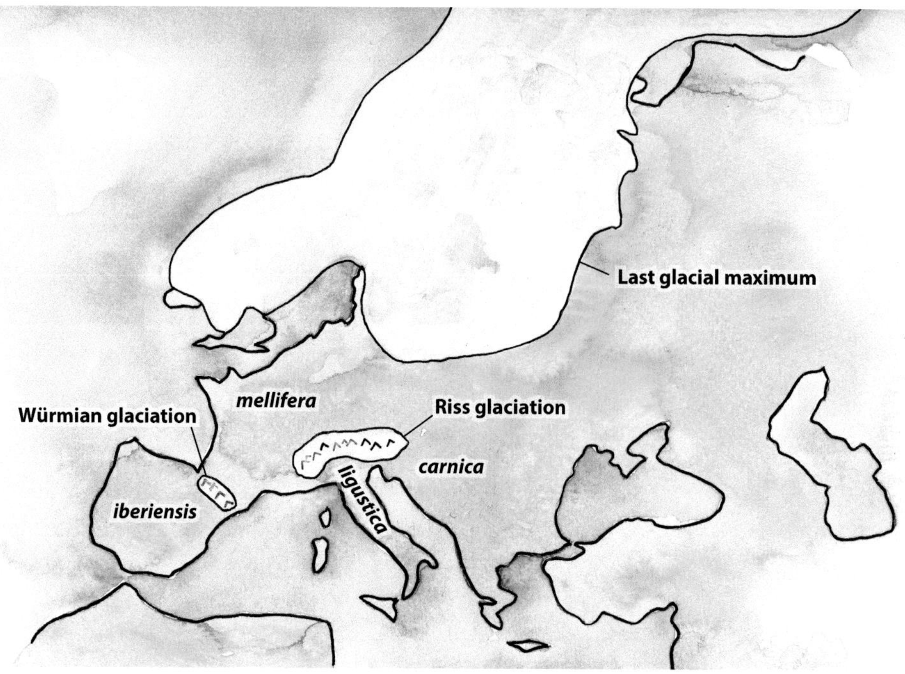

FIG. 22.9. Significant European Pleistocene glaciations: extent of last glacial maximum (23,000–19,000 years ago) (Ehlers and Gibbard, 2004; Mix et al., 2001; Svendsen et al., 2004), Riss glaciation over the Alps (200,000–160,000 years ago) (Antevs, 1929; Kukla, 2005), and Würmian glaciation over the Pyrenees (70,000–50,000 years ago) (Calvet et al., 2011). Map redrawn from Free World Maps—Atlas of the World (n.d.) and sources cited here. Italicized names indicate *A. mellifera* subspecies in their home natural ranges.

Europe from Africa across the Strait of Gibraltar whereas C entered Europe from the east [figures 22.8(A), 22.8(B), and 22.8(C)]. Thus, M's and C's modern European proximity is an accident of ancient migration patterns and no product of recent kinship.

There is evidence that however ancient the divergence of lineages, the radiation of subspecies within lineages is comparatively recent, specifically 38,000–13,000 years ago for M and 25,000 years ago for C (Wallberg et al., 2014), placing these events in the center of late-Pleistocene glacial and interglacial events.

Let's look first at M, specifically two of its subspecies, the northern European *A. m. mellifera* and Iberian honey bee, *A. m. iberiensis*. The Pyrenees Mountains run in a gentle northwest to southeast vector along the borders of present-day Spain and France, as well as the microstate of Andorra. The Pyrenees have long been considered an obstacle to gene flow between the M subspecies and therefore the traditional natural range boundary between the two (figure 22.9). Moreover, the Iberian Peninsula has long been considered a refugium to European honey bees (Ruttner, 1988) and many other animal groups (Hewitt, 1999) during times of glacial advance.

In the interest of studying gene sharing between populations of *A. m. mellifera* and *A. m. iberiensis*, Miguel et al. (2007) analyzed genetic variability in honey bee populations in Portugal, Spain, France, and Belgium using mitochondrial DNA (mtDNA) and microsatellites—very short sections of repeating DNA that occur at many sites across the honey bee's (and other organisms') genomes. "Microsats," as they are colloquially called, have high mutation rates, yet a disproportionately high number of them are "noncoding"—never translated into structures or behaviors. Being biologically neutral, neither selected for nor against, the numerous mutations in microsats tend to be conserved and inherited down successive generations, making them useful for evolutionary reconstructions.

Miguel et al. (2007) were able to identify two strong dynamics contributing to genetic differentiation for honey bees in western Europe: (1) simple distance in a southwest to northeast vector, and (2) the physical barrier of the Pyrenees. In other words, the greater the distance between any two honey bee populations, the greater their genetic differences, but additionally, these differences were categorically amplified depending on whether the population was located north or south of the mountains. Miguel et al. (2007) were able to identify similar patterns with genetic markers retained from an ancestral African A lineage, with a sharp decrease in A markers north of the Pyrenees compared with the south. This is consistent with the traditional interpretation that the M lineage migrated out of Africa across the strait of Gibraltar.

These findings support the Pyrenees as the traditional range boundary between *A. m. iberiensis* and *A. m. mellifera* (Ruttner, 1988) by highlighting the role of the Pyrenees as a physical barrier to gene flow between populations. However, Miguel et al. (2007) also found mtDNA evidence supporting the role of the Iberian Peninsula as a glacial refugium and offering clues to migration patterns of ancient populations as they recolonized Europe during warm interglacials.

Although genetic markers for the M lineage predictably predominated in all bee populations that Miguel et al. (2007) sampled (Portugal, Spain, France, and Belgium), the overall mtDNA diversity was significantly higher in the Iberian Peninsula than in France or Belgium; 14 out of 24 M markers were found exclusively in the Iberian Peninsula. Such loss of diversity north of the Pyrenees is a characteristic of recolonization events (Hewitt, 1999) as would be expected during warming post-glacial periods when small pioneering bee populations found their way around the western and eastern ends of the Pyrenees and re-entered present-day France. The populations that descend from such pioneers tend to be more genetically impoverished owing to at least two reasons: (1) the initially small number of founders fail to represent the full range of genetic resources of the parent population, and (2) risk of permanent gene loss is statistically higher in any small population, including pioneers, strictly from chance. This cause of genetic impoverishment is so common that it has a name—the "founder effect."

Miguel et al. (2007) showed that bee populations north and south of the Pyrenees at either end of the mountains share M markers; however, the markers at the western and eastern ends are different. Thus each "Pyrenean path" contributed differently to the post-glacial recolonization of northern Europe. For example, the "M8" marker is common in the north-central range of the mountains and along the Pyrenean Mediterranean coast but rare at the western end and the rest of Europe. At the western terminus of the mountains, genetic markers include many common throughout France and Belgium but absent from the Mediterranean region. Overall, it seems that the western path was the main conduit for recolonizing Europe with the M lineage. Wallberg et al. (2014) estimated the date of consolidation of the M lineage into *A. m. mellifera* (north of the Pyrenees) and *A. m. iberiensis* (south of the Pyrenees) at 38,000 years ago which falls between the retreat of the Würmian glaciation over the Pyrenees and the last continental glacial maximum (figure 22.9).

Now let's look at ancient formative events for the C lineage. For many beekeepers, the C lineage is the most interesting because it contains two of the most popular races for hobby and commercial beekeeping—the Italian bee *A. mellifera ligustica* and Carniolan bee *A. mellifera carnica*.

One of the most pertinent studies comes from Franck et al. (2000), who were interested in the extent to which the Italian Peninsula acted as a glacial refuge in a manner similar to Iberia. Capped as it is by the Alps mountain range and extending into more southerly latitudes, Italy has all the ecological ingredients for providing ancient geographic isolation and the chance for bees to genetically diverge. Franck et al. (2000) performed mtDNA and microsatellite analyses on populations of honey bees from Italy, the traditional range of *A. m. ligustica*. Additionally, they analyzed Carniolan bees *A. m. carnica* from Croatia as well as honey bees from diverse populations and lineages ranging from Guinea (A lineage), Morocco (A), France (M), Spain (M), Greece (C), and Lebanon (O).

In contrast to previous studies that placed Italian bees squarely in the C lineage, Franck et al. (2000) showed an unexpectedly high rate of M markers admixed with C across the *A. m. ligustica* populations of Italy. It was perhaps not surprising that they found French M genetic markers (M4) occurring at high frequency in northwest Italy bordering France; however, they also discovered a distinctly Italian M marker (M7) existing throughout the rest of Italy. Franck et al. (2000) interpreted these results as evidence of a hybrid origin for the Italian honey bee between lineages M and C. In an interesting twist, they suggested that the hybrid nature of *A. m. ligustica* has long been obscured by the fact that M markers are peculiarly absent from the main Italian queen production area in north central Italy (Forlì).

Franck et al. (2000) posited an historical scenario in which the Italian Peninsula was invaded from the west by M populations and from the east by C populations via the narrow land margins between the coast and either end of the Alps. Which came first? Based on the chronology of Wallberg et al. (2014) and figure 22.8(C), one could presume that M came first, and this is in fact the opinion of Franck et al. (2000), but for different reasons: they found the uniquely Italian M7 marker throughout the peninsula and concentrated in the south, whereas the uniquely Italian C marker (C1) was concentrated in northeast Italy where it would logically enter Italy from the natural range of its sister C subspecies *A. m. carnica*. They interpreted this pattern as evidence for an earlier western arrival (and saturation) of M with subsequent eastern invasion by C.

By comparing gene divergence rates between M and C markers in Italy with their neighboring counterparts in France and Croatia, Franck et al. (2000) were able to estimate the time at which bees on the Italian Peninsula were cut off from their neighbors east and west—190,000 years ago, which corresponds to the Riss glaciation over the Alps (figure 22.9). Trapped in Italy, the peninsular M and C populations began a process of hybridization that culminated in the subspecies known all over the world today as the Italian honey bee, *A. m. ligustica*. Those C populations restricted to the east of the Alps pursued an independent selection history that culminated in the modern Carniolan bee, *A. m. carnica*.

Finally, we must note that the Italian M7 marker has never yet been found in France; however, it is found at low rates, and in slightly altered form, across Spain. This is a ghost of post-glacial recolonization events and a reminder that Italy alongside Spain was an Ice Age refuge from which ancient *Apis mellifera* emerged to recolonize Europe.

Genes under Social Selection

The successful ecologic adaptations of *Apis mellifera* across such wide ranges of temperate and tropical latitudes as we have been discussing arose in no small part from benefits gained by increasing social sophistication.

Dogantzis et al. (2021), working with 18 native subspecies across Africa, Asia, and Europe, identified 145 genes indicating signal of independent selection. These genes were biased toward worker, not queen, traits. The adaptive radiation of *A. mellifera* was associated with genes governing development of wings, sensory organs, eyes, muscles, and appendages. There was active selection for genes regulating neural development, memory, task learning, and colony traits such as group defense, social immunity, and honey and royal jelly production.

Wallberg et al. (2014) identified positive selection for genes regulating vitellogenin, supporting a hypothesis that temperate bees increased vitellogenin storage capacity as an adaptation for overwintering survival. They found signal for a gene known to be involved in convergent eusocial evolution in bees. They found population differences in genes known to regulate sperm production and offered this as evidence for active sperm competition in African populations, whose queens express comparatively higher rates of polyandry.

Chen et al. (2016) compared the gene evolution of temperate *A. m. mellifera* and *A. m. sinisxinyuan* against tropical *A. m. scutellata* and found evidence for *selective sweep*—rapid population fixation of highly beneficial mutations associated, in this case, with surviving temperate climates. Temperate populations had higher gene enrichment for networks regulating vitellogenin, lipid storage, and worker and queen longevity, in genes regulating flight muscle involved in cluster heat generation, and in genes regulating organ size, which Chen et al. (2016) suggested underlies shifts in fat body deposition between winter and non-winter bees.

In general, in *Apis mellifera*, we witness a species with an extraordinary ancient genetic plasticity that enabled innovations in social organization, cognition, and physiology that resulted in, arguably, one of most ecologically successful extant species on Earth today.

Summary

Member species of the genus *Apis* are partitioned into three subgenera: the dwarf open nesters *Apis* (*Micrapis*), the giant open nesters *Apis* (*Megapis*), and the medium-sized cavity nesters *Apis* (*Apis*). The emergence of *Apis* at 38–28 million years ago falls within an Oligocene world with warm higher latitudes sustained from an earlier warm Eocene, conditions which are consistent with a long-standing presumption for warm-weather adaptations in *Apis* evolutionary biology. The natural range of the western honey bee *Apis mellifera* spreads across all of Africa, the Middle East, Europe north to the Arctic Circle, and almost a quarter of Asia, inside which no fewer than 32 subspecies have been recognized. These subspecies cluster naturally into five to eight lineages: the O group representing subspecies from the Middle East and western- to central Asia; the A group across all of Africa; the M group from northern and western Europe and eastward into central Asia; the C group from southern and eastern Europe; the Y group from eastern Africa and the southern Sinai Peninsula; the S lineage from Syria, the L lineage from Egypt; and the U lineage from Madagascar.

The M and C lineages advanced into Europe during the Pleistocene, a period encompassing no fewer than 20 major glaciation periods, the most extensive five to six occurring in the last 900,000 years. It was during this time that M and C populations began climate-driven ecological adaptations that resulted in the subspecies of *A. mellifera* used most commonly in beekeeping today. During their long-range expansion, populations of *A. mellifera* experienced social selection in gene sets biased toward worker over queen traits and regulating functions such as development of wings, sensory organs, eyes, appendages, neural apparatus, memory, task learning, nest defense, social immunity, and honey and royal jelly production. Populations expanding into temperate latitudes show social selection for gene networks affecting overwintering, including vitellogenin metabolism, lipid storage, worker and queen longevity, and in flight muscles involved in generating cluster heat.

PART IV

Beyond the Honey Bee Superorganism

CHAPTER 23

Who Gets to Be Queen?

WHEN IT COMES TO DRAMATIC plot devices, it's hard to beat royal succession, or at least when it goes bad. What would our literature, theater, and cinema be without it? We humans have shown a consistent and insatiable lust for power and influence and a willingness to fight for it. For much of our history, that power and influence has been nested in royalty, or in more recent centuries, vested in elected officials. One doesn't have to look far to find the same dynamics alive and well in our present political systems.

It's also not hard to read the same possibilities in the honey bee colony. The annual cycle of colony swarming makes annual requeening a near certainty, and that doesn't even count supersedures or emergency replacements. When order returns after a chaotic swarming season, we can expect one strong overwintered colony to have produced at most two new swarm colonies. Out of all available female larvae in the colony, only about 20 were reared to mature queen cells, of which only two (in our example) resulted in new successful queens. Why *those* two queens? What choices, if any, were being made by worker bees, and at how many critical junctures?

One thing that differs in the honey bee colony is the reward. The Big Prize is nothing so mundane as the right to wield authority over others. The currency that matters—the *only* currency that matters in natural selection—is the ability to pass one's genes on to the next generation. Moreover, in a genetic system as complicated as a honey bee colony, there can be varying degrees of success, and of honey bees, especially a worker, we can ask how *many* of her genes can she pass on? Which life strategy optimizes her gene transmission? As we learned in part I, workers can make choices that permit more, or fewer, of their genes to be directly passed along. And this raises another big difference between honey bee societies and our own: whereas our species seems to have no compunction against violent enforcement of self-interest, in honey bees, natural selection has moderated the most selfish impulses of workers so that the vast majority of them settle for less than optimum personal reproduction in favor of supporting their queen.

So yes, royal succession in the honey bee colony is messy, but there's also no getting around it that being queen is the biggest prize of all. She passes on 50% of her genes with every egg she lays—2,000 per day at peak season—with the whole colony's workforce mobilized to help her. Out of all the viable female larvae in a colony, what evolutionary constraints and opportunities tip the destiny of any of them into becoming the next queen?

The very reason we pose the question this way—who gets to be queen?—presupposes the possibility for disagreement among workers. Chapters 6 and 7 reviewed how the genetic structure of honey bees makes for some unusual degrees of relation in one colony. Male bees have only one set of chromosomes, not the diploid two, and this, coupled with the fact that the queen mates with 11–20 drones (see table 9.1) and stores and uses their sperm for the rest of her life, means that a normal bee colony is made up of numerous subfamilies, each with its own father. Workers in the same subfamily are supersisters and share 75% of their genes in common. Workers in different subfamilies are subsisters and share only 25% of their genes in common. Kin selection theory predicts that these "relatedness asymmetries" will lead to workers preferentially rearing their own supersister larvae into the next queen, if given the chance. Kin-favoring behavior like this is called *nepotism*, and nepotism in its various expressions has been one of the most appealing and intuitive predictions of kin selection theory. A literature search for the keywords "nepotism, *Apis mellifera*" yields no fewer than 628 hits.

All of which makes it seem a little odd for me to say that the intense search for nepotism has come up short. Nepotism is weakly evident, or missing altogether. At least when it comes to queen succession which, one would think, is the most likely place one would find it. Among those 628 papers, queen succession has been a prominent topic, and the weight of evidence is inescapable: nepotism plays a negligible to non-existent role; workers do not preferentially rear supersisters into the next queen.

To flesh this out a little better, let me list some of the ways nepotism *could* occur—and has *failed* to be convincingly demonstrated. Some of these studies are summarized and tabulated in an excellent review by Tarpy et al. (2004a). Bear in mind that there are opposing studies that do suggest evidence of nepotism, but these exceptions do not refute the majority evidence against it.

- When reciprocal transfers of young larvae are made between colonies of greater- or lesser-related worker nurses, the proportion of queens successfully reared by related workers is not higher than the proportion reared by non-related workers (Tarpy et al., 2004a).
- Colonies almost always rear more queens than they need, and the elimination of extras is an opportunity for nepotistic influence. This elimination is primarily accomplished by the young queens themselves as they fight to the death, but workers may aggressively interact with these queens during a duel. Gilley (2003) genotyped hundreds of workers involved in these duels and found that they were neither more nor less related to either queen than were background workers.
- During the queen elimination phase, some virgins are imprisoned inside their cells for up to a week beyond their normal emergence time. Workers cluster on the outside of the cell, feeding the young queen through slits she cuts in the cell wall—then patching them up again. If she manages to cut open her cell,

workers butt their heads against the opening to prevent her escaping while others repair the breach. Workers protect the imprisoned queen from "assassination" by other young queens. It is thought that this extra time allows imprisoned queens more time to mature, giving them an advantage over earlier queens who by now are weakened from previous duels. It is at the very least another opportunity for nepotism to be expressed. However, a controlled experiment was designed to detect genetic relationships between these "prison guard" workers and the imprisoned queens, and once again the data failed to show a convincing case for nepotism (Châline et al., 2005).

- The decision of swarming workers whether to join a departing queen is another opportunity for nepotism. Kin selection theory predicts that swarming workers will preferentially join swarms—or remain at the nest—depending on their relatedness to the associated queen. A genetic analysis of swarming bees and their queens was performed. The subfamily distributions were different between the mother colony and primary swarm and the mother colony and afterswarm, but the distributions were not biased in favor of the attending queens. There was no nepotism, but there was evidence that some subfamilies are more inclined to swarm than others (Kryger and Moritz, 1997).

The underwhelming performance of one of kin selection theory's most intuitive and attractive predictions is so egregious that it has fueled support for alternative models of social evolution. The strongest competitor is called group selection and posits the simple expression of "altruistic genes" that encourage the formation of groups (Nowak et al., 2010). This movement's champion is none other than E. O. Wilson, the founder of the science of sociobiology and author of numerous authoritative books on evolution, biodiversity conservation, and the interface of biology with human society. As fascinating as this debate is (and it *is* a debate, vigorously opposed by many), it is beyond our present purposes other than for me to point out this present beachhead of the controversy.

So what we have done so far is establish that queen succession is not driven by workers nepotistically favoring their supersister larvae. But what happens when we pose the question from the point of view of the larvae? Being rather passive creatures, afloat in their beds of royal jelly, it's hard to imagine larvae being a participant in this drama. But evidence is mounting that they are key players.

Evidence began trickling in once genotyping technology reached the point it could affordably and reliably determine relatedness between individuals. Investigators noticed that among all the subfamilies in a colony, some were overrepresented when it came to their members who became queens, and this preference had no association with the majority subfamily in the colony. In other words, it was not nepotism: the majority subfamily was not forcing its supersisters upon the rest of the colony. It was Moritz et al. (2005) who established that "honeybee queens are not reared at random but are preferentially reared from rare 'royal' subfamilies, which have extremely low frequencies in the colony's worker force but a high frequency in the queens reared." This study was performed with queen cells under emergency supersedure situations because under such conditions, it is workers who make the choice among available female larvae deposited in worker cells; under swarm conditions, it is the queen who determines larval fate because she deposits their eggs into queen cups (see figure 3.6).

When a subsequent study included queens reared under swarm conditions, it was confirmed that the preference for "royal" subfamilies was evident only when workers were the sole decision-makers—in a supersedure situation (Lattorff and Moritz, 2016). This is theoretically to be expected: in a swarm situation, the queen predetermines larval fate by depositing her eggs into queen cups, and because she is related equally to every egg she lays, she has no bias toward which one becomes the next queen.

If larvae are competing to become the next queen, by what mechanism could this happen? The most obvious explanation is pheromone signals that larvae use to solicit care and attention from nurse bees. Such pheromones are known to exist (Le Conte et al., 1995), and it is sensible to suppose that larvae from "royal" subfamilies are genetically primed to elicit this behavior from nurses. It remains to confirm whether such royals are, in fact, superior queens. Their existence is *a priori* evidence that they have weathered the test of natural selection; however, we have no evidence that queen quality is attached to competitive outcomes. In their review of the pertinent literature, Tarpy et al. (2004a, p. 520) concluded that "a queen's quality has little bearing on the outcomes of the queen elimination phase of queen replacement." Tarpy et al. (2004a, p. 513) also pointed out that "workers largely cooperate to raise queens of similar reproductive potential so that any queen is suitable to inherit the nest." Beekeeper experience seems to bear this out: with standard requeening practices, it is relatively easy to coerce workers into accepting any queen.

Let me summarize: There is no evidence that workers bias queen production in favor of their own supersisters. But there is evidence that some subfamilies are advantaged over others when it comes to their larvae becoming queens. The existence of such "royal" subfamilies is interesting in its own right; but it seems doubly interesting that these same subfamilies are *under*represented as workers. These subfamilies are really good at producing queens, but really bad at producing workers. And it seems triply interesting that these royal subfamilies are rare. What strange selection forces bring all this about?

One argument offers a mechanism for low worker representation in royal subfamilies: poor fertility in the drones that father these subfamilies (Lattorff and Moritz, 2016; Moritz et al., 2005). It's possible that the genes in a subfamily associated with high female fitness (royal larvae) are associated with low fitness in its males. If royal drones have little sperm, or highly infertile sperm, then royal workers may be rare for the simple reason that the whole subfamily is rare. If its few living representatives are overwhelmingly selected to become queens, this would explain the virtual absence of royal workers.

Next, let's face the fact that royal larvae competing for attention from nurse bees is bold-faced selfishness. After spending chapters reflecting on honey bee social evolution and the nuanced ways, including lack of nepotism, that worker selfishness is subjugated to the common good, we now see that the same rules don't apply to queens. A royal larva that dupes nurses into giving it preferential treatment stands to gain a genetic lottery prize for its selfish behavior. Such a potent mutation would spread rapidly in a population unless there were countervailing trade-offs that kept it at more stable levels.

What selective push-backs could accomplish this? Let's keep in mind that nurse workers seem incapable of judging queen quality as they rear new queens, and there is no reason to presume things are different in the case of royals. If so, then "royalty" may be evolutionarily uncoupled from "queen quality," and therefore at the colony level, royal genes are neutral, neither advantageous nor disadvantageous—an effect that limits their spread in the population.

But taking it one step further, there is actually a case to be made that royalty is maladaptive at the colony level (Lattorff and Moritz, 2016). In the Cape honey bee *Apis mellifera capensis*, we see a kind of social parasitism where some workers abandon nest duties, dupe other workers into giving them queen-like treatment, and lay their own eggs. Cape colonies with these "false queens" dwindle because their workers do a poor job rearing brood to adulthood. We have no evidence that "royal" queens (if I may be excused the redundancy) are similarly associated with poor colony performance, but their chicanery at the earliest larval stage bears the marks of incipient social parasitism which is known to be corrosive to colony integrity in other bees and social insects. At the very least, social parasitism is not cost-neutral, given the elaborate counter measures many social lineages have evolved to resist it (Grüter et al., 2018).

So where does this leave us? Who *does* get to be queen? We can confidently exclude nepotism: the majority superfamily does not force its supersister upon the rest of the colony. If queen succession is happening in a swarming context, virtually any subfamily in the nest has an equal chance of contributing the next queen. But if succession happens under supersedure, then significantly fewer subfamilies contribute the next queens, some of which subfamilies fit the pattern of being rare, poorly represented as workers, and thus "royal." Is "royal" synonymous with "parasitic?" We do not know if royals make poor queens, but based on experience with Cape honey bees and other social insects, selfish behavior in larvae is associated with poor colony performance. Presumably, royal queens and their progeny succeed frequently enough that natural selection keeps them from going extinct. Colony-level selection for social stability and against social parasites sustains a sort of equilibrium (Grüter et al., 2018).

Summary

Kin selection theory predicts that workers will rear their own supersister larvae into the next queens if given the chance. But the weight of experimental evidence refutes this: workers do not preferentially rear supersisters into the next queen. Instead, workers cooperate without bias to rear a queen from any suitable larva. Nevertheless, there is evidence that some subfamilies are advantaged over others when it comes to their larvae becoming queens. These so-called "royal" subfamilies are overrepresented as queens and underrepresented among workers, and evidence exists that their larvae excel at eliciting royal jelly from nurse bees.

Royal behavior is assumed to be a form of social parasitism because selfish behavior in larvae is associated with poor colony performance, and reproductive benefit is biased to the carrying patriline. It is normally held in check by the queen's habit of "choosing" the next queen without bias (she is related equally to every egg she lays) when she deposits eggs in queen cups, which workers preferentially rear into queens. But this advantage only exists when the queen "chooses" the next queen. In any situation in which workers "choose" the next queen, as for example in a secondary swarm, the risk of a royal takeover exists. These behaviors are evidence of the precarious stability that exists in the superorganism.

CHAPTER 24

Anarchy and Social Dissolution

Anarchy in Paradise

Having spent 23 chapters thinking together about the social evolution of the honey bee, it's worth pausing to reflect that the whole social enterprise rises or falls on the system's ability to rein in selfishness. E. O. Wilson, the "father of sociobiology" and Pulitzer-Prize–winning science writer, asserted that we humans are eusocial organisms like the honey bee, products of natural selection acting at two levels—the level of individual and the level of group. We need both: individual selection promotes fit and vigorous members; group selection promotes harmony and cooperation. But these two levels are unavoidably in tension; their interests only partly aligned. Selfish behavior may succeed in the short term and its genes spread within a group, but if those selfish genes undermine the competitiveness of the group, then ultimately, the group dies and those selfish genes go extinct. All things equal, a group of altruists will probably outlast a group of selfish narcissists. Or, as E. O. Wilson put it, ". . . individual selection promoted sin, while group selection promoted virtue (Wilson, 2014)."

Compromise between individual and group interests is necessary if lower levels of biological organization will ever coalesce to form higher ones, whether it's cells cooperating to form an organism, workers cooperating to form a colony, or humans cooperating to build a stable society. Wilson (2014, p. 33) again humanized the abstraction:

> So it came to pass that humans are forever conflicted by their prehistory of multilevel selection. They are suspended in unstable and constantly changing positions between the two extreme forces that created us. We are unlikely to yield completely to either force as the ideal solution to our social and political turmoil. To give in completely to the instinctual urgings born from individual selection would be to dissolve society. At the opposite extreme, to surrender to the urgings from group selection would turn us into angelic robots—the outsized equivalents of ants.

Cooperation is the answer, and it's a great understatement to say that cooperation is a winning strategy in Earth's natural history. Cooperation is the dynamic that leads to organisms, and organisms are the things life works with. Every one of our bodies is a testament to the success of organismality—the bundling together of lower biological entities into more-or-less cooperative and contiguous genomes. In our case, this means cells coalescing into tissues coalescing into organisms coalescing into superorganisms coalescing into communities . . . Where will it end? Organisms are the operative unit in biology; where there is life, there are organisms.

And for beekeepers, it should warm our hearts to know that in the honey bee colony, we see the very same evolutionary processes recapitulated that give rise to all organisms. Queller and Strassmann (2009) put it this way: "The evolution of organismality is a social process. All organisms originated from groups of simpler units that now show high cooperation among the parts and are nearly free of conflicts. We suggest that this near-unanimous cooperation be taken as the defining trait of organisms."

But notice that these authors called that cooperation "near-unanimous." As breathtaking as the scope of social evolution is, we cannot forget that evolution is optimization, not perfection. Among those coalesced genomes bundled into organisms such as ourselves, there are a few programmed to go rogue and rebel against the larger genome to which they are attached;

indeed, among the necessary steps toward organismality is the synchronous emergence of checks and balances to constrain this very thing (Grüter et al., 2018). But those constraints don't always work. If the constraints break down in metazoan organisms such as humans, then those rogue genomes—we call them cancer cells—reproduce unchecked with devastating results to the organism. In superorganisms such as a honey bee colony, there are also cancer cells, but these are reproductively active worker bees, called variously outlaws, social parasites, or anarchists—workers that evade the checks and balances that keep other workers passively content raising their mother's sons and daughters. Put another way, an anarchist worker is one that evades the colony's normal constraints against worker egg laying and produces sons.

The power of kin selection theory is its ability to predict reproductive and behavioral outcomes based on genetic relationships inside the colony. I have talked about these relationships throughout part I and summarized them figures 6.1 and 7.1. In general, theory always predicts that an individual will behave in a way that optimizes the transmission of its genes. Obviously, direct reproduction is ideal because (in queen honey bees and other diploid organisms) one gets to pass on 50% of one's genes. But in a social colony, there are constraints on a worker's direct reproduction, for good reason: workers that switch to egg laying invariably stop working on behalf of the colony (Delaplane and Harbo, 1987b; Hillesheim et al., 1989).

One of the constraints imposed on worker reproduction are pheromones produced by the queen and brood that suppress most workers' ovaries. Another constraint is practiced by the workers themselves against each other, so-called "enforced altruism" or mutual "policing"—the behavior by which workers eat each others' eggs (chapter 7). Figure 7.1 helps us understand the genetic rewards of worker mutual policing. A "law-abiding" worker would rather help her mother rear brothers (25% genes in common), half-sisters (25%), and especially an occasional supersister (75%) than help a half-sister rear a nephew with whom she shares only 12.5% genes in common. This is a genetic no-brainer, especially considering that raising nephews and rewarding her half-sister's anarchist behavior would lead to social disorder. Better that we all compromise a little and keep the colony intact. Or put a different way, ovary-inhibiting pheromones and worker mutual policing are the superorganism's evolved safeguards against cancer.

In normal European honey bee colonies, these constraints are powerful enough to keep worker ovary development rates at no more than 1% of the worker population (Visscher, 1996). But these constraints only make sense in the context of a properly functioning queen. In the event a queen is lost and workers are unable to replace her, not only is her pheromone-based suppression of worker ovaries removed, but so are the genetic incentives for law-abiding behavior. Faced with the choice of rearing nephews, workers now have every incentive to rear sons. Hence, queenless workers famously activate their ovaries, abandon nest duties, and lay eggs—with disastrous results for the colony. There's a reason beekeepers call hopelessly queenless colonies, well, hopeless: not only are workers unreplenished, but those that remain are either egg-layers themselves or too socially handicapped to sustain the brood nest.

Not all queenless workers join the egg-laying free-for-all; the propensity to activate one's ovaries is higher in young workers (Delaplane and Harbo, 1987a) and certain subfamily lines (Châline et al., 2002). But the ensuing social collapse is general: worker life spans decrease, colonies lose weight, and colony defensive responses decline (Delaplane and Harbo, 1987b).

Technically, the genetic argument for abandoning mutual policing applies every time there is queen succession; after all, with a new queen, the house bees are now producing nieces, the vast majority with whom they share only 12.5% genes in common (with a beekeeper-supplied queen, the relationship is zero). But in a normal succession, the interruption of queen pheromone signal is brief, or with brood pheromones, it's zero, leaving no time to permit activation of worker ovaries. Indeed, such a threat of "genetic mutiny" may explain the evolutionary persistence of pheromone-based suppression of worker ovaries. It acts as a redundant constraint on worker egg laying independent of policing.

So, we see that natural selection has, in the main, rewarded colony-level constraints that protect the integrity of the group. This is consistent with a vector in natural history toward organismality, the whole in favor to the part, or "virtue" over "sin" in the words of E. O. Wilson. But it is equally true that natural history documents that selfishness is powerfully adaptive. The persistence of cancers, whether organismal or superorganismal, is testimony that selfishness pays.

In fact, it pays so well that some honey bee worker lineages have figured out how to circumvent the colony's constraints, lay eggs, and rear them to functional drones, *even in the presence of a queen*. These so-called anarchist colonies persist with large numbers of workers sporting activated ovaries. In order to rear their "outlaw sons," (1) anarchist workers must be able to activate their ovaries in the presence of pheromonal inhibition, and (2) their eggs must escape policing. Each of these traits appears to be independent, and their occurrence in some, but not all, subfamilies in a colony underscores that anarchist behavior is under genetic control.

This was confirmed by experimenters who were able to selectively breed an anarchist line of honey bee (Oldroyd and Ratnieks, 2000). Although the pheromones of anarchist queens appeared normal in the sense that they repressed ovary development in anarchist workers (Hoover et al., 2005), the same was not true of anarchist brood (Oldroyd et al., 2001). And when it comes to policing, anarchist workers lay eggs that appear more acceptable to other workers, and the anarchists themselves are less likely to engage in policing behavior (Oldroyd and Ratnieks, 2000).

By dodging the colony's constraints and coopting the colony's brood-rearing machinery, these cheating anarchists win a sizeable genetic prize—the transmittal of 50% of their genes. The only thing that keeps it from being outright parasitism is the fact that these perpetrators perform their mischief in their own natal colonies. So, a natural question arises: why doesn't anarchy spread more generally in honey bee populations?

To date, the best mechanism limiting the spread of anarchy appears to be the fact that eggs of anarchist *queens* are removed at a significantly higher rate than those of non-anarchist queens (Beekman et al., 2007). The removal of queen-laid eggs constitutes a cost to *all* members of the colony, anarchist subfamilies and non-anarchist alike, hence supplying an evolutionary push-back against unchecked selfishness.

If this is true, then policing anarchist queens may constitute yet another, perhaps the latest, colony-level adaptive check on worker selfishness. If E. O.

Wilson was right that social evolution entails an eternal tension between the interests of self versus the group, then I can't help but cheer for the group. This bias for the beauty of coordinated groups, I think, is behind the sense of awe observers feel when witnessing a teeming colony of honey bees on a warm afternoon. Against such a wonder, a mob of self-serving lone rangers appears rather small and uninteresting.

But "uninteresting" is probably not the word that comes to mind for beekeepers who have to deal with an acute case of social parasitism happening in South Africa.

The Cape Bee: A Lesson in Social Dissolution

In the previous chapter and last section, I described two evolved threats to group cohesion: "royal" subfamilies whose female larvae coerce nurse bees into rearing them into queens, and "anarchistic" workers who evade the colony's normal checks on worker reproduction and lay eggs that successfully develop into males. Each is an example of individual selfishness at the expense of the group. Each is a ploy to win a larger slice of the Darwinian pie—more of one's genes passed on to the next generation. A "royal" female larva stands to win the biggest prize of all, namely, the chance to be queen, inherit the brood-rearing factory of her colony, and transmit 50% of her genes with every egg she lays. Meanwhile, "anarchistic" workers, although not as successful as royals, still get to pass on 50% of their genes with every outlaw son they produce. Each outcome is an improvement over the fate of a law-abiding worker who helps her mother produce siblings with whom she shares only ~27% of genes in common (chapter 7, equation 6).

These selfish gains come at great cost to the integrity of the colony. Workers who switch to egg laying invariably stop performing any work for the colony (Hillesheim et al., 1989). Beekeepers familiar with queenless colonies can attest how quickly a colony deteriorates when its workers give up working.

Anarchistic workers are to their colonies what cancer cells are to a metazoan organism such as humans. And just as higher metazoan bodies have evolved checks on the growth of cancer cells (Tollis et al., 2017), so the honey bee superorganism has evolved checks on selfish worker egg laying. These checks take the form of pheromones that queens and brood produce to inhibit worker ovary development along with the workers' practice of policing—eating each others' eggs.

When it comes to genetic motives for eating someone's eggs, kin selection theory offers explanations based on the degrees of relatedness existing in a colony. Figure 7.1 shows these relationships in a normal European colony. The bottom row, third generation, represents the progeny possibilities—all sons—for "outlaw" workers. Theory predicts that a worker would rather help her mother produce more siblings with whom she shares at least 25% of genes in common than help sisters produce nephews with whom she shares only 12.5% genes in common. Although the best pay-out for a worker is to produce a son with whom she shares 50% of her genes, if everyone did this, the colony would dissolve. So workers practice mutual policing, the tacit compromise being that passing along nearly 30% of one's genes is better than colony collapse and zero reproduction for everybody.

Here, I foreground the third member of our rogues' gallery, if you will, the Cape bee of South Africa, *Apis mellifera capensis*. Or more specifically, the opportunistic workers of *A. m. capensis*. This bee is a subspecies of *Apis mellifera* and cross-fertile with other subspecies. But the workers of *A. m. capensis* are well-known for their ability to produce diploid eggs produced without mating, a state known as *thelytoky*. This character is coded for by a single dominant allele under strong selection to maintain thelytoky in cape workers (Aumer et al., 2019). The resulting eggs are physiologically normal females, capable of being reared into workers or queens in the normal way depending on whether the larvae are fed royal jelly.

Cape workers produce these eggs by parthenogenesis. The meiosis that normally halves chromosome number during egg formation is altered, and the resulting egg not only contains a full dose of chromosomes, but those chromosomes are identical with the mother's. In other words, a Cape worker's daughter is a clone of her mother. This has drastic consequences on the relationships in the nest as well as the predictions of kin selection theory, as detailed in figure 24.1. A Cape worker is equally related to her subsister ($r = 0.25$) as to her subsister's clone daughter (0.25). The same holds true for a supersister (0.75) and the supersister's clone daughter (0.75). And finally,

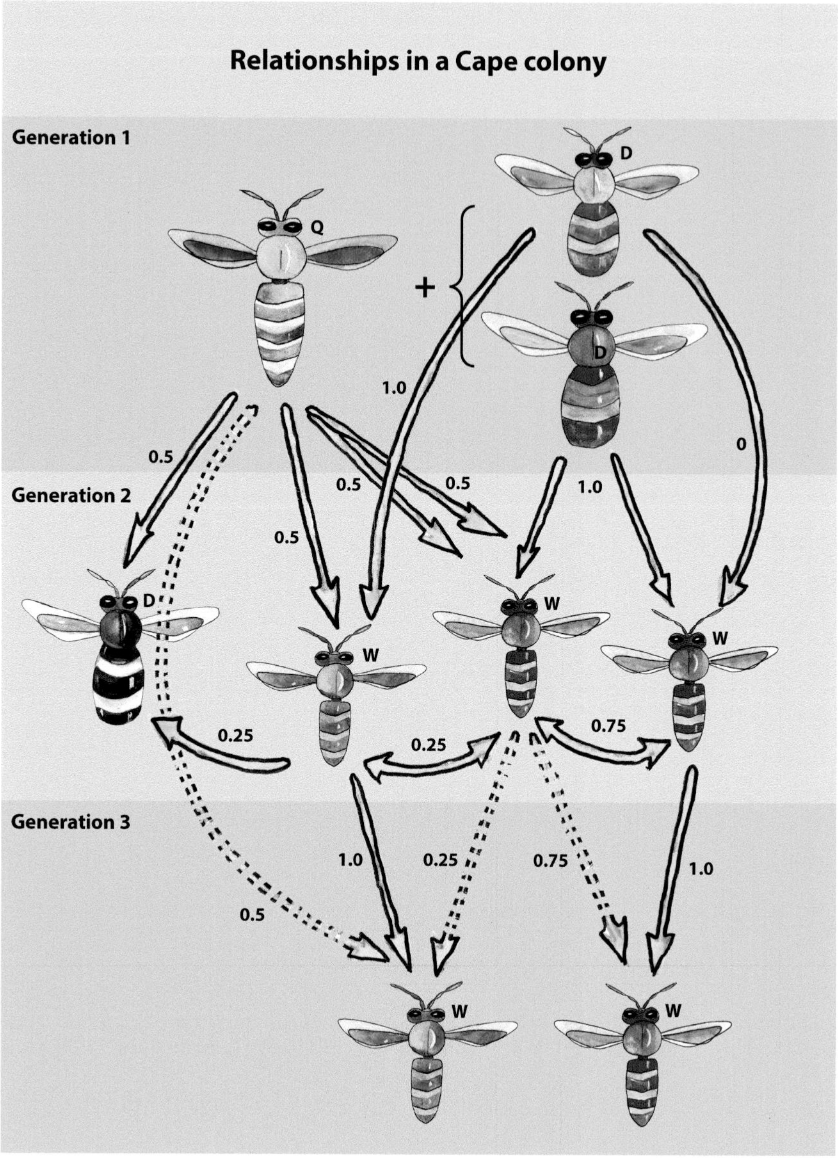

FIG. 24.1. Genetic relationships within a Cape bee colony. Hatched arrows show relatedness by indirect descent. Cape bee (*Apis mellifera capensis*) workers of South Africa, while sharing with worker honey bees of other subspecies the inability to mate, are able to produce diploid eggs by parthenogenesis. The resulting diploid egg is female, and like other subspecies of *A. mellifera*, the egg can be reared into either a worker or queen depending on its diet. Because these parthenogenetic female eggs derive entirely from the mother worker's genome, the daughter is a genetic clone of her mother (100% of genes in common, $r = 1.0$). Compared with a European colony, this dramatically alters the relationships of individuals within the colony; a Cape worker is as related to her sister (either subsister or supersister) as to that sister's daughter, and the queen is as related to her daughter as to her granddaughter.

the queen is equally related to her daughters (0.5) as to her daughters' daughters (0.5). In short, the genetic incentives for policing have fallen apart; with so much clonal inheritance it no longer "matters" as much to either the queen or a worker if other workers are producing daughters.

And indeed, controlled experiments performed by mixing Cape workers with workers of *Apis mellifera scutellata* showed that Cape workers can and do remove eggs laid by *scutellata* workers, but that neither *scutellata* nor Cape workers remove eggs laid by Cape workers (Calis et al., 2003; Martin et al., 2002a). In other words, Cape worker eggs are able to evade policing. The most plausible explanation for this is evolved chemical mimicry (Ratnieks, 1995) on the part of Cape worker eggs that makes them indistinguishable from queen-laid eggs.

But the absence of policing does not mean that there is no competition in the Cape bee nest. Far from it: the stakes are even higher, but this time kin selection theory is not as helpful because forces are in motion that favor direct, over indirect, inheritance.

Let's think for a moment about the stakes in this drama. In the case of a European anarchist worker, she stands to pass along 50% of her genes to every son (figure 7.1). In the case of a Cape worker, she stands to pass along 100% of her genes to every clonal daughter (figure 24.1). But because of significant life fate differences, these prizes are not equivalent, even beyond the obvious difference in gene numbers. The European worker's son has a realistic, albeit small, chance of mating with a new queen and contributing to the population gene pool. But if the Cape worker's daughter becomes just another worker, she's a virtual genetic dead-end: the only way for the Cape worker to win the genetic jackpot is if her daughter becomes the next queen. And because of clonal inheritance, the Cape worker who produces the next queen functionally *is* the next queen. Thus, for a Cape worker, the stakes are the highest they can possibly be: a dead-end versus the biggest prize of all.

More fundamentally however, the real possibility of direct clonal reproduction, passing along 100% of one's genes, means that altruism—all the work done to promote collateral transmission of one's genes through near kin, constituting nothing less than the hallmark of kin selection theory and bedrock of eusociality—begins to lose its luster. Instead, a kind of evolutionary arms race transpired in Cape workers grounded in selfish competition to become the next queen, a race conducted on two fronts, negative and positive. On the negative front, as group rewards for mutual policing faded, there was strong selection for one's eggs to not be singled out for policing: hence a rapid convergence of all subfamilies' eggs toward common indistinguishability from eggs of the queen. On the positive front, as rewards for direct clonal reproduction increased, there was strong selection for subfamilies whose workers could compete to become the next queen. And indeed, compared with European larvae, worker Cape larvae solicit and receive larger quantities of food from nurses (Allsopp et al., 2003), and Cape laying workers produce a pheromone bouquet that mimics that of the queen (Wossler, 2002). It's beginning to look like a return to Darwinian direct selection.

But not quite. It's a paean to the intricacies of natural history that complexification seems to win out over simplification. No such retreat to simpler life histories has been afforded the honey bee lineage: once complex eusociality is in place, there is no devolution to simpler modes of gene transmittal. Indeed, among the corbiculate bees (of which *Apis* is a part), termites, and ants, complex eusociality once gained has never been lost (Brady et al., 2006). Rather, in the face of competing modes of gene transmittal—direct individual transmission versus indirect transmission through near kin—we see (1) accommodation, (2) mild social parasitism, or (3) lethal social parasitism. The enigmatic Cape bee of South Africa demonstrates all three. There is also a fourth possibility, general social collapse, loss of the worker caste, and divergence into an obligately socially parasitic species. Although *A. m. capensis* has dodged that fate for the time being, there is evidence that it is experimenting with that option.

Let's first consider accommodation. It's worth stating the obvious that *Apis mellifera capensis* persists as a fully social species. Thus, it is self-evident that selfish worker egg-laying and social dissolution are not unrestrained. Second, it's worth recalling that the best prize for a laying Cape worker is to produce a queen daughter over a worker daughter. Thus, we can predict that Cape worker egg laying will be concentrated, if not restricted, to those times of year when the colony is raising queens. It's during swarm preparations that female larvae have the greatest chance of receiving royal jelly and the lavish nurse bee attention that will trigger their transformation into queens.

At first, experimental investigation seemed to bear this out: Cape workers produced about 31% of the workers during queen-rearing season but only 1.8% of the colony's workers during the rest of the year (Beekman et al., 2009). This had all the appearance of compromise, or more precisely, colony-level selective push-back in favor of group cohesion: by limiting her egg laying to swarm preparation season, a Cape worker stands to optimize her chance of producing a queen daughter; by withholding her egg laying the rest of the time, she participates in colony work and helps maintain colony integrity.

But later research seemed to discredit any idea of voluntary restraint on the part of Cape workers. Instead, a Cape worker is always ready to lay eggs, given the opportunity. When Cape colonies are dequeened, there is a cohort of workers who instantly lays eggs that become queen cells (Holmes et al., 2010); it's as if these workers live in a perpetual state of ovary activation or else are able to activate their ovaries extremely quickly.

The forces that keep these workers at bay are partly recognizable to beekeepers of European bees; others seem more unique to the Cape situation. We find, for example, that there is high genetic variation among Cape workers for worker reproduction (Moritz and Hillesheim, 1985); in other words, the ability of Cape workers to engage in reproductive behavior ranges from non-existent to potent—virulently so from a parasitic point of view. Second, guard bees are able to detect and deter some invading parasites (Wossler, 2002). House bees may show aggressive behavior toward them (Anderson, 1968), and *A. m. capensis* queens tend to have significantly higher queen mandibular pheromone levels than those expressed in other *A. mellifera* subspecies (Wossler, 2002). Given that colonies with high numbers of reproductively dominant workers express the typical malaise of poor brood production (Hillesheim et al., 1989), the existence of these mechanisms can be interpreted as colony-level adaptive constraints on selfish behavior—bids for group cohesion.

The second, and more insidious, outcome of competing modes of gene transmittal is social parasitism, and here we recall that parasitism by definition means exploiting another organism for reproductive advantage. The host, by definition, receives no benefit from the relationship. Social parasitism is common across many social insect groups, and the term applies when a reproductive individual gains access to a non-related colony and dupes its workers into rearing her eggs. In some groups, the parasitic species becomes so specialized that it loses its worker caste altogether.

There are mild forms of social parasitism that do not result in the death of the host colony, and predictably these milder forms predominate when the parasitism is within the same subspecies, as in the case of *A. mellifera capensis*. I say "predictably" because it is in the context of one's natal colony that selfishness and the beginnings of socially parasitic behavior can emerge, and only with later innovations can the parasitic behavior extend to near-neighbor colonies, to non-related colonies, or even to different subspecies—all the while during which the parasite may be diverging into a different species altogether. The Cape bee may in fact be modeling this phenomenon in real time (Neumann and Moritz, 2002).

One might say that, given the low rewards for policing and high rewards for directly and clonally engendering the next queen, that Cape workers are pre-adapted for parasitic behavior. Any situation that permits a worker to deposit a diploid egg is a statistical bid for a queen daughter, especially if it's queen-rearing season. But given that colonies can lose queens anytime, it's reasonable to predict that parasitic workers will increase their fitness if they can spread their eggs to more than one colony.

Observations and investigations bear this out. First of all, a state of queenlessness does indeed render Cape colonies more susceptible to invasion by alien Cape workers (Neumann and Hepburn, 2002), and second, some of these aliens are able to produce worker brood, sometimes at dominating rates. When investigators dequeened six Cape colonies and genotyped subsequent worker progeny, they found an average rate of alien-laid worker brood of 6.41%; however, in one colony, the rate soared to 62.5%, and even more curiously, this production was the work of just three alien workers (Härtel et al., 2006).

A subsequent study showed that not only do alien Cape workers gain access to dequeened Cape colonies and produce female brood, but that 5.3%–

20% of resulting queen cells are in fact the progeny of alien workers. Moreover, when the experimenters left the last round of queen cells in the test colonies, six cells successfully produced laying queens, four of which (67%) were progeny of alien workers (Holmes et al., 2010). Thus, it appears that not only is social parasitism naturally occurring in the Cape bee life cycle, but it is a complete and successful parasitic life cycle—a parasitic worker installing her clonal and usurping queen daughter into a non-natal colony who then propagates the parasitic line. Same-species social parasitism succeeds in *A. m. capensis.*

So let's take stock of our story so far. We've been talking about ways in which *A. m. capensis* deals with competing modes of gene transmittal—direct individual transmission versus indirect transmission through near-kin—and we have covered accommodation and mild social parasitism. Neither situation seems entirely stable; for each, the social cohesion of the group is sustained less by shared genetic interests and more by a suite of worker behaviors and queen pheromones that combine to resist the powerful incentive for direct reproduction that animates selfish (nestmate) or parasitic (non-nestmate) egg-laying workers.

And even though it is true that *A. m. capensis* persists as a eusocial species, within my professional memory and as recently as 1984, investigators were voicing concern for its imminent extinction, pointing to its vanishingly small natural range and morphological evidence of hybridization with neighboring races (Moritz and Kauhausen, 1984). These investigators were inclined to implicate migratory beekeeping practices for the shrinking range of *A. m. capensis*, but with our current understanding of the bees' biology, I can't help but wonder if any existential vulnerabilities of the Cape bee are better laid at the feet of the genetic instabilities introduced by parthenogenic clonal reproduction.

This leads us to the third possible outcome of competing modes of gene transmittal—lethal social parasitism—which, in a rather morbid manner may also constitute the salvation of *A. m. capensis*. For since that doleful 1984 paper warning of its imminent demise (Moritz and Kauhausen, 1984), the Cape bee has staged a remarkable comeback in the guise of a lethal social parasite, especially on its near-kin host *A. m. scutellata*. Hostile takeovers of

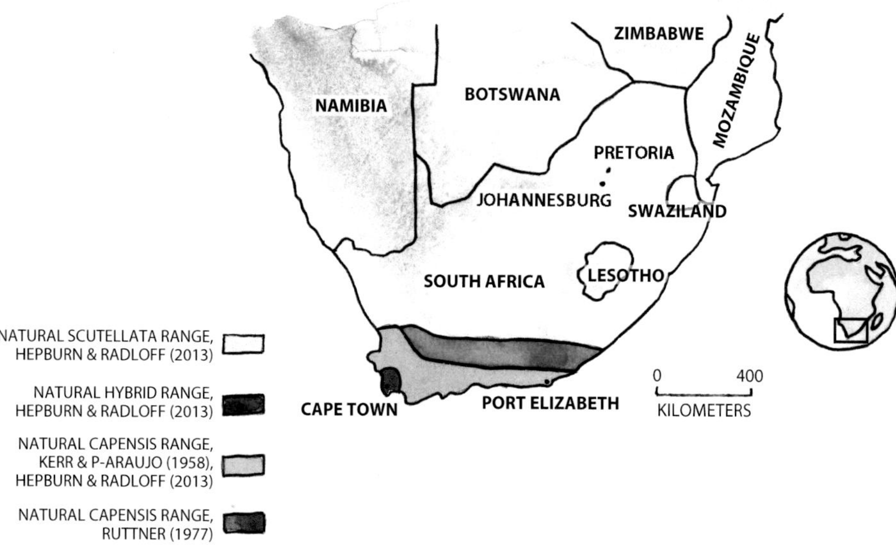

FIG. 24.2. Natural ranges of *Apis mellifera capensis*, *Apis mellifera scutellata*, and a naturally occurring hybrid zone between the two. Distributions shown here are amalgamated from various, sometimes conflicting, authorities (Hepburn and Radloff, 2013; Kerr and Portugal-Araújo 1958; Ruttner, 1977). The hybrid range according to Hepburn and Radloff extends to the western coast, whereas Kerr and Portugal-Araújo terminated the natural range of *A. m. capensis* about 200 kilometers west of Port Elizabeth.

A. m. capensis on other colonies have been known since the early twentieth century (Onions, 1912), but beginning in the early1990s, beekeepers in South Africa began migrating large numbers of *A. m. capensis* colonies north across a naturally-occurring hybrid zone between the two races (Hepburn and Crewe, 1991) into the natural range of *A. m. scutellata* (figure 24.2). Shortly thereafter, beekeepers began reporting large-scale decline of *scutellata* colonies, with symptoms typical of queenlessness and laying workers—poor brood production and inevitable collapse.

This so-called Capensis calamity (Allsopp, 1992) was shown to be the result of direct lethal social parasitism by Cape workers. Lacking the strong colony-level adaptations of *A. m. capensis* that constrain parasite colony entry and subsequent egg-laying, host *scutellata* colonies are highly susceptible to parasitic Cape workers who enter their hives, evade policing, and successfully rear workers [apparently not bothering to rear queens (Johannsmeier, 1983; Martin et al., 2002b)] who, in turn, drift to new host colonies and sustain the parasitic line. Inevitably, the host queen is killed, lost, or "ignored to death" (Neumann and Moritz, 2002), and the colony dies. There is reason to think this could be a long-term viable strategy for *A. m. capensis* because there are parasitic ant species that have lost the worker caste altogether as individuals compete to be queens. But when workerless parasites evolve in perennial social species, there appear to be fitness costs that render the parasitic populations small and geographically restricted (Nonacs and Tobin, 1992).

One more detail pushes the Capensis calamity story into the realm of the surreal, and that's the discovery that literally every single one of the billions of parasitic Cape workers that have propagated the crisis is each a clonal descendant of just one individual bee that lived in the early 1980s (Baudry et al., 2004). One could assume that this individual had an extraordinary mutation for extreme virulence, but Beekman and Oldroyd (2008) preferred to think that such virulence could fall within the range (albeit the extreme high end) of naturally occurring variation for reproductive competitiveness that is so strongly selected for in workers of *A. m. capensis*.

It is constructive to consider the evolutionary forces that engendered parthenogenic Cape workers in the first place. A human observer may wonder what benefits could possibly outweigh risks to social cohesion that seem implicit in the "parthenogenic option." The best answer evokes the striking geographic singularity that comprises the home range of *A. m. capensis*, and that is the southern cape of the African continent for which it is named—a region of extreme and unpredictable winds. If queen loss from high winds during mating flights was a real and constant risk to the *Apis mellifera* populations colonizing the region, then one can understand how diploid egg production by workers could be adaptive (Moritz, 1985a). It could have, in its earliest "lawful" stages, provided a colony the necessary female larvae for rearing replacement queens.

Beyond the Superorganism

I think it's unlikely in the long run that accommodating both direct individual gene transmission alongside indirect transmission through near-kin is a stable strategy. The two are essentially antagonistic. Parthenogenesis triggered strong competition among workers, uncoupled genetic interests among members of the colony, and provoked costly colony-level selective responses to constrain worker parasitism. If a modern observer reads a resuscitation of *A. m. capensis* in its current binge of lethal parasitism on *A. m. scutellata*, then one must remember that this is a recent situation made possible by human-assisted movement, not evidence of a long-term strategy. The conflict we see in *A. m. capensis* could be speciation in real time—social dissolution and the creative foment leading to a future "*Apis capensis*," a fully parasitic species in the fashion of workerless parasitic ants. It strikes me equally likely a portent of extinction. These and other dangers to social cohesion are the subject of Robin Moritz's and Robin Crewe's worthy 2018 book, *The Dark Side of the Hive*.

This brings us to the point where we can reflect on the implications of *A. m. capensis* for the whole project of social evolution. People sometimes ask me after a lecture on evolutionary transitions something to the effect, "If organisms are the accretion of formerly independent cells, and superorganisms the accretion of formerly independent organisms, then what comes after the superorganism?"

Ecosystems are the leading candidate for an organizational level above the superorganism. In fact, curiosity about evolved change at the community level has engendered the emerging field of *community and ecosystem genetics*, reviewed by Whitham et al. (2020). Study of the matter has always begun with a discussion on the appropriate building blocks for evolution. Lewontin (1970) stated that populations can evolve if they possess phenotypic variation, the variation is associated with fitness differences, and the fitness is heritable. But Lewontin went on to make clear that he doubted whether natural selection could act above the level of individual because higher organizational levels, communities and ecosystems, would be unlikely to exhibit heritability.

This objection was challenged by Swenson et al. (2000), whose experimental approach showed that heritable variation could be extended beyond single-species populations into entire ecosystems consisting of thousands of species. But this and subsequent work has been criticized for its overreliance on experimentally imposed community-level inheritance. Moreover, Lean et al. (2022) seemed skeptical that communities can rise to the level of evolutionary individuality, and they doubted the sufficiency of Lewontin (1970)'s criteria for accounting for evolution in a community context. They advocated instead approaches more like David Hull (1980)'s replicator-interactor framework and Richard Dawkins (1989)'s "extended phenotype" (chapter 2's "Multilevel Selection" section) that focused on the gene as the fundamental replicator and treated interactions between complex entities and their environment as the causal basis of differential reproduction in lower-level replicators.

Another consideration is the so-called ant supercolony that can contain multiple queens, multiple nests, freely-mixing worker populations numbering into the millions, and cover geographic ranges at a scale of kilometers, all contiguous and functionally one entity (Helanterä, 2022). Yet as impressive it is for its social complexity and ecologic dominance, the ant supercolony does not qualitatively rise above superorganism on a scale of evolutionary transitions (table 2.1). And in any case, for our candidate, *Apis mellifera*, I see no such novel options. As I suggested in chapter 9, the spatial discontinuity and volume limitations of a hollow tree compared with the infinitude that soil nesting afforded ants has selected against polygyny and constrained in bees anything like multi-queen supercolonies.

I tend to read in the biology of the South African subspecies *A. m. capensis* the limits of social organization for *Apis mellifera*. As I said in chapter 2's "Multilevel Selection" section, the superorganism is inherently less stable than the organism with the organism's developmental bottleneck of the single-celled zygote that renders each somatic cell a clone of one another. Superorganisms, by definition, are made up of units that are themselves organisms, not clones, so the possibility of worker conflict is acute and real. This conflict in the honey bee is constrained by a queen's mating strategies, her pheromones, and, crucially, workers' evolved behavior of policing one another.

But by evading her sister's policing efforts, the Cape bee strikes at the mechanistic tie that binds the superorganism together. In the case of the honey bee superorganism, its nesting biology precludes multi-queen supercolonies, and at the opposite pole, the Cape bee is demonstrating a viable path toward social parasitism. Although ants may be experimenting with higher expressions of sociality bordering on unity, it's hard for me to imagine organizational successions in the honey bee other than workerless parasites and social dissolution.

Summary

Eusociality involves natural selection acting at two levels—the level of individual and the level of group. These two levels are unavoidably in tension, their interests only partly aligned. Selfish behavior in the honey bee is manifested as workers abandoning colony work in favor of producing their own eggs. The colony normally constrains worker reproduction with pheromones and worker-on-worker "policing," which takes the form of eating each others' eggs. This chapter describes two situations by which workers "cheat," thereby evading the colony's constraints and successfully producing offspring of their own. Anarchy expresses in a colony when (1) one or more anarchist workers activates her ovaries in spite of the presence of pheromonal inhibition, and (2) her eggs escape policing.

A second situation is found in the Cape bee of South Africa, *Apis mellifera capensis*. Workers of this subspecies can produce diploid eggs without the need for mating. The resulting eggs are physiologically normal females, capable of being reared into workers or queens, and each is a clone of its worker mother. Under these circumstances, the genetic incentives for policing and altruism deteriorate and social dissolution ensues. Natural Cape bee colonies have evolved constraints that curb worker selfishness, allowing the eusocial state to persist, but when Cape bees were moved into the same range as their near-relative *A. m. scutellata*, they expressed social parasitism at high rates. Cape workers drift into *A. m. scutellata* colonies, evade policing, and successfully rear workers who, in turn, drift to new host colonies and sustain the parasitic line. Inevitably, the host queen is killed, lost, or "ignored to death," and the colony dies.

CHAPTER 25

The Eusocial Primate

IT HAS BEEN MY GOAL in this book to showcase the breadth and ambit of social evolution, its generality as an explanation for organismal evolution, its association and synergy with emergent biological order, and the warrants it declares for cooperation, connectivity, and solidarity in the history and practice of life. In its most charismatic exemplar, the honey bee, we witness a species in whose shadow our own has evolved. Our relationship is ancient and mutually formative to the point I consider it viable to ask whether human-on-honey bee predation fueled the evolution of early *Homo* brains (discussed in chapter 1's "Contemporaneous Relationships between *Apis mellifera* and *Homo sapiens*" section).

But the connections don't stop there. In the social structures of modern human groups, more than one thinker has seen parallels to the kinds of social structures we have been studying in this book. There are credible arguments to be made that to the representative species of eusociality found among the honey bees, other insects, crustaceans, and vertebrates may be added our own, *Homo sapiens*.

To refresh, in 1971, E. O. Wilson summarized his enduring three criteria for eusociality ("Degrees and Qualities of Sociality" section in chapter 5).

A conspecific group must express all of these at some point in its life cycle: (1) cooperative brood care, (2) reproductive division of labor (some members reproduce; some don't), and (3) overlapping generations. In an influential commentary, Foster and Ratnieks (2005) applied Wilson's criteria to vertebrates, paying special attention to the qualities of a non-reproducing caste that helps care for close relatives. They saw parallels with menopause in middle-aged women. This rapid and irreversible onset of sterility has been met with two fitness-based explanations. One is a "mother effect," which explains mid-life female sterility as an adaptation freeing mothers from late-life child-bearing, enabling them to take better care of existing children. The second is a "grandmother effect," which sees mid-life sterility as an adaptation freeing grandmothers to help their reproductive daughters (Shanley and Kirkwood, 2001).

An analysis of pre-modern (eighteenth and nineteenth century) demographic data from Canada and Finland affirmed that post-reproductive grandmothers enhanced reproductive fitness of their offspring by enabling them to procreate earlier, more often, and more successfully (Lahdenperä et al., 2004). By foregrounding sterile human grandmothers "staying at the nest" to help their reproductive daughters, Foster and Ratnieks (2005) showed that humans fulfill all three of Wilson's criteria for eusociality. It is noteworthy that Wilson himself is on record supporting the inclusion of humans in the eusocial roster (Nowak et al., 2010).

The parallels are not exact of course. Human grandmothers are helping rear grandchildren ($r = 0.25$), not siblings ($r = 0.50$); the "castes" are not permanent, but rather human females cycling from a reproductive to non-reproductive phase; and the state of eusociality is not obligatory in that humans are capable of living and reproducing as solitaries. But given that Wilson's criteria have long accepted variations along a continuum of social expression (table 5.1), these peculiarities do not present fatal obstacles as long as fitness benefits can be shown.

Admission of humans into the eusociality club has stimulated supporting lines of creative scholarly thinking. Hardisty and Cassill (2010) saw another commonality shared across eusocial species—the need for polyphasic sleep patterns to care for offspring around the clock. In contrast to monophasic sleep—one extended sleep period per 24-hour day, polyphasic sleep is accumulated over numerous short naps throughout a day, meaning that in a social context, enough workers are awake at any moment to handle any task required. Hardisty and Cassill (2010) cited examples of polyphasic sleep from the fire ant *Solenopsis invicta* and naked mole rats and extended the function of polyphasic sleep to humans, who have innovated artificial heat and lighting to extend our work days and, if needed, can access family members or hired help to provide constant care to our young.

But a widening circle of creative discussion is taking place on the role and ongoing effects of social selection on such patently human institutions and qualities as empathy, music, a sense of equity, economics, politics, group allegiance, morality, and religion (Ebstein et al., 2010; Feierman, 2016). The idea of religion is treated in a dedicated chapter in one of E. O. Wilson's final books, his 2014 *The Meaning of Human Existence*. Religion, saw Wilson, is foremost an engine for the formation and maintenance of the group. Given that we inherited our genes for selfish direct fitness (our "sins") from our solitary forebears, it was for the sake of maintaining the group, and maintaining our place in the group that we evolved our capacities for altruism, cooperation, and citizenship (our "virtues"). To be human means to live in tension between these moral imperatives, to weigh every moment of every day our selfish needs and desires against our group reputation—in short, to be under simultaneous self- and group-selection. Social psychologist Jonathan Haidt borrowed directly from Wilson in the former's book *The Righteous Mind: Why Good People Are Divided by Politics and Religion* (Haidt, 2012) where, in a chapter entitled "Religion is a Team Sport," he unpacks the power of religion to discourage cheats and foster communities with shared trust and agreed-upon narratives on origins, morals, and meaning. For the eusocial human, we call that group-level adaptation; for the honey bee, it's called colony-level adaptation, but it's the same thing.

It is worth driving home the point that the outcomes of positive social selection accrue to efficiency gains in cooperation, and better cooperation accrues to better fitness of the group. There is no better example of this than language—both bee and human. Language is manifestly a socially evolved behavior. Social selection may have engendered the honey bee dance recruitment language, but evolution did not reward dancers who performed the most beautiful, original, or athletic dances; it rewarded the colony whose dancers and recruits used the encoded information to cooperatively store enough food for winter.

In the case of humans, gifted speakers may win individual fitness if they reverse the opinion of a hostile jury or woo a fertile partner with love sonnets. But larger than that, it is through and because of language that *Homo sapiens* realized the features that set it apart from all other biological life: its unmatched capacity for marshaling cooperation at a scale of populations and across populations.

The historian Yuval Noah Harari unpacked this uniquely human superpower in his books *Sapiens: A Brief History of Humankind* (Harari, 2015) and *Homo Deus: A Brief History of Tomorrow* (Harari, 2017). It is our socially evolved powers of language, abstraction, and story-telling that enabled us to export religious beliefs, values, and origin myths outside our own family units. Evolution may not have directly rewarded the most entertaining storyteller, but evolution rewarded his tribe of 500, unified in their certainty of the Great Lion God's favor, who drove out the disorganized bands of tens or dozens *Homo neanderthalensis* occupying the rich valley bottom.

Such dark scenarios are indeed plausible explanations for what Coolidge and Wynn (2018, p. 221) have called "the wooly mammoth in the room," the fact that all hominin taxa, except for our own, have gone extinct. The historian Harari (2015, 2017) highlighted this sober fact alongside other hallmarks of human enterprise—governments, economies, whole civilizations—as testimony to the world-shaping power of *Homo sapiens* cooperation. Has such a pitch of cooperation been adaptive? Most certainly so, at least in the short span of our species's existence. But the more salient question upon which our future hangs is whether *H. sapiens* can extend its powers of cooperation into ecosystem scales (chapter 24's "Beyond the Superorganism" section) and choose to live sustainably alongside its sibling species on this beautiful, precious world.

Haidt and Harari are among a widening community of scholars who are frankly embracing social evolution as the long-neglected window into understanding the deepest compulsions and desires of *Homo sapiens*, understanding, if you will, human nature as it really is—biological life as it really is.

Throughout this book, I have used the honey bee as the carrier of this larger story. I think they have been sufficient to the task. Through them, we have witnessed the tectonic forces, the *social* forces, that push smaller units of individuality into larger, more inclusive units, forces that are general to biology not just particular to the oddball social insects. Through the honey bee, we have celebrated the marvels of emergent order that fortifies the species's foothold on its habitat and inclines it toward higher expressions of social sophistication.

Equally, through the honey bee, we have been alerted to vulnerabilities in the superorganismal project and warned that solitary opportunism is corrosive to group stability. When it comes to descent with modification from natural selection, the honey bee has shown us that Charles Darwin's Great Idea is not license for dystopian selfishness. It carries in its message license for cooperation, altruism, and conflict resolution—all that is best in us.

REFERENCES

Abbot, P., J. Abe, J. Alcock, S. Alizon, J. Alpedrinha, M. Andersson, J.-B. André, et al. 2011. "Inclusive Fitness Theory and Eusociality." *Nature* 471 (7339): E1–4. https://doi.org/10.1038/nature09831.

Abrams, J., and G. C. Eickwort. 1981. "Nest Switching and Guarding by the Communal Sweat Bee *Agapostemon virescens* (Hymenoptera, Halictidae)." *Insectes Sociaux* 28 (2): 105–116. https://doi.org/10.1007/bf02223699.

Adams, J., E. D. Rothman, W. E. Kerr, and Z. L. Paulino. 1977. "Estimation of the Number of Sex Alleles and Queen Matings from Diploid Male Frequencies in a Population of *Apis mellifera*." *Genetics* 86 (3): 583–596. https://doi.org/10.1093/genetics/86.3.583.

Al Ghzawi, A. Al-M., S. Zaitoun, and H. K. Shannag. 2009. "Management of *Braula orientalis* Örösi (Diptera: Braulidae) in Honeybee Colonies with Tobacco Smoke under Semiarid Conditions." *Entomological Research* 39 (3): 168–174. https://doi.org/10.1111/j.1748-5967.2009.00215.x.

Al-Khafaji, K., S. Tuljapurkar, J. R. Carey, and R. E. Page. 2009. "Hierarchical Demography: A General Approach with an Application to Honey Bees." *Ecology* 90 (2): 556–566. https://doi.org/10.1890/08-0402.1.

Albalat, R., and C. Cañestro. 2016. "Evolution by Gene Loss." *Nature Reviews Genetics* 17 (7): 379–391. https://doi.org/10.1038/nrg.2016.39.

Alburaki, M., B. Bertrand, H. Legout, S. Moulin, A. Alburaki, W. S. Sheppard, and L. Garnery. 2013. "A Fifth Major Genetic Group among Honeybees Revealed in Syria." *BMC Genomic Data* 14 (1): 117. https://doi.org/10.1186/1471-2156-14-117.

Alexander, B. A. 1991. "Phylogenetic Analysis of the Genus *Apis* (Hymenoptera: Apidae)." *Annals of the Entomological Society of America* 84 (2): 137–149. https://doi.org/10.1093/aesa/84.2.137.

Alexander, R. D. 1974. "The Evolution of Social Behavior." *Annual Review of Ecology and Systematics* 5 (1): 325–383. https://doi.org/10.1146/annurev.es.05.110174.001545.

Alford, D. 1975. *Bumblebees*. London: Davis-Poynter.

Allen, M. D. 1956. "The Behaviour of Honeybees Preparing to Swarm." *British Journal of Animal Behaviour* 4 (1): 14–22. https://doi.org/10.1016/s0950-5601(56)80011-7.

Allen, M. D. 1959. "Respiration Rates of Worker Honeybees of Different Ages and at Different Temperatures." *Journal of Experimental Biology* 36 (1): 92–101. https://doi.org/10.1242/jeb.36.1.92..

Allsopp, M. H. 1992. "The Capensis Calamity." *South African Bee Journal* 64, 52–55.

Allsopp, M. H., J. N. M. Calis, and W. J. Boot. 2003. "Differential Feeding of Worker Larvae Affects Caste Characters in the Cape Honeybee, *Apis mellifera capensis*." *Behavioral Ecology and Sociobiology* 54 (6): 555–561. https://doi.org/10.1007/s00265-003-0666-4.

Almeida, E. a. B., S. Bossert, B. N. Danforth, D. S. Porto, F. V. Freitas, C. C. Davis, E. A. Murray, et al. 2023. "The Evolutionary History of Bees in Time and Space." *Current Biology* 33 (16): 3409–3422.e6. https://doi.org/10.1016/j.cub.2023.07.005.

Alpedrinha, J., A. Gardner, and S. A. West. 2014. "Haplodiploidy and the Evolution of Eusociality: Worker Revolution." *American Naturalist* 184 (3): 303–317. https://doi.org/10.1086/677283.

Alpedrinha, J., S. A. West, and A. Gardner. 2013. "Haplodiploidy and the Evolution of Eusociality: Worker Reproduction." *American Naturalist* 182 (4): 421–438. https://doi.org/10.1086/671994.

Amdam, G. V., K. Norberg, M. K. Fondrk, and R. E. Page. 2004a. "Reproductive Ground Plan May Mediate Colony-Level Selection Effects on Individual Foraging Behavior in Honey Bees." *Proceedings of the National Academy of Sciences of the United States of America* 101 (31): 11350–11355. https://doi.org/10.1073/pnas.0403073101.

Amdam, G. V., Z. L. P. Simões, A. Hagen, K. Norberg, K. H. Schrøder, Ø. Mikkelsen, T. B. L. Kirkwood, and S. W. Omholt. 2004b. "Hormonal Control of the Yolk Precursor Vitellogenin Regulates Immune Function and Longevity in

Honeybees." *Experimental Gerontology* 39 (5): 767–773. https://doi.org/10.1016/j.exger.2004.02.010.

Amdam, G. V., K. Norberg, S. W. Omholt, P. Kryger, A. P. Lourenço, M. M. G. Bitondi, and Z. L. P. Simões. 2005. "Higher Vitellogenin Concentrations in Honey Bee Workers May Be an Adaptation to Life in Temperate Climates." *Insectes Sociaux* 52 (4): 316–319. https://doi.org/10.1007/s00040-005-0812-2

Amdam, G. V., A. Csondes, M. K. Fondrk, and R. E. Page. 2006. "Complex Social Behaviour Derived from Maternal Reproductive Traits." *Nature* 439 (7072): 76–78. https://doi.org/10.1038/nature04340.

Amdam, G. V., and S. W. Omholt. 2002. "The Regulatory Anatomy of Honeybee Lifespan." *Journal of Theoretical Biology* 216 (2): 209–228. https://doi.org/10.1006/jtbi.2002.2545.

Amdam, G. V., and S. W. Omholt. 2003. "The Hive Bee to Forager Transition in Honeybee Colonies: The Double Repressor Hypothesis." *Journal of Theoretical Biology* 223 (4): 451–464. https://doi.org/10.1016/s0022-5193(03)00121-8.

Anderson, R. 1968. "The Effect of Queen Loss on Colonies of *Apis mellifera capensis*." *South African Journal of Agricultural Science* 11, 383–388.

Anderson, R. M., and R. M. May. 1981. "The Population Dynamics of Microparasites and Their Invertebrate Hosts." *Philosophical Transactions of the Royal Society of London* 291 (1054): 451–524. https://doi.org/10.1098/rstb.1981.0005.

Antevs, E. 1929. "Maps of the Pleistocene Glaciations." *Bulletin of the Geological Society of America* 40, 631–720.

Arechavaleta-Velasco, M. E., G. J. Hunt, and C. Emore. 2003. "Quantitative Trait Loci That Influence the Expression of Guarding and Stinging Behaviors of Individual Honey Bees." *Behavior Genetics* 33, 357–364. https://doi.org/10.1023/A:1023458827643.

Arias, M. C., and W. S. Sheppard. 1996. "Molecular Phylogenetics of Honey Bee Subspecies (*Apis mellifera* L.) Inferred from Mitochondrial DNA Sequence." *Molecular Phylogenetics and Evolution* 5 (3): 557–566. https://doi.org/10.1006/mpev.1996.0050.

Aron, S., L. De Menten, D. R. Van Bockstaele, S. M. Blank, and Y. Roisin. 2005. "When Hymenopteran Males Reinvented Diploidy." *Current Biology* 15 (9): 824–827. https://doi.org/10.1016/j.cub.2005.03.017.

Asencot, M., and Y. Lensky. 1976. "The Effect of Sugars and Juvenile Hormone on the Differentiation of the Female Honeybee Larvae (*Apis mellifera* L.) to Queens." *Life Sciences* 18 (7): 693–699. https://doi.org/10.1016/0024-3205(76)90180-6.

Asencot, M., and Y. Lensky. 1985. "The Phagostimulatory Effect of Sugars on the Induction of 'Queenliness' in Female Honeybee (*Apis mellifera* L.) Larvae."

Comparative Biochemistry and Physiology Part A: Physiology 81 (1): 203–208. https://doi.org/10.1016/0300-9629(85)90289-0.

Asencot, M., and Y. Lensky. 1988. "The Effect of Soluble Sugars in Stored Royal Jelly on the Differentiation of Female Honeybee (*Apis mellifera* L.) Larvae to Queens." *Insect Biochemistry* 18 (2): 127–133. https://doi.org/10.1016/0020-1790(88)90016-9.

Aumer, D., E. Stolle, M. H. Allsopp, F. N. Mumoki, C. W. Werner Pirk, and R. F. A. Moritz. 2019. "A Single SNP Turns a Social Honey Bee (*Apis mellifera*) Worker into a Selfish Parasite." *Molecular Biology and Evolution* 36 (3): 516–526. https://doi.org/10.1093/molbev/msy232.

Avalos, J., H. Rosero, G. Maldonado, and F. J. Reynaldi. 2019. "Honey Bee Louse (*Braula schmitzi*) as a Honey Bee Virus Vector?" *Journal of Apicultural Research* 58 (3): 427–429. https://doi.org/10.1080/00218839.2019.1565726.

Avise, J. C., J. Arnold, R. Ball, E. Bermingham, T. Lamb, J. E. Neigel, C. A. Reeb, and N. C. Saunders. 1987. "Intraspecific Phylogeography: The Mitochondrial DNA Bridge between Population Genetics and Systematics." *Annual Review of Ecology and Systematics* 18 (1): 489–522. https://doi.org/10.1146/annurev.es.18.110187.002421.

Baer, B., J. Collins, K. Maalaps, and S. P. a. Den Boer. 2016. "Sperm Use Economy of Honeybee (*Apis mellifera*) Queens." *Ecology and Evolution* 6 (9): 2877–2885. https://doi.org/10.1002/ece3.2075.

Baer, B., E. D. Morgan, and P. Schmid-Hempel. 2001. "A Nonspecific Fatty Acid within the Bumblebee Mating Plug Prevents Females from Remating." *Proceedings of the National Academy of Sciences of the United States of America* 98 (7): 3926–3928. https://doi.org/10.1073/pnas.061027998.

Baer, B., and P. Schmid-Hempel. 2001. "Unexpected Consequences of Polyandry for Parasitism and Fitness in the Bumblebee, *Bombus terrestris*." *Evolution* 55 (8): 1639–1643. https://doi.org/10.1111/j.0014-3820.2001.tb00683.x.

Bailey, L. 1958. "Wild Honeybees and Disease." *Bee World* 39, 92–95.

Bakhtin, Y., M. I. Katsnelson, Y. I. Wolf, and E. V. Koonin. 2021. "Evolution in the Weak-mutation Limit: Stasis Periods Punctuated by Fast Transitions between Saddle Points on the Fitness Landscape." *Proceedings of the National Academy of Sciences of the United States of America* 118 (4) e2015665118. https://doi.org/10.1073/pnas.2015665118.

Bale, J. S., and S. A. L. Hayward. 2010. "Insect Overwintering in a Changing Climate." *Journal of Experimental Biology* 213 (6): 980–994. https://doi.org/10.1242/jeb.037911.

Bang, C., K. Weidenbach, T. Gutsmann, H. Heine, and R. A. Schmitz. 2014. "The Intestinal Archaea *Methanosphaera stadtmanae* and *Methanobrevibacter smithii*

Activate Human Dendritic Cells." *PLOS ONE* 9 (6): e99411. https://doi.org/10.1371/journal.pone.0099411.

Baracchi, David, and Stefano Turillazzi. 2010. "Differences in Venom and Cuticular Peptides in Individuals of *Apis mellifera* (Hymenoptera: Apidae) Determined by MALDI-TOF MS." *Journal of Insect Physiology* 56 (4): 366–375. https://doi.org/10.1016/j.jinsphys.2009.11.013.

Barchuk, A. R., R. Maleszka, and Z. L. P. Simões. 2004. "*Apis mellifera* Ultraspiracle: cDNA Sequence and Rapid Up-Regulation by Juvenile Hormone." *Insect Molecular Biology* 13 (5): 459–467. https://doi.org/10.1111/j.0962-1075.2004.00506.x

Baudry, E., P. Kryger, M. H. Allsopp, N. Kœniger, D. Vautrin, F. Mougel, J.-M. Cornuet, and M. Solignac. 2004. "Whole-Genome Scan in Thelytokous-Laying Workers of the Cape Honeybee (*Apis mellifera capensis*): Central Fusion, Reduced Recombination Rates and Centromere Mapping Using Half-Tetrad Analysis." *Genetics* 167 (1): 243–252. https://doi.org/10.1534/genetics.167.1.243.

Baudry, E., M. Solignac, L. Garnery, M. Gries, J. M. Cornuet, and N. Kœniger. 1998. "Relatedness among Honeybees (*Apis mellifera*) of a Drone Congregation." *Proceedings of the Royal Society B: Biological Sciences* 265 (1409): 2009–2014. https://doi.org/10.1098/rspb.1998.0533.

Baum, J., J. P. St George, and K. McCall. 2005. "Programmed Cell Death in the Germline." *Seminars in Cell & Developmental Biology* 16 (2): 245–259. https://doi.org/10.1016/j.semcdb.2004.12.008.

Beekman, M., M. H. Allsopp, L. A. Jordan, J. Lim, and B. P. Oldroyd. 2009. "A Quantitative Study of Worker Reproduction in Queenright Colonies of the Cape Honey Bee, *Apis mellifera capensis*." *Molecular Ecology* 18 (12): 2722–2727. https://doi.org/10.1111/j.1365-294x.2009.04224.x.

Beekman, M., S. J. Martin, F. P. Drijfhout, and B. P. Oldroyd. 2007. "Higher Removal Rate of Eggs Laid by Anarchistic Queens—A Cost of Anarchy?" *Behavioral Ecology and Sociobiology* 61 (12): 1847–1853. https://doi.org/10.1007/s00265-007-0424-0.

Beekman, M., and B. P. Oldroyd. 2008. "When Workers Disunite: Intraspecific Parasitism by Eusocial Bees." *Annual Review of Entomology* 53 (1): 19–37. https://doi.org/10.1146/annurev.ento.53.103106.093515.

Berg, S., N. Koeniger, G. Koeniger, and S. Fuchs. 1997. "Body Size and Reproductive Success of Drones (*Apis mellifera L*)." *Apidologie* 28 (6): 449–460. https://doi.org/10.1051/apido:19970611.

Berger, S. L., T. Kouzarides, R. Shiekhattar, and A. Shilatifard. 2009. "An Operational Definition of Epigenetics." *Genes & Development* 23 (7): 781–783. https://doi.org/10.1101/gad.1787609.

Bergmann, C. 1847. Über die Verhältnisse der Wärmeökonomie der Thiere zu ihrer Grösse. *Göttinger Studien* 3, 595–708.

Berry, J. A., and K. S. Delaplane. 2001. "Effects of Comb Age on Honey Bee Colony Growth and Brood Survivorship." *Journal of Apicultural Research* 40 (1): 3–8. https://doi.org/10.1080/00218839.2001.11101042.

Betts, H. C., M. N. Puttick, J. W. Clark, T. A. Williams, P. C. J. Donoghue, and D. Pisani. 2018. "Integrated Genomic and Fossil Evidence Illuminates Life's Early Evolution and Eukaryote Origin." *Nature Ecology and Evolution* 2 (10): 1556–1562. https://doi.org/10.1038/s41559-018-0644-x.

Beye, M., I. Gattermeier, M. Hasselmann, T. Gempe, M. Schioett, J. F. Baines, D. I. Schlipalius, et al. 2006. "Exceptionally High Levels of Recombination across the Honey Bee Genome." *Genome Research* 16 (11): 1339–1344. https://doi.org/10.1101/gr.5680406.

Biesmeijer, J. C., and H. De Vries. 2001. "Exploration and Exploitation of Food Sources by Social Insect Colonies: A Revision of the Scout-Recruit Concept." *Behavioral Ecology and Sociobiology* 49 (2–3): 89–99. https://doi.org/10.1007/s002650000289.

Blumberg, M. S. 2005. *Basic Instinct: The Genesis of Behavior.* New York: Thunder's Mouth Press.

Bonduriansky, R. 2011. "Sexual Selection and Conflict as Engines of Ecological Diversification." *American Naturalist* 178 (6): 729–745. https://doi.org/10.1086/662665.

Bonner, J. T. 1988. *The Evolution of Complexity by Means of Natural Selection.* Princeton, NJ: Princeton University Press.

Bonoan, R. E., P. M. I. Feliciano, J. Chang, and P. T. Starks. 2020. "Social Benefits Require a Community: The Influence of Colony Size on Behavioral Immunity in Honey Bees." *Apidologie* 51 (5): 701–709. https://doi.org/10.1007/s13592-020-00754-5.

Boomsma, J. J., B. Baer, and J. Heinze. 2005. "The Evolution of Male Traits in Social Insects." *Annual Review of Entomology* 50 (1): 395–420. https://doi.org/10.1146/annurev.ento.50.071803.130416.

Boomsma, J. J., and F. L. W. Ratnieks. 1996. "Paternity in Eusocial Hymenoptera." *Philosophical Transactions of the Royal Society B* 351 (1342): 947–975. https://doi.org/10.1098/rstb.1996.0087.

Boots, M., and M. Begon. 1993. "Trade-Offs with Resistance to a Granulosis Virus in the Indian Meal Moth, Examined by a Laboratory Evolution Experiment." *Functional Ecology* 7 (5): 528. https://doi.org/10.2307/2390128.

Boraas, M. E., D. B. Seale, and J. E. Boxhorn. 1998. "Phagotrophy by a Flagellate Selects for Colonial Prey: A Possible Origin of Multicellularity." *Evolutionary Ecology* 12 (2): 153–164. https://doi.org/10.1023/a:1006527528063.

Borowiec, M. L., C. S. Moreau, and C. Rabeling. 2021. Ants: Phylogeny and Classification. In *Encyclopedia of Social Insects*, edited by C. Starr. Cham, Switzerland: Springer Nature.

Bosch, T. C. G. 2009. "Hydra and the Evolution of Stem Cells." *BioEssays* 31 (4): 478–486. https://doi.org/10.1002/bies.200800183.

Bossert, S., E. A. Murray, E. A. B. Almeida, S. G. Brady, B. B. Blaimer, and B. N. Danforth. 2019. "Combining Transcriptomes and Ultraconserved Elements to Illuminate the Phylogeny of Apidae." *Molecular Phylogenetics and Evolution* 130 (January): 121–131. https://doi.org/10.1016/j.ympev.2018.10.012.

Bourke, A. 1999. "Colony Size, Social Complexity and Reproductive Conflict in Social Insects." *Journal of Evolutionary Biology* 12 (2): 245–257. https://doi.org /10.1046/j.1420-9101.1999.00028.x

Bourke, A. F. 1988. "Worker Reproduction in the Higher Eusocial Hymenoptera." *Quarterly Review of Biology* 63 (3): 291–311. https://doi.org/10.1086/415930.

Bourke, A. F. 1994. "Worker Matricide in Social Bees and Wasps." *Journal of Theoretical Biology* 167 (3): 283–292. https://doi.org/10.1006/jtbi.1994.1070.

Bourke, A. F. 2011. *Principles of Social Evolution*. Oxford, UK: Oxford University Press.

Bourke, A. F. 2015. "Sex Investment Ratios in Eusocial Hymenoptera Support Inclusive Fitness Theory." *Journal of Evolutionary Biology* 28 (11): 2106–2111. https://doi.org/10.1111/jeb.12710.

Bourke, A. F., and N. R. Franks. 1995. *Social Evolution in Ants*. Princeton, NJ: Princeton University Press.

Boyce, M. S., and C. M. Perrins. 1987. "Optimizing Great Tit Clutch Size in a Fluctuating Environment." *Ecology* 68 (1): 142–153. https://doi.org/10.2307 /1938814.

Bozina, K. 1961. "How Long Does the Queen Live?" [in Russian]. *Pchelovodstvo* 38: 13.

Brady, S. G., S. D. Sipes, A. M. Pearson, and B. N. Danforth. 2006. "Recent and Simultaneous Origins of Eusociality in Halictid Bees." *Proceedings of the Royal Society B: Biological Sciences* 273 (1594): 1643–1649. https://doi.org/10.1098/rspb .2006.3496.

Branstetter, M. G., B. N. Danforth, J. P. Pitts, B. C. Faircloth, P. S. Ward, M. L. Buffington, M. W. Gates, R. R. Kula, and S. G. Brady. 2017. "Phylogenomic Insights into the Evolution of Stinging Wasps and the Origins of Ants and Bees." *Current Biology* 27 (7): 1019–1025. https://doi.org/10.1016/j.cub.2017.03.027.

Breed, M. D. 1976. "The Evolution of Social Behavior in Primitively Social Bees: A Multivariate Analysis." *Evolution* 30 (2): 234–240. https://doi.org/10.1111/j.1558 -5646.1976.tb00906.x.

Breed, M. D., E. Guzmán-Novoa, and G. J. Hunt. 2004. "Defensive Behavior of Honey Bees: Organization, Genetics, and Comparisons with Other Bees."

Annual Review of Entomology 49 (1): 271–298. https://doi.org/10.1146/annurev.ento.49.061802.123155.

Breed, M. D., G. E. Robinson, and R. E. Page. 1990. "Division of Labor during Honey Bee Colony Defense." *Behavioral Ecology and Sociobiology* 27 (6): 395–401. https://doi.org/10.1007/bf00164065.

Breed, M. D., and K. B. Rogers. 1991. "The Behavioral Genetics of Colony Defense in Honeybees: Genetic Variability for Guarding Behavior." *Behavior Genetics* 21 (3): 295–303. https://doi.org/10.1007/bf01065821.

Breed, M. D., J. M. Silverman, and W. J. Bell. 1978. "Agonistic Behavior, Social Interactions, and Behavioral Specialization in a Primitively Eusocial Bee." *Insectes Sociaux* 25 (4): 351–364. https://doi.org/10.1007/bf02224299.

Breed, M. D., T. Smith, and A. Torres. 1992. "Role of Guard Honey Bees (Hymenoptera: Apidae) in Nestmate Discrimination and Replacement of Removed Guards." *Annals of the Entomological Society of America* 85 (5): 633–637. https://doi.org/10.1093/aesa/85.5.633.

Brown, M. J. F., and P. Schmid-Hempel. 2003. "The Evolution of Female Multiple Mating in Social Hymenoptera." *Evolution* 57 (9): 2067–2081. https://doi.org/10.1111/j.0014-3820.2003.tb00386.x.

Brückner, D. 1978. "Why Are There Inbreeding Effects in Haplo-Diploid Systems?" *Evolution* 32 (2): 456–458. https://doi.org/10.1111/j.1558-5646.1978.tb00661.x.

Bucholz, S., M. O. Schäfer, S. Spiewok, J. S. Pettis, M. Duncan, W. Ritter, R. Spooner-Hart, and P. Neumann. 2008. "Alternative Food Sources of *Aethina tumida* (Coleoptera: Nitidulidae)." *Journal of Apicultural Research* 47 (3): 202–209. https://doi.org/10.1080/00218839.2008.11101460.

Bull, J. J., I. J. Molineux, and W. R. Rice. 1991. "Selection of Benevolence in a Host–Parasite System." *Evolution* 45 (4): 875–882. https://doi.org/10.1111/j.1558-5646.1991.tb04356.x

Bull, N. J., A. C. Mibus, Y. Norimatsu, B. L. Jarmyn, and M. P. Schwarz. 1998. "Giving Your Daughters the Edge: Bequeathing Reproductive Dominance in a Primitively Social Bee." *Proceedings of the Royal Society B: Biological Sciences* 265 (1404): 1411–1415. https://doi.org/10.1098/rspb.1998.0450.

Butler, C. 1609. *The Feminine Monarchie.* Oxford: Joseph Barnes.

Buttstedt, A., C. Ihling, M. Pietzsch, and R. F. A. Moritz. 2016. "Royalactin Is Not a Royal Making of a Queen." *Nature* 537 (7621): E10–E12. https://doi.org/10.1038/nature19349.

Buttstedt, A., C. I. Mureşan, H. Lilie, G. Hause, C. Ihling, S.-H. Schulze, M. Pietzsch, and R. F. A. Moritz. 2018. "How Honeybees Defy Gravity with Royal Jelly to Raise Queens." *Current Biology* 28 (7): 1095–1100.e3. https://doi.org/10.1016/j.cub.2018.02.022.

Byers, G. W. 1959. "An Unusual Nest of the Honey Bee." *Journal of the Kansas Entomological Society* 32: 46–48.

Calder, W. A. 1984. *Size, Function, and Life History*. Cambridge, MA: Harvard University Press.

Calderone, N. W., and R. E. Page. 1991. "Evolutionary Genetics of Division of Labor in Colonies of the Honey Bee (*Apis mellifera*)." *American Naturalist* 138 (1): 69–92. https://doi.org/10.1086/285205.

Calis, J. N., W. J. Boot, and M. H. Allsopp. 2003. "Cape Honeybees: Crucial Steps Leading to Social Parasitism." In *Proceedings of the Section Experimental and Applied Entomology, Netherlands Entomological Society*, 39–44.

Calvet, M., M. Delmas, Y. Gunnell, R. Braucher, and D. Bourlès. 2011. "Recent Advances in Research on Quaternary Glaciations in the Pyrenees." *Developments in Quaternary Science* 15: 127–139. https://doi.org/10.1016/b978-0-444-53447-7.00011-8.

Camargo, J. M., and S. R. Pedro. 2003. "Meliponini neotropicais: o gênero Partamona Schwarz, 1939 (Hymenoptera, Apidae, Apinae)-bionomia e biogeografia." *Revista Brasileira de Entomologia* 47: 311–372.

Camazine, S. 1991. "Self-Organizing Pattern Formation on the Combs of Honey Bee Colonies." *Behavioral Ecology and Sociobiology* 28 (1): 61–76. https://doi.org/10.1007/bf00172140.

Camazine, S., J.-L. Deneubourg, N. R. Franks, J. Sneyd, G. Theraula, and E. Bonabeau. 2001. *Self-Organization in Biological Systems*. Princeton, NJ: Princeton University Press.

Cameron, S. A. 1993. "Multiple Origins of Advanced Eusociality in Bees Inferred from Mitochondrial DNA Sequences." *Proceedings of the National Academy of Sciences of the United States of America* 90 (18): 8687–8691. https://doi.org/10.1073/pnas.90.18.8687.

Cane, J., and J. Payne. 1990. "Native Bee Pollinates Rabbiteye Blueberry." *Alabama Agricultural Experiment Station* 37: 4.

Cappa, F., C. Bruschini, I. Protti, S. Turillazzi, and R. Cervo. 2016. "Bee Guards Detect Foreign Foragers with Cuticular Chemical Profiles Altered by Phoretic *Varroa* Mites." *Journal of Apicultural Research* 55: 268–277.

Cardinal, S., and B. N. Danforth. 2011. "The Antiquity and Evolutionary History of Social Behavior in Bees." *PLOS ONE* 6 (6): e21086. https://doi.org/10.1371/journal.pone.0021086.

Cardinal, S., and L. Packer. 2007. "Phylogenetic Analysis of the Corbiculate Apinae Based on Morphology of the Sting Apparatus (Hymenoptera: Apidae)." *Cladistics* 23 (2): 99–118. https://doi.org/10.1111/j.1096-0031.2006.00137.x.

Cardinal, S., J. Straka, and B. N. Danforth. 2010. "Comprehensive Phylogeny of Apid Bees Reveals the Evolutionary Origins and Antiquity of Cleptoparasitism." *Proceedings of the National Academy of Sciences of the United States of America* 107 (37): 16207–16211. https://doi.org/10.1073/pnas.1006299107.

Carr, S. M. 2023. "Multiple Mitogenomes Indicate Things Fall Apart with Out of Africa or Asia Hypotheses for the Phylogeographic Evolution of Honey Bees (*Apis mellifera*)." *Scientific Reports* 13 (1). https://doi.org/10.1038/s41598-023-35937-4.

Cavaliere, V., C. Taddei, and G. Gargiulo. 1998. "Apoptosis of Nurse Cells at the Late Stages of Oogenesis of Drosophila Melanogaster." *Development Genes and Evolution* 208 (2): 106–112. https://doi.org/10.1007/s004270050160.

Châline, N., S. F. Martin, and F. L. W. Ratnieks. 2004. "Absence of Nepotism Toward Imprisoned Young Queens During Swarming in the Honey Bee." *Behavioral Ecology* 16 (2): 403–409. https://doi.org/10.1093/beheco/ari003.

Châline, N., F. L. W. Ratnieks, and T. Burke. 2002. "Anarchy in the UK: Detailed Genetic Analysis of Worker Reproduction in a Naturally Occurring British Anarchistic Honeybee, *Apis mellifera*, Colony Using DNA Microsatellites." *Molecular Ecology* 11 (9): 1795–1803. https://doi.org/10.1046/j.1365-294x.2000.01569.x.

Chambers, J., M. B. Quinlan, A. Evans, and R. J. Quinlan. 2020. "Dog-Human Coevolution: Cross-Cultural Analysis of Multiple Hypotheses." *Journal of Ethnobiology* 40 (4): 414–433. https://doi.org/10.2993/0278-0771-40.4.414.

Chantawannakul, P., L. De Guzman, J. Li, and G. R. Williams. 2015. "Parasites, Pathogens, and Pests of Honeybees in Asia." *Apidologie* 47 (3): 301–324. https://doi.org/10.1007/s13592-015-0407-5.

Chapman, N. C., B. P. Oldroyd, and W. O. H. Hughes. 2007. "Differential Responses of Honeybee (*Apis mellifera*) Patrilines to Changes in Stimuli for the Generalist Tasks of Nursing and Foraging." *Behavioral Ecology and Sociobiology* 61 (8): 1185–1194. https://doi.org/10.1007/s00265-006-0348-0.

Chen, C., Z. Liu, Q. Pan, C. Xiao, H. Wang, H. Guo, S. Liu, et al. 2016. "Genomic Analyses Reveal Demographic History and Temperate Adaptation of the Newly Discovered Honey Bee Subspecies *Apis mellifera sinisxinyuann*. ssp." *Molecular Biology and Evolution* 33 (5): 1337–1348. https://doi.org/10.1093/molbev/msw017.

Chéron, B., A. L. Cronin, C. Doums, P. Fédérici, C. Haussy, C. Tirard, and T. Monnin. 2011. "Unequal Resource Allocation among Colonies Produced by Fission in the Ant *Cataglyphis cursor*." *Ecology* 92 (7): 1448–1458. https://doi.org/10.1890/10-2347.1.

Chittka, L., and J. E. Niven. 2009. "Are Bigger Brains Better?" *Current Biology* 19 (21): R995–R1008. https://doi.org/10.1016/j.cub.2009.08.023.

Cho, S., H. Zhi, D. R. Green, D. R. Smith, and J. Zhang. 2006. "Evolution of the Complementary Sex-Determination Gene of Honey Bees: Balancing Selection and Trans-Species Polymorphisms." *Genome Research* 16 (11): 1366–1375. https://doi.org/10.1101/gr.4695306.

Choe, J. C., and B. J. Crespi. 1997. *The Evolution of Social Behaviour in Insects and Arachnids*. Cambridge: Cambridge University Press.

Cockell, C. S., and P. A. Bland. 2005. "The Evolutionary and Ecological Benefits of Asteroid and Comet Impacts." *Trends in Ecology and Evolution* 20 (4): 175–179. https://doi.org/10.1016/j.tree.2005.01.009.

Cole, B. J. 1983. "Multiple Mating and the Evolution of Social Behavior in the Hymenoptera." *Behavioral Ecology and Sociobiology* 12 (3): 191–201. https://doi.org/10.1007/bf00290771.

Colgan, D. J., and W. F. Ponder. 2000. "Incipient Speciation in Aquatic Snails in an Arid-Zone Spring Complex." *Biological Journal of the Linnean Society* 71 (4): 625–641. https://doi.org/10.1111/j.1095-8312.2000.tb01282.x.

Collins, A. M., T. E. Rinderer, J. R. Harbo, and M. A. Brown. 1984. "Heritabilities and Correlations for Several Characters in the Honey Bee." *Journal of Heredity* 75 (2): 135–140. https://doi.org/10.1093/oxfordjournals.jhered.a109888.

Collins, A. M. 1979. "Genetics of the Response of the Honeybee to an Alarm Chemical, Isopentyl Acetate." *Journal of Apicultural Research* 18 (4): 285–291. https://doi.org/10.1080/00218839.1979.11099984.

Cook, C. N., and M. D. Breed. 2013. "Social Context Influences the Initiation and Threshold of Thermoregulatory Behaviour in Honeybees." *Animal Behaviour* 86 (2): 323–329. https://doi.org/10.1016/j.anbehav.2013.05.021.

Coolidge, F. L., and T. G. Wynn. 2018. *The Rise of Homo sapiens: The Evolution of Modern Thinking*. Oxford: Oxford University Press.

Cornell, D. 2017. "Derrida's Negotiations as a Technique of Liberation." *Discourse: Journal for Theoretical Studies in Media and Culture* 39 (2): 195–215. https://doi.org/10.13110/discourse.39.2.0195.

Cotter, S. C., and R. M. Kilner. 2010. "Personal Immunity versus Social Immunity." *Behavioral Ecology* 21 (4): 663–668. https://doi.org/10.1093/beheco/arq070.

Coulson, T., S. Tuljapurkar, and D. Z. Childs. 2010. "Using Evolutionary Demography to Link Life History Theory, Quantitative Genetics and Population Ecology." *Journal of Animal Ecology* 79 (6): 1226–1240. https://doi.org/10.1111/j.1365-2656.2010.01734.x.

Couzin, I. D., J. Krause, N. R. Franks, and S. A. Levin. 2005. "Effective Leadership and Decision-Making in Animal Groups on the Move." *Nature* 433 (7025): 513–516. https://doi.org/10.1038/nature03236.

Craig, R. E. 1979. "Parental Manipulation, Kin Selection, and the Evolution of Altruism." *Evolution* 33 (1 Part 2): 319–334. https://doi.org/10.1111/j.1558-5646.1979.tb04685.x.

Crane, E. 1990. *Bees and Beekeeping: Science, Practice and World Resources*. Oxford,: Heinemann.

Crane, E. 1999. *The World History of Beekeeping and Honey Hunting*. New York: Routledge.

Cremer, S., S. A. O. Armitage, and P. Schmid-Hempel. 2007. "Social Immunity." *Current Biology* 17 (16): R693–702. https://doi.org/10.1016/j.cub.2007.06.008.

Cremer, S., and M. Sixt. 2008. "Analogies in the Evolution of Individual and Social Immunity." *Philosophical Transactions of the Royal Society B* 364 (1513): 129–142. https://doi.org/10.1098/rstb.2008.0166.

Crespi, B. J., and D. Yanega. 1995. "The Definition of Eusociality." *Behavioral Ecology* 6 (1): 109–115. https://doi.org/10.1093/beheco/6.1.109.

Cridland, J, M., N. D. Tsutsui, and S. R. Ramírez. 2017. "The Complex Demographic History and Evolutionary Origin of the Western Honey Bee, *Apis mellifera*." *Genome Biology and Evolution* 9 (2): 457–472. https://doi.org/10.1093/gbe/evx009.

Cronin, A. L., M. Molet, C. Doums, T. Monnin, and C. Peeters. 2013. "Recurrent Evolution of Dependent Colony Foundation across Eusocial Insects." *Annual Review of Entomology* 58 (1): 37–55. https://doi.org/10.1146/annurev-ento-120811-153643.

Crozier, R. H. 2008. "Advanced Eusociality, Kin Selection and Male Haploidy." *Australian Journal of Entomology* 47 (1): 2–8. https://doi.org/10.1111/j.1440-6055.2007.00621.x.

Crozier, R. H., and R. E. Page. 1985. "On Being the Right Size: Male Contributions and Multiple Mating in Social Hymenoptera." *Behavioral Ecology and Sociobiology* 18 (2): 105–115. https://doi.org/10.1007/bf00299039.

Crozier, R. H., and E. J. Fjerdingstad. 2001. "Polyandry in Social Hymenoptera—Disunity in Diversity?" *Annales Zoologici Fennici*, 38 (3–4): 267–285. https://www.jstor.org/stable/23735845.

Cruz-Landim, C. D., R. D. Reginato, and V. L. Imperatriz-Fonseca. 1998. "Variation on Ovariole Number in Meliponinae (Hymenoptera, Apidae) Queen's Ovaries, with Comments on Ovary Development and Caste Differentiation." *Papeis Avulsos de Zoologia* 40 (18): 289–296. https://www.revistas.usp.br/paz/article/view/211652/193882.

Danforth, B. N. 2001. "Evolution of Sociality in a Primitively Eusocial Lineage of Bees." *Proceedings of the National Academy of Sciences of the United States of America* 99 (1): 286–290. https://doi.org/10.1073/pnas.012387999.

Danforth, B. N., R. L. Minckley, and J. L. Neff. 2019. *The Solitary Bees: Biology, Evolution, Conservation*. Princeton, NJ: Princeton University Press.

Danforth, B. N., J. A. Neff, and P. Barretto-Ko. 1996. "Nestmate Relatedness in a Communal Bee, *Perdita texana* (Hymenoptera: Andrenidae), Based on DNA Fingerprinting." *Evolution* 50 (1): 276. https://doi.org/10.2307/2410799.

Darwin, C. 1859. *On the Origin of Species*. New York: Random House.

Da Silva, J. 2021. "Life History and the Transitions to Eusociality in the Hymenoptera." *Frontiers in Ecology and Evolution* 9 (December). https://doi.org/10.3389/fevo.2021.727124.

Dawkins, R. 1976. *The Selfish Gene*. Oxford, UK: Oxford University Press.

Dawkins, R. 1989. *The Selfish Gene*, 2nd ed. Oxford: Oxford University Press.

De Camargo, Aparecida. 1972. "Mating of the Social Bee *Melipona quadrifasciata* under Controlled Conditions (Hymenoptera, Apidae)." *Journal of the Kansas Entomological Society* 45 (4): 520–23. https://digitalcommons.usu.edu/bee_lab_bu/93/.

De Castro, J. B., and M. Martinón-Torres. 2022. "The Origin of the *Homo sapiens* Lineage: When and Where?" *Quaternary International* 634: 1–13.

De Miranda, J. R., and I. Fries. 2008. "Venereal and Vertical Transmission of Deformed Wing Virus in Honeybees (*Apis mellifera* L.)." *Journal of Invertebrate Pathology* 98 (2): 184–189. https://doi.org/10.1016/j.jip.2008.02.004.

De Morgan, A. 1915. *A Budget of Paradoxes*. London: Open Court.

De Souza, D. A., Y. Wang, O. Kaftanoğlu, D. De Jong, G. V. Amdam, L. S. Gonçalves, and T. M. Francoy. 2015. "Morphometric Identification of Queens, Workers and Intermediates in In Vitro Reared Honey Bees (*Apis mellifera*)." *PLOS ONE* 10 (4): e0123663. https://doi.org/10.1371/journal.pone.0123663.

Dedej, S., and K. S. Delaplane. 2003. "Honey Bee (Hymenoptera: Apidae) Pollination of Rabbiteye Blueberry *Vaccinium Ashei* Var. 'Climax' Is Pollinator Density-Dependent." *Journal of Economic Entomology* 96 (4): 1215–1220. https://doi.org/10.1603/0022-0493-96.4.1215.

Dedej, S., K. Hartfelder, P. Aumeier, P. Rosenkranz, and W. Engels. 1998. "Caste Determination Is a Sequential Process: Effect of Larval Age at Grafting on Ovariole Number, Hind Leg Size and Cephalic Volatiles in the Honey Bee (*Apis mellifera carnica*)." *Journal of Apicultural Research* 37 (3): 183–190. https://doi.org/10.1080/00218839.1998.11100970.

Degrandi-Hoffman, G., A. Collins, J. H. Martin, J. O. Schmidt, and H. G. Spangler. 1998. "Nest Defense Behavior in Colonies from Crosses between Africanized and European Honey Bees (*Apis mellifera* L.) (Hymenoptera: Apidae)." *Journal of Insect Behavior* 11: 37–45. https://doi.org/10.1023/A:1020862432087.

Delaplane, K. S. 2007. *First Lessons in Beekeeping*. Hamilton, IL: Dadant & Sons.

Delaplane, K. S. 2015. "Management for Honey Production." In *The Hive and the Honey Bee, revised ed.*, edited by J. Graham. Hamilton, IL: Dadant & Sons.

Delaplane, K. S. 2021. *Crop Pollination by Bees: Evolution, Ecology, Conservation, and Management*. Oxfordshire: CABI.

Delaplane, K. S., K. Given, J. Menz, and D. A. Delaney. 2021. "Colony Fitness Increases in the Honey Bee at Queen Mating Frequencies Higher than Genetic Diversity Asymptote." *Behavioral Ecology and Sociobiology* 75 (9): 126. https://doi.org/10.1007/s00265-021-03065-6.

Delaplane, K. S., K. Hagan, K. Vogel, and L. Bartlett. 2024. "Mechanisms for Polyandry Evolution in a Complex Social Bee." *Behavioral Ecology and Sociobiology* 78(3). http://doi.org/10.1007/s00265-024-03450-x.

Delaplane, K. S., and J. R. Harbo. 1987a. "Drone Production by Young versus Old Worker Honeybees in Queenless Colonies." *Apidologie* 18 (2): 115–120. https://doi.org/10.1051/apido:19870201.

Delaplane, K. S., and Harbo, J. R. 1987b. "Effect of Queenlessness on Worker Survival, Honey Gain and Defence Behaviour in Honeybees." *Journal of Apicultural Research* 26 (1): 37–42. https://doi.org/10.1080/00218839.1987.11100732.

Delaplane, K. S., and W. M. Hood. 1999. "Economic Threshold for *Varroa jacobsoni* Oud. in the Southeastern USA." *Apidologie* 30 (5): 383–395. https://doi.org/10.1051/apido:19990504.

Delaplane, K. S., and P. P La Fage. 1989. "Foraging Tenacity of *Reticulitermes flavipes* and *Coptotermes formosanus* (Isoptera: Rhinotermitidae)." *Sociobiology* 16 (2): 183–189. https://pascal-francis.inist.fr/vibad/index.php?action=getRecordDetail&idt=6900111.

Delaplane, K. S., S. Pietravalle, M. Brown, and G. E. Budge. 2015. "Honey Bee Colonies Headed by Hyperpolyandrous Queens Have Improved Brood Rearing Efficiency and Lower Infestation Rates of Parasitic Varroa Mites." *PLOS ONE* 10 (12): e0142985. https://doi.org/10.1371/journal.pone.0142985.

Desai, S. D., and R. W. Currie. 2015. "Genetic Diversity within Honey Bee Colonies Affects Pathogen Load and Relative Virus Levels in Honey Bees, *Apis mellifera* L." *Behavioral Ecology and Sociobiology* 69 (9): 1527–1541. https://doi.org/10.1007/s00265-015-1965-2.

Dethlefsen, L., M. J. McFall-Ngai, and D. A. Relman. 2007. "An Ecological and Evolutionary Perspective on Human–Microbe Mutualism and Disease." *Nature* 449 (7164): 811–818. https://doi.org/10.1038/nature06245.

Dogantzis, K. A., B. A. Harpur, A. Rodrigues, L. Beani, A. L. Toth, and A. Zayed. 2018. "Insects with Similar Social Complexity Show Convergent Patterns of

Adaptive Molecular Evolution." *Scientific Reports* 8 (1): 10388. https://doi.org/10.1038/s41598-018-28489-5.

Dogantzis, K. A., T. Tiwari, I. M. Conflitti, A. Dey, H. M. Patch, E. Muli, L. Garnery, et al. 2021. "Thrice Out of Asia and the Adaptive Radiation of the Western Honey Bee." *Science Advances* 7 (49). https://doi.org/10.1126/sciadv.abj2151.

Dogantzis, K. A., and A. Zayed. 2019. "Recent Advances in Population and Quantitative Genomics of Honey Bees." *Current Opinion in Insect Science* 31 (February): 93–98. https://doi.org/10.1016/j.cois.2018.11.010.

Döke, M. A., C. M. McGrady, M. Otieno, C. M. Grozinger, and M. Frazier. 2019. "Colony Size, Rather Than Geographic Origin of Stocks, Predicts Overwintering Success in Honey Bees (Hymenoptera: *Apidae*) in the Northeastern United States." *Journal of Economic Entomology* 112: 525–533.

Donaldson-Matasci, M. C., G. DeGrandi-Hoffman, and A. Dornhaus. 2013. "Bigger Is Better: Honeybee Colonies as Distributed Information-Gathering Systems." *Animal Behaviour* 85 (3): 585–592. https://doi.org/10.1016/j.anbehav.2012.12.020.

Doolittle, W. F. 2020. "Evolution: Two Domains of Life or Three?" *Current Biology* 30 (4): R177–R179. https://doi.org/10.1016/j.cub.2020.01.010.

Dornhaus, A., and L. Chittka. 2004. "Why Do Honey Bees Dance?" *Behavioral Ecology and Sociobiology* 55 (4): 395–401. https://doi.org/10.1007/s00265-003-0726-9.

Dressler, R. L. 1982. "Biology of the Orchid Bees (Euglossini)." *Annual Review of Ecology and Systematics* 13 (1): 373–394. https://doi.org/10.1146/annurev.es.13.110182.002105.

Dunn, T. 2003. "When to Bee Social: Interactions among Environmental Constraints, Incentives, Guarding, and Relatedness in a Facultatively Social Carpenter Bee." *Behavioral Ecology* 14 (3): 417–424. https://doi.org/10.1093/beheco/14.3.417.

Dupraw, E. J. 1961. "A Unique Hatching Process in the Honeybee." *Transactions of the American Microscopical Society* 80 (2): 185. https://doi.org/10.2307/3223908.

Dyer, F. C. 2002. "The Biology of the Dance Language." *Annual Review of Entomology* 47 (1): 917–949. https://doi.org/10.1146/annurev.ento.47.091201.145306.

Dyer, F. C., and T. D. Seeley. 1987. "Interspecific Comparisons of Endothermy in Honey-Bees (*Apis*): Deviations From the Expected Size-Related Patterns." *Journal of Experimental Biology* 127 (1): 1–26. https://doi.org/10.1242/jeb.127.1.1.

Dyer, F. C., and T. D. Seeley. 1991. "Nesting Behavior and the Evolution of Worker Tempo in Four Honey Bee Species." *Ecology* 72 (1): 156–170. https://doi.org/10.2307/1938911.

Dyer, F. C., and T. D. Seeley. 1994. "Colony Migration in the Tropical Honey bee *Apis dorsata* F. (Hymenoptera: Apidae)." *Insectes Sociaux* 41 (2): 129–140. https://doi.org/10.1007/bf01240473.

Dzierzon, J. 1845. "Gutachten über die von Herrn Direktor Stöhr im ersten und zweiten Kapitel des General-Gutachtens aufgestellten Fragen." *Eichstädter Bienenzeitung,* 1, 119–121.

Eberhard, M. J. W. 1975. "The Evolution of Social Behavior by Kin Selection." *Quarterly Review of Biology* 50 (1): 1–33. https://doi.org/10.1086/408298.

Ebstein, R. P., S. Israel, S. H. Chew, S. Zhong, and A. Knafo. 2010. "Genetics of Human Social Behavior." *Neuron* 65 (6): 831–844. https://doi.org/10.1016/j.neuron.2010.02.020.

Eckholm, B. J., K. E. Anderson, M. Weiss, and G. DeGrandi-Hoffman. 2010. "Intracolonial Genetic Diversity in Honeybee (*Apis mellifera*) Colonies Increases Pollen Foraging Efficiency." *Behavioral Ecology and Sociobiology* 65 (5): 1037–1044. https://doi.org/10.1007/s00265-010-1108-8.

Eckholm, B. J., M.-B. Huang, K. E. Anderson, B. M. Mott, and G. DeGrandi-Hoffman. 2014. "Honey Bee (*Apis mellifera*) Intracolonial Genetic Diversity Influences Worker Nutritional Status." *Apidologie* 46 (2): 150–163. https://doi.org/10.1007/s13592-014-0311-4.

Egley, R. L., and M. D. Breed. 2012. "The Fanner Honey Bee: Behavioral Variability and Environmental Cues in Workers Performing a Specialized Task." *Journal of Insect Behavior* 26 (2): 238–45. https://doi.org/10.1007/s10905-012-9357-1.

Ehlers, J., and P. L. Gibbard. 2004. *Quaternary Glaciations-Extent and Chronology: Part I: Europe.* Amsterdam, Netherlands: Elsevier.

Ehlers, J., and P. L. Gibbard. 2007. "The Extent and Chronology of Cenozoic Global Glaciation." *Quaternary International* 164–165 (April): 6–20. https://doi.org/10.1016/j.quaint.2006.10.008.

El-Niweiri, M. a. A., and R. F. A. Moritz. 2011. "Mating in the Rain? Climatic Variance for Polyandry in the Honeybee (*Apis mellifera jemenitica*)." *Population Ecology* 53 (3): 421–427. https://doi.org/10.1007/s10144-011-0271-8.

Elliott, D. K., and J. D. Nations. 1998. "Bee Burrows in the Late Cretaceous (Late Cenomanian) Dakota Formation, Northeastern Arizona." *Ichnos* 5 (4): 243–253. https://doi.org/10.1080/10420949809386423.

Ellis, J., R. Hepburn, K. S. Delaplane, P. Neumann, and P. J. Elzen. 2003. "The Effects of Adult Small Hive Beetles, *Aethina tumida* (Coleoptera: Nitidulidae), on Nests and Flight Activity of Cape and European Honey Bees (*Apis mellifera*)." *Apidologie* 34 (4): 399–408. https://doi.org/10.1051/apido:2003038.

Ellis, J., P. Neumann, R. Hepburn, and P. J. Elzen. 2002. "Longevity and Reproductive Success of *Aethina tumida* (Coleoptera: Nitidulidae) Fed Different Natural

Diets." *Journal of Economic Entomology* 95 (5): 902–907. https://doi.org/10.1093/jee/95.5.902.

Elsik, C. G., K. C. Worley, A. Bennett, M. Beye, F. Cámara, C. P. Childers, D. C. De Graaf, et al. 2014. "Finding the Missing Honey Bee Genes: Lessons Learned From a Genome Upgrade." *BMC Genomics* 15 (1): 86. https://doi.org/10.1186/1471-2164-15-86.

Engel, M. S. 1999. "The Taxonomy of Recent and Fossil Honey Bees (Hymenoptera: Apidae; *Apis*)." *Journal of Hymenoptera Research* 8 (2): 165–196. http://biostor.org/reference/28973.

Engel, M. S. 2001. "A Monograph of the Baltic Amber Bees and Evolution of the Apoidea (Hymenoptera)." *Bulletin of the American Museum of Natural History* 259 (April): 1–192. https://doi.org/10.1206/0003-0090(2001)259.

Engel, M., I. Hinojosa- Díaz, and A. Rasnitsyn. 2009. "A honey bee from the Miocene of Nevada and the biogeography of *Apis* (Hymenoptera: Apidae: Apini)." *Proceedings of the California Academy of Sciences*. 60: 23–38.

Engels, E., and W. Engels. 1988. "Age-Dependent Queen Attractiveness for Drones and Mating in the Stingless Bee, *Scaptotrigona postica*." *Journal of Apicultural Research* 27 (1): 3–8. https://doi.org/10.1080/00218839.1988.11100773.

Engels, W., H. Kaatz, A. Zillikens, Z. P. Simões, A. Trube, R. Braun, and F. Dittrich. 1990. "Honey Bee Reproduction: Vitellogenin and Caste-Specific Regulation of Fertility." In *Advances in Invertebrate Reproduction*, edited by M. Hashi and O. Yamashita. Amsterdam, Netherlands: Elsevier.

Erwin, D. H., M. Laflamme, S. M. Tweedt, E. A. Sperling, D. Pisani, and K. J. Peterson. 2011. "The Cambrian Conundrum: Early Divergence and Later Ecological Success in the Early History of Animals." *Science* 334 (6059): 1091–1097. https://doi.org/10.1126/science.1206375.

Estoup, A., L. Garnery, M. Solignac, and J.-M. Cornuet. 1995. "Microsatellite Variation in Honey Bee (*Apis mellifera* L.) Populations: Hierarchical Genetic Structure and Test of the Infinite Allele and Stepwise Mutation Models." *Genetics* 140 (2): 679–695. https://doi.org/10.1093/genetics/140.2.679.

Estoup, A., M. Solignac, and J.-M. Cornuet. 1994. "Precise Assessment of the Number of Patrilines and of Genetic Relatedness in Honeybee Colonies." *Proceedings of the Royal Society B: Biological Sciences* 258 (1351): 1–7. https://doi.org/10.1098/rspb.1994.0133.

Evans, J. D., K. Aronstein, Y. P. Chen, C. Hétru, I. Jl, H. Jiang, M. R. Kanost, G. Thompson, Z. Zou, and D. Hultmark. 2006. "Immune Pathways and Defence Mechanisms in Honey Bees *Apis mellifera*." *Insect Molecular Biology* 15 (5): 645–656. https://doi.org/10.1111/j.1365-2583.2006.00682.x.

Evans, J., and J. S. Pettis. 2005. "Colony-Level Impacts of Immune Responsiveness in Honey Bees, *Apis mellifera*." *Evolution* 59 (10): 2270–2274. https://doi.org/10.1111/j.0014-3820.2005.tb00935.x.

Evans, J., J. S. Pettis, and H. Shimanuki. 2000. "Mitochondrial DNA Relationships in an Emergent Pest of Honey Bees: *Aethina tumida* (Coleoptera: Nitidulidae) from the United States and Africa." *Annals of the Entomological Society of America* 93 (3): 415–20. https://doi.org/10.1603/0013-8746(2000)093.

Farrar, C. 1937. "The Influence of Colony Populations on Honey Production." *Journal of Agricultural Research* 54 (12): 945–954. https://projects.sare.org/wp-content/uploads/1937-CL-Farrar-Article.pdf.

Fefferman, N. H., and P. T. Starks. 2006. "A Modeling Approach to Swarming in Honey Bees (*Apis mellifera*)." *Insectes Sociaux* 53 (1): 37–45. https://doi.org/10.1007/s00040-005-0833-x.

Feierman, J. R. 2016. "Religion's Possible Role in Facilitating Eusocial Human Societies. A Behavioral Biology (Ethological) Perspective." *Studia Humana* 5 (4): 5–33. https://doi.org/10.1515/sh-2016-0021.

Field, J. 1992. "Intraspecific Parasitism as an Alternative Reproductive Tactic in Nest-Building Wasps and Bees." *Biological Reviews* 67 (1): 79–126. https://doi.org/10.1111/j.1469-185x.1992.tb01659.x.

Fisher, R. A. 1930. *The Genetical Theory of Natural Selection*, New York: Dover.

Fisher, R. M., C. K. Cornwallis, and S. A. West. 2013. "Group Formation, Relatedness, and the Evolution of Multicellularity." *Current Biology* 23 (12): 1120–1125. https://doi.org/10.1016/j.cub.2013.05.004.

Fittkau, E. J., and H. Klinge. 1973. "On Biomass and Trophic Structure of the Central Amazonian Rain Forest Ecosystem." *Biotropica* 5 (1): 2. https://doi.org/10.2307/2989676.

Fletcher, D., and G. Tribe. 1977. "Swarming Potential of the African Bee, *Apis mellifera adansonii* L." In *African Bees: Taxonomy, Biology and Economic Use*, edited by D. Fletcher. Pretoria, South Africa: Apimondia.

Fletcher, D. 1978. "The African Bee, *Apis mellifera adansonii*, in Africa." *Annual Review of Entomology* 23 (1): 151–171. https://doi.org/10.1146/annurev.en.23.010178.001055.

Fluri, P., H. Wille, L. Gerig, and M. Lüscher. 1977. "Juvenile Hormone, Vitellogenin and Haemocyte Composition in Winter Worker Honeybees (*Apis mellifera*)." *Experientia* 33 (9): 1240–1241. https://doi.org/10.1007/bf01922354.

Forêt, S., K. W. Wanner, and R. Maleszka. 2007. "Chemosensory Proteins in the Honey Bee: Insights From the Annotated Genome, Comparative Analyses and Expressional Profiling." *Insect Biochemistry and Molecular Biology* 37 (1): 19–28. https://doi.org/10.1016/j.ibmb.2006.09.009.

Foster, K. R., and F. L. W. Ratnieks. 2005. "A New Eusocial Vertebrate?" *Trends in Ecology and Evolution* 20 (7): 363–364. https://doi.org/10.1016/j.tree.2005.05.005.

Fouks, B., P. Brand, H. N. Nguyen, J. Herman, F. Cámara, D. Ence, D. E. Hagen, et al. 2021. "The Genomic Basis of Evolutionary Differentiation among Honey Bees." *Genome Research* 31 (7): 1203–1215. https://doi.org/10.1101/gr.272310.120.

Franck, P., L. Garnery, G. Celebrano, M. Solignac, and J.-M. Cornuet. 2000. "Hybrid Origins of Honeybees From Italy (*Apis mellifera ligustica*) and Sicily (*A. m. sicula*)." *Molecular Ecology* 9 (7): 907–921. https://doi.org/10.1046/j.1365-294x.2000.00945.x.

Franck, P., L. Garnery, A. Loiseau, B. P. Oldroyd, H. R. Hepburn, M. Solignac, and J.-M. Cornuet. 2001. "Genetic Diversity of the Honeybee in Africa: Microsatellite and Mitochondrial Data." *Heredity* 86 (4): 420–430. https://doi.org/10.1046/j.1365-2540.2001.00842.x.

Franck, P., M. Solignac, D. Vautrin, J.-M. Cornuet, G. Koeniger, and N. Kœniger. 2002. "Sperm Competition and Last-Male Precedence in the Honeybee." *Animal Behaviour* 64 (3): 503–509. https://doi.org/10.1006/anbe.2002.3078.

Free, J. B., and P. A. Racey. 1968. "The Effect of the Size of Honeybee Colonies on Food Consumption, Brood Rearing and the Longevity of the Bees During Winter." *Entomologia Experimentalis Et Applicata* 11 (2): 241–249. https://doi.org/10.1111/j.1570-7458.1968.tb02048.x.

Free, J. B., and Y. Spencer-Booth. 1960. "Chill-Coma and Cold Death Temperatures of *Apis mellifera*." *Entomologia Experimentalis Et Applicata* 3 (3): 222–230. https://doi.org/10.1111/j.1570-7458.1960.tb00451.x.

Free, J. B., and I. H. Williams. 1974. "Factors Determining Food Storage and Brood Rearing in Honeybee (*Apis mellifera L.*) Comb." *Journal of Entomology* 49 (1): 47–63. https://doi.org/10.1111/j.1365-3032.1974.tb00067.x.

Free World Maps—Atlas of the World. n.d. *Europe*. Accessed October 4, 2023. http://www.freeworldmaps.net/europe/europe-map.jpg.

Fuchs, S., and R. F. A. Moritz. 1999. "Evolution of Extreme Polyandry in the Honeybee *Apis mellifera L.*" *Behavioral Ecology and Sociobiology* 45 (3–4): 269–275. https://doi.org/10.1007/s002650050561.

Gadagkar, R.. 1990. "Evolution of Eusociality: The Advantage of Assured Fitness Returns." *Philosophical Transactions of the Royal Society B* 329 (1252): 17–25. https://doi.org/10.1098/rstb.1990.0146.

Gandon, S, M. Van Baalen, and V. Jansen. 2002. "The Evolution of Parasite Virulence, Superinfection, and Host Resistance." *American Naturalist* 159 (6): 658–169. https://doi.org/10.1086/339993.

Gardner, A., J. Alpedrinha, and S. A. West. 2012. "Haplodiploidy and the Evolution of Eusociality: Split Sex Ratios." *American Naturalist* 179 (2): 240–256. https://doi.org/10.1086/663683.

Gardner, A., and A. Grafen. 2009. "Capturing the Superorganism: A Formal Theory of Group Adaptation." *Journal of Evolutionary Biology* 22 (4): 659–671. https://doi.org/10.1111/j.1420-9101.2008.01681.x.

Gardner, K. E., T. D. Seeley, and N. Calderone, 2006. *The Round and Waggle Dances of the Honey Bee: One Dance or Two?* Washington, DC: International Union for the Study of Social Insects.

Garnery, L., J.-M. Cornuet, and M. Solignac. 1992. "Evolutionary History of the Honey Bee *Apis mellifera* Inferred From Mitochondrial DNA Analysis." *Molecular Ecology* 1 (3): 145–154. https://doi.org/10.1111/j.1365-294x.1992.tb00170.x.

Garnery, L., D. Vautrin, J.-M. Cornuet, and M. Solignac. 1991. "Phylogenetic Relationships in the Genus *Apis* Inferred From Mitochondrial DNA Sequence Data." *Apidologie* 22 (1): 87–92. https://doi.org/10.1051/apido:19910111.

Gary, N. E. 1962. "Chemical Mating Attractants in the Queen Honey Bee." *Science* 136 (3518): 773–774. https://doi.org/10.1126/science.136.3518.773.

Gates, B. N. 1914. *The Temperature of the Bee Colony*. Washington, DC: US Deptartment of Agriculture.

Gempe, T., S. Stach, K. Bienefeld, and M. Beye. 2012. "Mixing of Honeybees With Different Genotypes Affects Individual Worker Behavior and Transcription of Genes in the Neuronal Substrate." *PLOS ONE* 7 (2): e31653. https://doi.org/10.1371/journal.pone.0031653.

Genise, J. F., J. C. Sciutto, J. H. Laza, M. G. González, and E. S. Bellosi. 2002. "Fossil Bee Nests, Coleopteran Pupal Chambers and Tuffaceous Paleosols From the Late Cretaceous Laguna Palacios Formation, Central Patagonia (Argentina)." *Palaeogeography, Palaeoclimatology, Palaeoecology* 177 (3–4): 215–235. https://doi.org/10.1016/s0031-0182(01)00333-9.

Gilbert, J. A., and C. Henry. 2015. "Predicting Ecosystem Emergent Properties at Multiple Scales." *Environmental Microbiology Reports* 7 (1): 20–22. https://doi.org/10.1111/1758-2229.12258.

Gillespie, J. H. 1977. "Natural Selection for Variances in Offspring Numbers: A New Evolutionary Principle." *American Naturalist* 111 (981): 1010–1014. https://doi.org/10.1086/283230.

Gilley, D. C. 1998. "The Identity of Nest-Site Scouts in Honey Bee Swarms." *Apidologie* 29 (3): 229–240. https://doi.org/10.1051/apido:19980303.

Gilley, D. C. 2003. "Absence of Nepotism in the Harassment of Duelling Queens by Honeybee Workers." *Proceedings of the Royal Society B: Biological Sciences* 270 (1528): 2045–2049. https://doi.org/10.1098/rspb.2003.2461.

Gilley, D. C., and D. R. Tarpy. 2005. "Three Mechanisms of Queen Elimination in Swarming Honey Bee Colonies." *Apidologie* 36 (3): 461–474. https://doi.org/10.1051/apido:2005033.

Gilliam, M., and S. Taber. 1991. “Diseases, Pests, and Normal Microflora of Honeybees, *Apis mellifera*, from Feral Colonies.” *Journal of Invertebrate Pathology* 58 (2): 286–289. https://doi.org/10.1016/0022-2011(91)90077-4.

Girard, M. B., H. R. Mattila, and T. D. Seeley. 2010. “Recruitment-Dance Signals Draw Larger Audiences When Honey Bee Colonies Have Multiple Patrilines.” *Insectes Sociaux* 58 (1): 77–86. https://doi.org/10.1007/s00040-010-0118-x.

Giray, T., and G. E. Robinson. 1996. “Common Endocrine and Genetic Mechanisms of Behavioral Development in Male and Worker Honey Bees and the Evolution of Division of Labor.” *Proceedings of the National Academy of Sciences of the United States of America* 93 (21): 11718–11722. https://doi.org/10.1073/pnas.93.21.11718.

Glen, W. 1990. “What Killed the Dinosaurs?” *American Scientist* 78 (4): 354–370. https://www.jstor.org/stable/29774121.

Godfray, H. C. J., and A. Grafen. 1988. “Unmatedness and the Evolution of Eusociality.” *American Naturalist* 131 (2): 303–305. https://doi.org/10.1086/284791.

Goetze, G., and B. Bessling. 1959. Die Wirkung verschiedener Fütterung der Honigbiene auf Wachserzeugung und Bautätigkeit. *Z Bienenforsch* 4: 202–209.

Goren-Inbar, N., N. Alperson, M. E. Kislev, O. Simchoni, Y. Melamed, A. Ben-Nun, and E. Werker. 2004. “Evidence of Hominin Control of Fire at Gesher Benot Ya’Aqov, Israel.” *Science* 304 (5671): 725–727. https://doi.org/10.1126/science.1095443.

Gould, J. L., J. L. Kirschvink, and K. S. Deffeyes. 1978. “Bees Have Magnetic Remanence.” *Science* 201 (4360): 1026–1028. https://doi.org/10.1126/science.201.4360.1026.

Graham, S., M. R. Myerscough, J. C. Jones, and B. P. Oldroyd. 2006. “Modelling the Role of Intracolonial Genetic Diversity on Regulation of Brood Temperature in Honey Bee (*Apis mellifera* L.) Colonies.” *Insectes Sociaux* 53 (2): 226–232. https://doi.org/10.1007/s00040-005-0862-5.

Grassé, P.-P. 1959. “La reconstruction du nid et les coordinations interindividuelles chez *Bellicositermes natalensis* et *Cubitermes* sp. la théorie de la stigmergie: Essai d’interprétation du comportement des termites constructeurs.” *Insectes Sociaux* 6: 41–80.

Grimaldi, D. and M. S. Engel. 2005. *Evolution of the Insects,* New York: Cambridge University Press.

Grimaldi, D. A., and B. A. Underwood. 1986. “*Megabraula*, a New Genus for Two New Species of Braulidae (Diptera), and a Discussion of Braulid Evolution.” *Systematic Entomology* 11 (4): 427–438. https://doi.org/10.1111/j.1365-3113.1986.tb00534.x.

Grosberg, R. K., and R. R. Strathmann. 2007. "The Evolution of Multicellularity: A Minor Major Transition?" *Annual Review of Ecology, Evolution, and Systematics* 38 (1): 621–654. https://doi.org/10.1146/annurev.ecolsys.36.102403.114735.

Grozinger, C. M., Y. Fan, S. E Hoover, and M. L. Winston. 2007. "Genome-wide Analysis Reveals Differences in Brain Gene Expression Patterns Associated with Caste and Reproductive Status in Honey Bees (*Apis mellifera*)." *Molecular Ecology* 16 (22): 4837–4848. https://doi.org/10.1111/j.1365-294x.2007.03545.x.

Grozinger, C. M., J. Richards, and H. R. Mattila. 2013. "From Molecules to Societies: Mechanisms Regulating Swarming Behavior in Honey Bees (*Apis* Spp.)." *Apidologie* 45 (3): 327–346. https://doi.org/10.1007/s13592-013-0253-2.

Grüter, C., E. Jongepier, and S. Foitzik. 2018. "Insect Societies Fight Back: The Evolution of Defensive Traits Against Social Parasites." *Philosophical Transactions of the Royal Society B* 373 (1751): 20170200. https://doi.org/10.1098/rstb.2017.0200.

Guidugli, K. R., A. M. Do Nascimento, G. V. Amdam, A. R. Barchuk, S. W. Omholt, Z. L. P. Simões, and K. Hartfelder. 2005. "Vitellogenin Regulates Hormonal Dynamics in the Worker Caste of a Eusocial Insect." *FEBS Letters* 579 (22): 4961–4965. https://doi.org/10.1016/j.febslet.2005.07.085.

Guzmán-Novoa, E., B. Emsen, P. Unger, L. G. Espinosa-Montaño, and T. Petukhova. 2012. "Genotypic Variability and Relationships Between Mite Infestation Levels, Mite Damage, Grooming Intensity, and Removal of Varroa Destructor Mites in Selected Strains of Worker Honey Bees (*Apis mellifera* L.)." *Journal of Invertebrate Pathology* 110 (3): 314–320. https://doi.org/10.1016/j.jip.2012.03.020.

Guzmán-Novoa, E., G. J. Hunt, J. L. Uribe-Rubio, and D. Prieto-Merlos. 2004. "Genotypic Effects of Honey Bee (*Apis mellifera*) Defensive Behavior at the Individual and Colony Levels: The Relationship of Guarding, Pursuing and Stinging." *Apidologie* 35 (1): 15–24. https://doi.org/10.1051/apido:2003061.

Guzmán-Novoa, E., and R. E. Page. 1999. "Selective Breeding of Honey Bees (Hymenoptera: Apidae) in Africanized Areas." *Journal of Economic Entomology* 92 (3): 521–525. https://doi.org/10.1093/jee/92.3.521.

Gwynne, D. T. 1984. "Male Mating Effort, Confidence of Paternity, and Insect Sperm Competition." In *Sperm Competition and the Evolution of Animal Mating Systems*, edited by R. Smith. New York: Academic Press.

Haberl, M., and D. Tautz. 1998. "Sperm Usage in Honey Bees." *Behavioral Ecology and Sociobiology* 42 (4): 247–255. https://doi.org/10.1007/s002650050436.

Haidt, J. 2012. *The Righteous Mind: Why Good People Are Divided by Politics and Religion*. New York: Vintage.

Haldane, J. B. S. 1932. *The Causes of Evolution*. London: Longmans, Green & Co.

Hales, T. C. 2001. "The Honeycomb Conjecture." *Discrete and Computational Geometry* 25 (1): 1–22. https://doi.org/10.1007/s004540010071.

Hamilton, W. D. 1964. "The Genetical Evolution of Social Behaviour. I." *Journal of Theoretical Biology* 7 (1): 1–16. https://doi.org/10.1016/0022-5193(64)90038-4.

Hamilton, W. D. 1964. "The Genetical Evolution of Social Behaviour. II." *Journal of Theoretical Biology* 7 (1): 17–52. https://doi.org/10.1016/0022-5193(64)90039-6.

Hamilton, W. D. 1972. "Altruism and Related Phenomena, Mainly in Social Insects." *Annual Review of Ecology and Systematics* 3 (1): 193–232. https://doi.org/10.1146/annurev.es.03.110172.001205.

Hamilton, W. D. 1976. *Gamblers Since Life Began: Barnacles, Aphids, Elms.* New York: Stony Brook Foundation, Inc.

Han, F., A. Wållberg, and M. T. Webster. 2012. "From Where Did the Western Honeybee (*Apis mellifera*) Originate?" *Ecology and Evolution* 2 (8): 1949–1957. https://doi.org/10.1002/ece3.312

Harari, Y. 2015. *Sapiens: A Brief History of Humankind.* New York: HarperCollins.

Harari, Y. 2017. *Homo Deus: A Brief History of Tomorrow.* New York: HarperCollins.

Harbo, J. R. 1979. "The Rate of Depletion of Spermatozoa in the Queen Honeybee Spermatheca." *Journal of Apicultural Research* 18 (3): 204–207. https://doi.org/10.1080/00218839.1979.11099969.

Harbo, J. R. 1986. "Oviposition Rates of Instrumentally Inseminated and Naturally Mated Queen Honey Bees (Hymenoptera: Apidae)." *Annals of the Entomological Society of America* 79 (1): 112–115. https://doi.org/10.1093/aesa/79.1.112.

Harbo, J. R. 1990. "Artificial Mixing of Spermatozoa From Honeybees and Evidence for Sperm Competition." *Journal of Apicultural Research* 29 (3): 151–158. https://doi.org/10.1080/00218839.1990.11101212.

Harbo, J. R., and J. W. Harris. 2009. "Responses to *Varroa* by Honey Bees With Different Levels of *Varroa* Sensitive Hygiene." *Journal of Apicultural Research* 48 (3): 156–161. https://doi.org/10.3896/ibra.1.48.3.02.

Hardisty, B. E., and D. L. Cassill. 2009. "Extending Eusociality to Include Vertebrate Family Units." *Biology and Philosophy* 25 (3): 437–440. https://doi.org/10.1007/s10539-009-9176-8.

Harpur, B. A., C. F. Kent, D. Molodtsova, J. M. D. Lebon, A. S. Alqarni, A. A. Owayss, and A. Zayed. 2014. "Population Genomics of the Honey Bee Reveals Strong Signatures of Positive Selection on Worker Traits." *Proceedings of the National Academy of Sciences of the United States of America* 111 (7): 2614–2619. https://doi.org/10.1073/pnas.1315506111.

Harris, J. W., R. G. Danka, and J. D. Villa. 2010. "Honey Bees (Hymenoptera: Apidae) with the Trait of *Varroa* Sensitive Hygiene Remove Brood With All

Reproductive Stages of *Varroa* Mites (Mesostigmata: Varroidae)." *Annals of the Entomological Society of America* 103 (2): 146–152. https://doi.org/10.1603/an09138.

Harris, J. W., and J. Woodring. 1999. "Effects of Dietary Precursors to Biogenic Amines on the Behavioural Response From Groups of Caged Worker Honey Bees (*Apis mellifera*) to the Alarm Pheromone Component Isopentyl Acetate." *Physiological Entomology* 24 (3): 285–291. https://doi.org/10.1046/j.1365-3032.1999.00141.x.

Härtel, S., P. Neumann, F. S. Raassen, R. F. A. Moritz, and H. R. Hepburn. 2006. "Social Parasitism by Cape Honeybee Workers in Colonies of Their Own Subspecies (*Apis mellifera capensis* Esch.)." *Insectes Sociaux* 53 (2): 183–193. https://doi.org/10.1007/s00040-005-0857-2.

Hartfelder, K. 2000. "Insect Juvenile Hormone: From 'Status Quo' to High Society." *Brazilian Journal of Medical and Biological Research* 33 (2): 157–177. https://doi.org/10.1590/s0100-879x2000000200003.

Hartfelder, K., and H. Rembold. 1991. "Caste-Specific Modulation of Juvenile Hormone III Content and Ecdysteroid Titer in Postembryonic Development of the Stingless Bee, *Scaptotrigona postica depilis*." *Journal of Comparative Physiology B-Biochemical Systemic and Environmental Physiology* 160 (6): 617–620. https://doi.org/10.1007/bf00571258.

Hasiotis, S. T. 2003. "Complex Ichnofossils of Solitary and Social Soil Organisms: Understanding Their Evolution and Roles in Terrestrial Paleoecosystems." *Palaeogeography, Palaeoclimatology, Palaeoecology* 192 (1–4): 259–320. https://doi.org/10.1016/s0031-0182(02)00689-2.

Haydak, M. H. 1958. "Do the Nurse Honey Bees Recognize the Sex of the Larvae?" *Science* 127 (3306): 1113. https://doi.org/10.1126/science.127.3306.1113.

Heidinger, I. M. M., M. D. Meixner, S. Berg, and R. Büchler. 2014. "Observation of the Mating Behavior of Honey Bee (*Apis mellifera* L.) Queens Using Radio-Frequency Identification (RFID): Factors Influencing the Duration and Frequency of Nuptial Flights." *Insects* 5 (3): 513–27. https://doi.org/10.3390/insects5030513.

Heinrich, B. 1993. *The Hot-Blooded Insects: Strategies and Mechanisms of Thermoregulation.* Cambridge, MA: Harvard University Press.

Helanterä, H. 2022. "Supercolonies of Ants (Hymenoptera: Formicidae): Ecological Patterns, Behavioural Processes and Their Implications for Social Evolution." *Myrmecological News* 32: 1–22. https://doi.org/10.25849/myrmecol.news_032:001.

Hellmich, R. L., J. M. Kulinčević, and W. C. Rothenbuhler. 1985. "Selection for High and Low Pollenhoarding Honey Bees." *Journal of Heredity* 76 (3): 155–158. https://doi.org/10.1093/oxfordjournals.jhered.a110056.

Henshilwood, C. S., F. D'Errico, R. Yates, Z. Jacobs, C. Tribolo, G.A.T. Duller, N. Mercier, et al. 2002. "Emergence of Modern Human Behavior: Middle Stone Age Engravings From South Africa." *Science* 295 (5558): 1278–1280. https://doi.org/10.1126/science.1067575.

Hepburn, H. 2006. "Absconding, Migration and Swarming in Honeybees: An Ecological and Evolutionary Perspective." In *Life Cycles in Social Insects: Behaviour, Ecology and Evolution*, edited by V. Kipyatkov. St. Petersburg, Russia: St. Petersburg University Press.

Hepburn, H. R., and R. M. Crewe. 1991. "Portrait of the Cape Honeybee, *Apis mellifera capensis*." *Apidologie* 22 (6): 567–580. https://doi.org/10.1051/apido:19910601.

Hepburn, H., C. Pirk, and O. Duangphakdee. 2014. *Honeybee Nests: Composition, Structure, Function,* Berlin, Germany: Springer.

Hepburn, H. R., and S. E. Radloff. 2013. *Honeybees of Africa*. Berlin, Germany: Springer.

Hepburn, H. R., D. R. Smith, S. E. Radloff, and G. W. Otis. 2001. "Infraspecific Categories of *Apis cerana*: Morphometric, Allozymal and mtDNA Diversity." *Apidologie* 32 (1): 3–23. https://doi.org/10.1051/apido:2001108.

Herbeck, Y. E., R. Gulevich, D. V. Shepeleva, and V. Grinevich. 2017. "Oxytocin: Coevolution of Human and Domesticated Animals." *Russian Journal of Genetics: Applied Research* 7 (3): 235–242. https://doi.org/10.1134/s2079059717030042.

Hermann, H. R. 1971. "Sting Autotomy, a Defensive Mechanism in Certain Social Hymenoptera." *Insectes Sociaux* 18 (2): 111–120. https://doi.org/10.1007/bf02223116.

Hewitt, G. M. 1999. "Post-Glacial Re-colonization of European Biota." *Biological Journal of the Linnean Society* 68 (1–2): 87–112. https://doi.org/10.1111/j.1095-8312.1999.tb01160.x.

Hillesheim, E., N. Kœniger, and R. F. A. Moritz. 1989. "Colony Performance in Honeybees (*Apis mellifera capensis* Esch.) Depends on the Proportion of Subordinate and Dominant Workers." *Behavioral Ecology and Sociobiology* 24 (5): 291–296. https://doi.org/10.1007/bf00290905.

Hoffmann, J. A. 2003. "The Immune Response of Drosophila." *Nature* 426 (6962): 33–38. https://doi.org/10.1038/nature02021.

Hogendoorn, K., and H. H. W. Velthuis. 1995. "The Role of Young Guards in *Xylocopa pubescens*." *Insectes Sociaux* 42 (4): 427–48. https://doi.org/10.1007/bf01242171.

Hogendoorn, K., and H. H. W. Velthuis. 1993. "The Sociality of *Xylocopa pubescens*: Does a Helper Really Help?" *Behavioral Ecology and Sociobiology* 32 (4). https://doi.org/10.1007/bf00166514.

Holland, J. H. 2014. *Complexity: A Very Short Introduction*. Oxford, UK: Oxford University Press.

Holmes, M. J., B. P. Oldroyd, M. H. Allsopp, J. Lim, T. C. Wossler, and M. Beekman. 2010. "Maternity of Emergency Queens in the Cape Honey Bee, *Apis mellifera capensis*." *Molecular Ecology* 19 (13): 2792–2799. https://doi.org/10.1111/j.1365-294x.2010.04683.x.

Hood, W. M. 2004. "The Small Hive Beetle, *Aethina tumida*: A Review." *Bee World* 85 (3): 51–59. https://doi.org/10.1080/0005772x.2004.11099624.

Hoover, S. E, B. P. Oldroyd, T. C. Wossler, and M. L. Winston. 2005. "Anarchistic Queen Honey Bees Have Normal Queen Mandibular Pheromones." *Insectes Sociaux* 52 (1): 6–10. https://doi.org/10.1007/s00040-004-0772-y.

Hoover, S. E., C. I. Keeling, M. L. Winston, and K. N. Slessor. 2003. "The Effect of Queen Pheromones on Worker Honey Bee Ovary Development." *Science of Nature* 90 (10): 477–480. https://doi.org/10.1007/s00114-003-0462-z.

Huang, Z., and G. E. Robinson. 1992. "Honeybee Colony Integration: Worker-Worker Interactions Mediate Hormonally Regulated Plasticity in Division of Labor." *Proceedings of the National Academy of Sciences of the United States of America* 89 (24): 11726–11729. https://doi.org/10.1073/pnas.89.24.11726.

Huang, Z., and G. E. Robinson. 1996. "Regulation of Honey Bee Division of Labor by Colony Age Demography." *Behavioral Ecology and Sociobiology* 39 (3): 147–158. https://doi.org/10.1007/s002650050276.

Huang, Z., G. E. Robinson, and D. W. Borst. 1994. "Physiological Correlates of Division of Labor Among Similarly Aged Honey Bees." *Journal of Comparative Physiology A-Neuroethology Sensory Neural and Behavioral Physiology* 174 (6): 731–739. https://doi.org/10.1007/bf00192722.

Hughes, D., N. E. Pierce, and J. J. Boomsma. 2008a. "Social Insect Symbionts: Evolution in Homeostatic Fortresses." *Trends in Ecology and Evolution* 23 (12): 672–677. https://doi.org/10.1016/j.tree.2008.07.011.

Hughes, W. O. H., B. P. Oldroyd, M. Beekman, and F. L. W. Ratnieks. 2008b. "Ancestral Monogamy Shows Kin Selection Is Key to the Evolution of Eusociality." *Science* 320 (5880): 1213–1216. https://doi.org/10.1126/science.1156108.

Hughes, W. O. H., F. L. W. Ratnieks, and B. P. Oldroyd. 2008c. "Multiple Paternity or Multiple Queens: Two Routes to Greater Intracolonial Genetic Diversity in the Eusocial Hymenoptera." *Journal of Evolutionary Biology* 21 (4): 1090–1095. https://doi.org/10.1111/j.1420-9101.2008.01532.x.

Hull, D. L. 1980. "Individuality and Selection." *Annual Review of Ecology and Systematics* 11 (1): 311–332. https://doi.org/10.1146/annurev.es.11.110180.001523.

Hunt, G. J., E. Guzmán-Novoa, M. K. Fondrk, and R. E. Page. 1998. "Quantitative Trait Loci for Honey Bee Stinging Behavior and Body Size." *Genetics* 148 (3): 1203–1213. https://doi.org/10.1093/genetics/148.3.1203.

Huxley, J. S. 1912. *The Individual in the Animal Kingdom*. Cambridge, UK: Cambridge University Press.

Ichikawa, M. 1981. "Ecological and sociological importance of honey to the Mbuti nest hunters, Eastern Zaire." *African Study Monographs* 1: 55–68.

Inoue, T., S. F. Sakagami, S. Salmah, and S. Yamane. 1984. "The Process of Colony Multiplication in the Sumatran Stingless Bee *Trigona* (*Tetragonula*) Laeviceps." *Biotropica* 16 (2): 100. https://doi.org/10.2307/2387841.

Jaffé, R., V. Dietemann, M. H. Allsopp, C. Costa, R. M. Crewe, R. Dall'Olio, P. De La Rúa, et al. 2010. "Estimating the Density of Honeybee Colonies Across Their Natural Range to Fill the Gap in Pollinator Decline Censuses." *Conservation Biology* 24 (2): 583–593. https://doi.org/10.1111/j.1523-1739.2009.01331.x.

Jaffé, R., V. Dietemann, R. M. Crewe, and R. F. A. Moritz. 2009. "Temporal Variation in the Genetic Structure of a Drone Congregation Area: An Insight into the Population Dynamics of Wild African Honeybees (*Apis mellifera scutellata*)." *Molecular Ecology* 18 (7): 1511–1522. https://doi.org/10.1111/j.1365-294x.2009.04143.x.

Jaffé, R., and R. F. A. Moritz. 2009. "Mating Flights Select for Symmetry in Honeybee Drones (*Apis mellifera*)." *Science of Nature* 97 (3): 337–343. https://doi.org/10.1007/s00114-009-0638-2.

Janson, S., M. Middendorf, and M. Beekman. 2005. "Honeybee Swarms: How Do Scouts Guide a Swarm of Uninformed Bees?" *Animal Behaviour* 70 (2): 349–358. https://doi.org/10.1016/j.anbehav.2004.10.018.

Jeanne, R. L. 2013. "The Evolution of the Organization of Work in Social Insects." *Monitore Zoologico Italiano-Italian Journal of Zoology* 20 (2): 119–133. https://www.tandfonline.com/doi/abs/10.1080/00269786.1986.10736494.

Jeanson, R., and J. H. Fewell. 2008. "Influence of the Social Context on Division of Labor in Ant Foundress Associations." *Behavioral Ecology* 19 (3): 567–574. https://doi.org/10.1093/beheco/arn018.

Ji, Y. 2021. "The Geographical Origin, Refugia, and Diversification of Honey Bees (*Apis* Spp.) Based on Biogeography and Niche Modeling." *Apidologie* 52 (2): 367–377. https://doi.org/10.1007/s13592-020-00826-6.

Johannsmeier, M. 1983. "Experiences with the Cape bee in the Transvaal." *South African Bee Journal* 55: 130–138.

Johansson, T., and Johansson, M. 1967. "Lorenzo L. Langstroth and the Bee Space." *Bee World* 48: 133–143.

Johnson, B. R. 2003. "Organization of Work in the Honeybee: A Compromise Between Division of Labour and Behavioural Flexibility." *Proceedings of the Royal Society B: Biological Sciences* 270 (1511): 147–152. https://doi.org/10.1098/rspb.2002.2207.

Johnson, B. R. 2008. "Within-Nest Temporal Polyethism in the Honey Bee." *Behavioral Ecology and Sociobiology* 62 (5): 777–784. https://doi.org/10.1007/s00265-007-0503-2.

Johnson, B. R. 2009. "Pattern Formation on the Combs of Honeybees: Increasing Fitness by Coupling Self-organization With Templates." *Proceedings of the Royal Society B: Biological Sciences* 276 (1655): 255–261. https://doi.org/10.1098/rspb.2008.0793.

Johnson, B. R., and J. C. Nieh. 2010. "Modeling the Adaptive Role of Negative Signaling in Honey Bee Intraspecific Competition." *Journal of Insect Behavior* 23 (6): 459–471. https://doi.org/10.1007/s10905-010-9229-5.

Johnson, B. R., and N. D. Tsutsui. 2011. "Taxonomically Restricted Genes Are Associated With the Evolution of Sociality in the Honey Bee." *BMC Genomics* 12 (1). https://doi.org/10.1186/1471-2164-12-164.

Jones, J. C., M. R. Myerscough, S. Graham, and B. P. Oldroyd. 2004. "Honey Bee Nest Thermoregulation: Diversity Promotes Stability." *Science* 305 (5682): 402–404. https://doi.org/10.1126/science.1096340.

Jones, J. C., P. Nanork, and B. P. Oldroyd. 2006. "The Role of Genetic Diversity in Nest Cooling in a Wild Honey Bee, *Apis florea*." *Journal of Comparative Physiology A-Neuroethology Sensory Neural and Behavioral Physiology* 193 (2): 159–165. https://doi.org/10.1007/s00359-006-0176-8.

Kaiser, D. 2001. "Building a Multicellular Organism." *Annual Review of Genetics* 35 (1): 103–123. https://doi.org/10.1146/annurev.genet.35.102401.090145.

Kaiser, W. 1988. "Busy Bees Need Rest, Too." *Journal of Comparative Physiology A-Neuroethology Sensory Neural and Behavioral Physiology* 163 (5): 565–584. https://doi.org/10.1007/bf00603841.

Kajobe, R., and D. W. Roubik. 2006. "Honey-Making Bee Colony Abundance and Predation by Apes and Humans in a Uganda Forest Reserve." *Biotropica* 38 (2): 210–218. https://doi.org/10.1111/j.1744-7429.2006.00126.x.

Kawamura, M. 2011. "Royalactin Induces Queen Differentiation in Honeybees." *Nature* 473 (7348): 478–483. https://doi.org/10.1038/nature10093.

Kamakura, M. 2016. "Kamakura Replies." *Nature* 537 (7621): E13. https://doi.org/10.1038/nature19350.

Kapheim, K. M., H. Pan, L. Cai, S. L. Salzberg, D. Puiu, T. Magoč, H. M. Robertson, et al. 2015. "Genomic Signatures of Evolutionary Transitions From Solitary to

Group Living." *Science* 348 (6239): 1139–1143. https://doi.org/10.1126/science.aaa4788.

Karihaloo, B. L., K. Zhang, and J. Wang. 2013. "Honeybee Combs: How the Circular Cells Transform Into Rounded Hexagons." *Journal of the Royal Society Interface* 10 (86): 20130299. https://doi.org/10.1098/rsif.2013.0299.

Kaspar, R. E., C. N. Cook, and M. D. Breed. 2018. "Experienced Individuals Influence the Thermoregulatory Fanning Behaviour in Honey Bee Colonies." *Animal Behaviour* 142 (August): 69–76. https://doi.org/10.1016/j.anbehav.2018.06.004.

Kaspari, M., and E. L. Vargo. 1995. "Colony Size as a Buffer against Seasonality: Bergmann's Rule in Social Insects." *American Naturalist,* 145 (4): 610–632. https://www.jstor.org/stable/2462971.

Kastberger, G., E. Schmelzer, and I. Kranner. 2008. "Social Waves in Giant Honeybees Repel Hornets." *PLOS ONE* 3 (9): e3141. https://doi.org/10.1371/journal.pone.0003141.

Katsnelson, M. I., Y. I. Wolf, and E. V. Koonin. 2019. "On the Feasibility of Saltational Evolution." *Proceedings of the National Academy of Sciences of the United States of America* 116 (42): 21068–21075. https://doi.org/10.1073/pnas.1909031116.

Kawakita, A., J. S. Ascher, T. Sota, M. Katô, and D. W. Roubik. 2008. "Phylogenetic Analysis of the Corbiculate Bee Tribes Based on 12 Nuclear Protein-Coding Genes (Hymenoptera: Apoidea: Apidae)." *Apidologie* 39 (1): 163–175. https://doi.org/10.1051/apido:2007046.

Kay, T., L. Keller, and L. Lehmann. 2020. "The Evolution of Altruism and the Serial Rediscovery of the Role of Relatedness." *Proceedings of the National Academy of Sciences of the United States of America* 117 (46): 28894–28898. https://doi.org/10.1073/pnas.2013596117.

Keller, I., P. Fluri, and A. Imdorf. 2005. "Pollen Nutrition and Colony Development in Honey Bees: Part 1." *Bee World* 86 (1): 3–10. https://doi.org/10.1080/0005772x.2005.11099641.

Keller, L.. 1997. "Indiscriminate Altruism: Unduly Nice Parents and Siblings." *Trends in Ecology and Evolution* 12 (3): 99–103. https://doi.org/10.1016/s0169-5347(96)10065-3.

Keller, L., and M. Genoud. 1997. "Extraordinary Lifespans in Ants: A Test of Evolutionary Theories of Ageing." *Nature* 389 (6654): 958–960. https://doi.org/10.1038/40130.

Kerr, W., and V. Portugal-Araújo. 1958. "Racas de abelhas de Africa." *Garcia De Orta Serie de Zoologia* 6: 53–59.

Kerr, W. E., S. D. L. Del Rio, and B. Md. 1982. The southern limits of the distribution of the Africanized honey bee in South America. *American Bee Journal* 122: 196–198.

Khoury, D. S., M. R. Myerscough, and A. B. Barron. 2011. "A Quantitative Model of Honey Bee Colony Population Dynamics." *PLOS ONE* 6 (4): e18491. https://doi.org/10.1371/journal.pone.0018491.

Kirchner, W. H., and C. Dreller. 1993. "Acoustical Signals in the Dance Language of the Giant Honeybee, *Apis dorsata*." *Behavioral Ecology and Sociobiology* 33 (2): 67–72. https://doi.org/10.1007/bf00171657.

Kirkwood, T. B. L., and M. R. Rose. 1991. "Evolution of Senescence: Late Survival Sacrificed for Reproduction." *Philosophical Transactions of the Royal Society B* 332 (1262): 15–24. https://doi.org/10.1098/rstb.1991.0028.

Kistner, D. H. 1982. "The social insects' bestiary." *Social Insects* 3: 1–244.

Knoll, S., W. Pinna, A. Varcasia, A. Scala, and M. G. Cappai. 2020. "The Honey Bee (*Apis mellifera* L., 1758) and the Seasonal Adaptation of Productions. Highlights on Summer to Winter Transition and Back to Summer Metabolic Activity. A Review." *Livestock Science* 235 (May): 104011. https://doi.org/10.1016/j.livsci.2020.104011.

Kocher, S. D., and R. J. Paxton. 2014. "Comparative Methods Offer Powerful Insights into Social Evolution in Bees." *Apidologie* 45 (3): 289–305. https://doi.org/10.1007/s13592-014-0268-3.

Kocher, S. D., D. R. Tarpy, and C. M. Grozinger. 2010. "The Effects of Mating and Instrumental Insemination on Queen Honey Bee Flight Behaviour and Gene Expression." *Insect Molecular Biology* 19 (2): 153–162. https://doi.org/10.1111/j.1365-2583.2009.00965.

Kochin, B. F., J. J. Bull, and R. Antia. 2010. "Parasite Evolution and Life History Theory." *PLOS Biology* 8 (10): e1000524. https://doi.org/10.1371/journal.pbio.1000524.

Kodai, T., K. Umebayashi, T. Nakatani, K. Ishiyama, and N. Noda. 2007. "Compositions of Royal Jelly II. Organic Acid Glycosides and Sterols of the Royal Jelly of Honeybees (*Apis mellifera*)." *Chemical & Pharmaceutical Bulletin* 55 (10): 1528–1531. https://doi.org/10.1248/cpb.55.1528.

Koenig, J. P., G. M. Boush, and E. H. Erickson. 1986. "Effect of Type of Brood Comb on Chalk Brood Disease in Honeybee Colonies." *Journal of Apicultural Research* 25 (1): 58–62. https://doi.org/10.1080/00218839.1986.11100694.

Koeniger, G. 1988. "Mating behavior of honey bees." In *Africanized Honey Bees and Bee Mites*, edited by G. R. Needham, R. E. Page, M. Delfinado-Baker, and C. E. Bowman. New York: Halstead Press.

Koeniger, G. 1990. “The Role of the Mating Sign in Honey Bees, *Apis mellifera* L.: Does It Hinder or Promote Multiple Mating?” *Animal Behaviour* 39 (3): 444–449. https://doi.org/10.1016/s0003-3472(05)80407-5.

Koeniger, G., N. Kœniger, M. Mardan, G. W. Otis, and S. Wongsiri. 1991. “Comparative Anatomy of Male Genital Organs in the Genus *Apis*.” *Apidologie* 22 (5): 539–552. https://doi.org/10.1051/apido:19910507.

Koeniger, G., N. Koeniger, M. Mardan, R. W. K. Punchihewa, and G. W. Otis. 1990. “Numbers of Spermatozoa in Queens and Drones Indicate Multiple Mating of Queens in *Apis andreniformis* and *Apis dorsata*.” *Apidologie* 21 (4): 281–286. https://doi.org/10.1051/apido:19900402.

Koeniger, G., N. Kœniger, H. Pechhacker, F. Ruttner, and S. Berg. 1989a. “Assortative Mating in a Mixed Population of European Honeybees, *Apis mellifera ligustica* and *Apis mellifera carnica*.” *Insectes Sociaux* 36 (2): 129–138. https://doi.org/10.1007/bf02225908.

Kœniger, N., G. Koeniger, and S. Wongsiri. 1989b. “Mating and Sperm Transfer in *Apis florea*.” *Apidologie* 20 (5): 413–418. https://doi.org/10.1051/apido:19890506.

Koeniger, G., N. Kœniger, and S. Tingek. 1994. “Mating Flights, Number of Spermatozoa, Sperm Transfer and Degree of Polyandry in *Apis koschevnikovi* (Buttel-Reepen, 1906).” *Apidologie* 25 (2): 224–238. https://doi.org/10.1051/apido:19940209.

Koeniger, G., N. Kœniger, S. Tingek, and A. Kelitu. 2000. “Mating Flights and Sperm Transfer in the Dwarf Honeybee *Apis andreniformis* (Smith, 1858).” *Apidologie* 31 (2): 301–311. https://doi.org/10.1051/apido:2000124..

Koeniger, N. 1970. “Factors Determining the Laying of Drone and Worker Eggs by the Queen Honeybee.” *Bee World* 51: 166–169.

Kœniger, N., and G. Koeniger. 1991. “An Evolutionary Approach to Mating Behaviour and Drone Copulatory Organs in *Apis*.” *Apidologie* 22 (6): 581–590. https://doi.org/10.1051/apido:19910602.

Kœniger, N., and G. Koeniger. 2000. “Reproductive Isolation Among Species of the Genus *Apis*.” *Apidologie* 31 (2): 313–339. https://doi.org/10.1051/apido:2000125.

Kœniger, N., and G. Koeniger. 2007. “Mating Flight Duration of *Apis mellifera* Queens: As Short as Possible, as Long as Necessary.” *Apidologie* 38 (6): 606–611. https://doi.org/10.1051/apido:2007060

Kœniger, N., G. Koeniger, M. Gries, and S. Tingek. 2005. “Drone Competition at Drone Congregation Areas in Four *Apis* Species.” *Apidologie* 36 (2): 211–221. https://doi.org/10.1051/apido:2005011.

Kœniger, N., G. Koeniger, R. K. W. Punchihewa, M. Fabritius, and M. Fabritius. 1982. “Observations and Experiments on Dance Communication in *Apis florea* in Sri Lanka." *Journal of Apicultural Research* 21 (1): 45–52. https://doi.org/10.1080/00218839.1982.11100515.

Koeniger, N., G. Koeniger, and D. Smith. 2011. "Phylogeny of the genus *Apis*." In *Honeybees of Asia*. Berlin, Germany: Springer.

Kojima, J.-I. 1996. "Colony Cycle of an Australian Swarm-Founding Paper Wasp, *Ropalidia romandi* (Hymenoptera: Vespidae)." *Insectes Sociaux* 43 (4): 411–420. https://doi.org/10.1007/bf01258413.

Koschwanez, J. H., K. R. Foster, and A. W. Murray. 2013. "Improved Use of a Public Good Selects for the Evolution of Undifferentiated Multicellularity." *eLife* 2 (April). https://doi.org/10.7554/elife.00367.

Kotthoff, U., T. Wappler, and M. S. Engel. 2013. "Greater Past Disparity and Diversity Hints at Ancient Migrations of European Honey Bee Lineages Into Africa and Asia." *Journal of Biogeography* 40 (10): 1832–1838. https://doi.org/10.1111/jbi.12151.

Kraus, F. B., P. Neumann, and R. F. A. Moritz. 2005. "Genetic Variance of Mating Frequency in the Honeybee (*Apis mellifera* L.)." *Insectes Sociaux* 52 (1): 1–5. https://doi.org/10.1007/s00040-004-0766-9.

Kraus, F. B., P. Neumann, J. Van Praagh, and R. F. A. Moritz. 2003. "Sperm Limitation and the Evolution of Extreme Polyandry in Honeybees (*Apis mellifera* L.)." *Behavioral Ecology and Sociobiology* 55 (5): 494–501. https://doi.org/10.1007/s00265-003-0706-0.

Kritsky, G. 2015. *The Tears of Re: Beekeeping in Ancient Egypt*. New York: Oxford University Press.

Kritsky, G. 2020. "Revisiting Prehistoric Honey Hunting at Bicorp, Spain." *American Bee Journal* 160: 1247–1250.

Kryger, P., and R. F. A. Moritz. 1997. "Lack of Kin Recognition in Swarming Honeybees (*Apis mellifera*)." *Behavioral Ecology and Sociobiology* 40 (4): 271–276. https://doi.org/10.1007/s002650050342.

Kühnholz, S., and T. D. Seeley. 1997. "The Control of Water Collection in Honey Bee Colonies." *Behavioral Ecology and Sociobiology* 41 (6): 407–422. https://doi.org/10.1007/s002650050402.

Kukla, G. 2005. "Saalian Supercycle, Mindel/Riss Interglacial and Milankovitch's Dating." *Quaternary Science Reviews* 24 (14–15): 1573–1583. https://doi.org/10.1016/j.quascirev.2004.08.023.

Kukuk, P., and G. Eickwort. 1987. "Alternative Social Structures in Halictine Bees." *Chemistry and Biology of Social Insects* 555–556.

Lahdenperä, M., V. Lummaa, S. Helle, M. A. Tremblay, and A. F. Russell. 2004. "Fitness Benefits of Prolonged Post-reproductive Lifespan in Women." *Nature* 428 (6979): 178–181. https://doi.org/10.1038/nature02367.

Landim, C. D. C.. 1963. "Evaluation of the Wax and Scent Glands in the Apinae (Hymenoptera: Apidae)." *Journal of the New York Entomological Society* 71 (January): 2–13. https://digitalcommons.usu.edu/bee_lab_co/2/.

Lattorff, H. M. G., and R. F. A. Moritz. 2016. "Context Dependent Bias in Honeybee Queen Selection: Swarm Versus Emergency Queens." *Behavioral Ecology and Sociobiology* 70 (8): 1411–1417. https://doi.org/10.1007/s00265-016-2151-x.

Le Conte, Y., L. Sréng, N. Sacher, J. Trouiller, G. Dusticier, and S. Poitout. 1994. "Chemical Recognition of Queen Cells by Honey Bee Workers *Apis mellifera* (Hymenoptera: Apidae)." *Chemoecology* 5–6 (1): 6–12. https://doi.org/10.1007/bf01259967.

Leakey, R. E., and A. Walker. 1985. "Further Hominids From the Plio-Pleistocene of Koobi Fora, Kenya." *American Journal of Biological Anthropology* 67 (2): 135–163. https://doi.org/10.1002/ajpa.1330670209.

Lean, C. H., W. F. Doolittle, and J. P. Bielawski. 2022. "Community-Level Evolutionary Processes: Linking Community Genetics With Replicator-Interactor Theory." *Proceedings of the National Academy of Sciences of the United States of America* 119 (46). https://doi.org/10.1073/pnas.2202538119.

LeBoeuf, A. C., and C. M. Grozinger. 2014. "Me and We: The Interplay Between Individual and Group Behavioral Variation in Social Collectives." *Current Opinion in Insect Science* 5 (November): 16–24. https://doi.org/10.1016/j.cois.2014.09.010.

Lechner, S., L. Ferretti, C. Schöning, W. Kinuthia, D. Willemsen, and M. Hasselmann. 2013. "Nucleotide Variability at Its Limit? Insights Into the Number and Evolutionary Dynamics of the Sex-Determining Specificities of the Honey Bee *Apis mellifera*." *Molecular Biology and Evolution* 31 (2): 272–287. https://doi.org/10.1093/molbev/mst207.

LeDoux, M. N, M. L. Winston, H. Higo, C. I. Keeling, K. N. Slessor, and Y. Le Conte. 2001. "Queen and Pheromonal Factors Influencing Comb Construction by Simulated Honey Bee (*Apis mellifera* L.) Swarms." *Insectes Sociaux* 48 (1): 14–20. https://doi.org/10.1007/pl00001738.

Lee, G. M., M. J. F. Brown, and B. P. Oldroyd. 2012. "Inbred and Outbred Honey Bees (*Apis mellifera*) Have Similar Innate Immune Responses." *Insectes Sociaux* 60 (1): 97–102. https://doi.org/10.1007/s00040-012-0271-5.

Lee, P. C., and M. L. Winston. 1987. "Effects of Reproductive Timing and Colony Size on the Survival, Offspring Colony Size and Drone Production in the Honey Bee (*Apis mellifera*)." *Ecological Entomology* 12 (2): 187–195. https://doi.org/10.1111/j.1365-2311.1987.tb00997.x.

Leibniz, G. 1996. *Leibniz: New Essays on Human Understanding*. Cambridge, UK: Cambridge University Press.

Leimar, O., K. Hartfelder, M. D. Laubichler, and R. E. Page. 2012. "Development and Evolution of Caste Dimorphism in Honeybees–A Modeling Approach." *Ecology and Evolution* 2 (12): 3098–3109. https://doi.org/10.1002/ece3.414.

Lensky, Y., and Y. Slabezki. 1981. "The Inhibiting Effect of the Queen Bee (*Apis mellifera* L.) Foot-Print Pheromone on the Construction of Swarming Queen Cups." *Journal of Insect Physiology* 27 (5): 313–323. https://doi.org/10.1016/0022-1910(81)90077-9.

Leonard, W. R. 2002. "Food for Thought. Dietary Change Was a Driving Force in Human Evolution." *Scientific American* 287 (6): 106–116. https://pubmed.ncbi.nlm.nih.gov/12469653.

Léoncini, I., Y. Le Conte, G. Costagliola, E. Plettner, A. L. Toth, M. Wang, H. Zhi, et al. 2004. "Regulation of Behavioral Maturation by a Primer Pheromone Produced by Adult Worker Honey Bees." *Proceedings of the National Academy of Sciences of the United States of America* 101 (50): 17559–17564. https://doi.org/10.1073/pnas.0407652101.

Leonhardt, S. D., F. Menzel, V. Nehring, and T. Schmitt. 2016. "Ecology and Evolution of Communication in Social Insects." *Cell* 164 (6): 1277–1287. https://doi.org/10.1016/j.cell.2016.01.035.

Lewis, L. A., and S. Schneider. 2008. "'Migration Dances' in Swarming Colonies of the Honey Bee, *Apis mellifera*." *Apidologie* 39 (3): 354–361. https://doi.org/10.1051/apido:2008018..

Lewontin, R. C. 1970. "The Units of Selection." *Annual Review of Ecology and Systematics* 1 (1): 1–18. https://doi.org/10.1146/annurev.es.01.110170.000245.

Liersch, S., and P. Schmid-Hempel. 1998. "Genetic Variation Within Social Insect Colonies Reduces Parasite Load." *Proceedings of the Royal Society B: Biological Sciences* 265 (1392): 221–225. https://doi.org/10.1098/rspb.1998.0285.

Lindauer, M. 1954. "Temperaturregulierung und wasserhaushalt im bienenstaat." *Journal of Comparative Physiology A: Neuroethology, Sensory, Neural, and Behavioral Physiology* 36: 391–432.

Lindauer, M. 1957. "Communication in Swarm-Bees Searching for a New Home." *Nature* 179: 63–66.

Lindauer, M. 1961. *Communication among Social Bees*. Cambridge, MA: Harvard University Press.

Linksvayer, T. A., J. H. Fewell, J. Gadau, and M. D. Laubichler. 2012. "Developmental Evolution in Social Insects: Regulatory Networks From Genes to Societies." *Journal of Experimental Zoology Part B: Molecular and Developmental Evolution* 318 (3): 159–169. https://doi.org/10.1002/jez.b.22001.

Linksvayer, T. A., O. Kaftanoğlu, E. Akyol, S. Blatch, G. V. Amdam, and R. E. Page. 2011. "Larval and Nurse Worker Control of Developmental Plasticity and the Evolution of Honey Bee Queen-worker Dimorphism." *Journal of Evolutionary Biology* 24 (9): 1939–1948. https://doi.org/10.1111/j.1420-9101.2011.02331.x.

List, C., C. Elsholtz, and T. D. Seeley. 2008. "Independence and Interdependence in Collective Decision Making: An Agent-based Model of Nest-Site Choice by Honeybee Swarms." *Philosophical Transactions of the Royal Society B* 364 (1518): 755–762. https://doi.org/10.1098/rstb.2008.0277.

Liu, Z., M. Pagani, D. Zinniker, R. M. DeConto, M. Huber, H. Brinkhuis, S. R. Shah, R. M. Leckie, and A. Pearson. 2009. "Global Cooling During the Eocene-Oligocene Climate Transition." *Science* 323 (5918): 1187–1190. https://doi.org/10.1126/science.1166368.

Livina, V., F. Kwasniok, G. Lohmann, J. W. Kantelhardt, and T. Lenton. 2011. "Changing Climate States and Stability: From Pliocene to Present." *Climate Dynamics* 37 (11–12): 2437–2453. https://doi.org/10.1007/s00382-010-0980-2.

Loftus, J., M. L. Smith, and T. D. Seeley. 2016. "How Honey Bee Colonies Survive in the Wild: Testing the Importance of Small Nests and Frequent Swarming." *PLOS ONE* 11 (3): e0150362. https://doi.org/10.1371/journal.pone.0150362.

Loper, G. M., W. W. Wolf, and O. R. Taylor Jr. 1992. "Honey Bee Drone Flyways and Congregation Areas—Radar Observations." *Journal of the Kansas Entomological Society* 65 (3): 223–230. https://europepmc.org/article/AGR/IND93006951.

MacArthur, R. H., and E. O. Wilson. 1967. *The Theory of Island Biogeography*. Princeton, NJ: Princeton University Press.

Mahfooz, N., N. Turchyn, M. Mihajlovic, S. Hrycaj, and A. Popadić. 2007. "Ubx Regulates Differential Enlargement and Diversification of Insect Hind Legs." *PLOS ONE* 2 (9): e866. https://doi.org/10.1371/journal.pone.0000866.

Makinson, J. C., B. P. Oldroyd, T. M. Schaerf, W. Wattanachaiyingcharoen, and M. Beekman. 2010. "Moving Home: Nest-Site Selection in the Red Dwarf Honeybee (*Apis florea*)." *Behavioral Ecology and Sociobiology* 65 (5): 945–958. https://doi.org/10.1007/s00265-010-1095-9.

Maleszka, R. 2018. "Beyond Royalactin and a Master Inducer Explanation of Phenotypic Plasticity in Honey Bees." *Communications Biology* 1 (1). https://doi.org/10.1038/s42003-017-0004-4.

Margulis, L., and M. Chapman. 2010. *Kingdoms and Domains: An Illustrated Guide to the Phyla of Life on Earth*. Amsterdam, Netherlands: Elsevier.

Markwick, P. 2007. "The Palaeogeographic and Palaeoclimatic Significance of Climate Proxies for Data-Model Comparisons." In *Deep-Time Perspectives on Climate Change: Marrying the Signal from Computer Models and Biological Proxies*, edited by M. Williams, A. Haywood, F. Gregory, and D. Schmidt. London: Geological Society.

Marlowe, F. W., J. C. Berbesque, B. M. Wood, A. N. Crittenden, C. Porter, and A. Mabulla. 2014. "Honey, Hadza, Hunter-Gatherers, and Human Evolution." *Journal of Human Evolution* 71 (June): 119–128. https://doi.org/10.1016/j.jhevol.2014.03.006.

Martin, S. J., and J. Bayfield. 2014. "Is the Bee Louse *Braula coeca* (Diptera) Using Chemical Camouflage to Survive within Honeybee Colonies?" *Chemoecology* 24: 165–169. https://doi.org/10.1007/s00049-014-0158-1.

Martin, S. J., M. Beekman, T. C. Wossler, and F. L. W. Ratnieks. 2002a. "Parasitic Cape Honeybee Workers, *Apis mellifera capensis*, Evade Policing." *Nature* 415 (6868): 163–165. https://doi.org/10.1038/415163a.

Martin, S. J., T. C. Wossler, and P. Kryger. 2002b. "Usurpation of African *Apis mellifera scutellata* Colonies by Parasitic *Apis mellifera capensis* Workers." *Apidologie* 33 (2): 215–232. https://doi.org/10.1051/apido:2002003.

Martins, A. C., and G. a. R. Melo. 2015. "The New World Oil-Collecting Bees *Centris* and *Epicharis* (Hymenoptera, Apidae): Molecular Phylogeny and Biogeographic History." *Zoologica Scripta* 45 (1): 22–33. https://doi.org/10.1111/zsc.12133.

Martins, A. C., G. a. R. Melo, and S. S. Renner. 2014. "The Corbiculate Bees Arose From New World Oil-Collecting Bees: Implications for the Origin of Pollen Baskets." *Molecular Phylogenetics and Evolution* 80 (November): 88–94. https://doi.org/10.1016/j.ympev.2014.07.003.

Mattila, H. R., K. M. Burke, and T. D. Seeley. 2008. "Genetic Diversity Within Honeybee Colonies Increases Signal Production by Waggle-Dancing Foragers." *Proceedings of the Royal Society B: Biological Sciences* 275 (1636): 809–816. https://doi.org/10.1098/rspb.2007.1620.

Mattila, H. R., and T. D. Seeley. 2007. "Genetic Diversity in Honey Bee Colonies Enhances Productivity and Fitness." *Science* 317 (5836): 362–364. https://doi.org/10.1126/science.1143046.

Mattila, H. R., and T. D. Seeley. 2010. "Promiscuous Honeybee Queens Generate Colonies With a Critical Minority of Waggle-Dancing Foragers." *Behavioral Ecology and Sociobiology* 64 (5): 875–889. https://doi.org/10.1007/s00265-010-0904-5.

Mattila, H. R., and T. D. Seeley. 2011. "Does a Polyandrous Honeybee Queen Improve Through Patriline Diversity the Activity of Her Colony's Scouting Foragers?" *Behavioral Ecology and Sociobiology* 65 (4): 799–811. https://doi.org/10.1007/s00265-010-1083-0.

Mattila, H. R., and T. D. Seeley. 2013. "Extreme Polyandry Improves a Honey Bee Colony's Ability to Track Dynamic Foraging Opportunities via Greater Activity of Inspecting Bees." *Apidologie* 45 (3): 347–363. https://doi.org/10.1007/s13592-013-0252-3.

Maynard Smith, J. 1988. "Evolutionary Progress and Levels of Selection." In *Evolutionary Progress*, edited by M. H. Nitecki. Chicago: University of Chicago Press.

Maynard Smith, J. M., and E. Szathmáry. 1995. *The Major Transitions in Evolution*. Oxford, UK: Oxford University Press.

McDougall, I., F. H. Brown, and J. G. Fleagle. 2005. "Stratigraphic Placement and Age of Modern Humans from Kibish, Ethiopia." *Nature* 433 (7027): 733–736. https://doi.org/10.1038/nature03258.

Medved, V., H. Zhi, and A. Popadić. 2014. "Ubxpromotes Corbicular Development in *Apis mellifera*." *Biology Letters* 10 (1): 20131021. https://doi.org/10.1098/rsbl.2013.1021.

Mencken, H. L. 1982. *A Mencken Chrestomathy,* New York: Vintage Books.

Menzel, J. G., H. Wunderer, and D. G. Stavenga. 1991. "Functional Morphology of the Divided Compound Eye of the Honeybee Drone (*Apis mellifera*)." *Tissue & Cell* 23 (4): 525–535. https://doi.org/10.1016/0040-8166(91)90010-q.

Meunier, J. 2015. "Social Immunity and the Evolution of Group Living in Insects." *Philosophical Transactions of the Royal Society B* 370 (1669): 20140102. https://doi.org/10.1098/rstb.2014.0102.

Michelsen, A., B. B. Andersen, J. Storm, W. H. Kirchner, and M. Lindauer. 1992. "How Honeybees Perceive Communication Dances, Studied by Means of a Mechanical Model." *Behavioral Ecology and Sociobiology* 30: 143–150.

Michelsen, A., W. H. Kirchner, and M. Lindauer. 1986. "Sound and Vibrational Signals in the Dance Language of the Honeybee, *Apis mellifera*." *Behavioral Ecology and Sociobiology* 18 (3): 207–212. https://doi.org/10.1007/bf00290824.

Michener, C. D. 1944. Comparative External Morphology, Phylogeny, and a Classification of the Bees (Hymenoptera). *Bulletin of the American Natural History Museum* 82 (6): 151–326.

Michener, C. D. 1969. "Comparative Social Behavior of Bees." *Annual Review of Entomology* 14: 299–342.

Michener, C. D. 1974. *The Social Behavior of the Bees: A Comparative Study*. Cambridge, MA: Harvard University Press.

Michener, C. D. 1979. "Biogeography of the Bees." *Annals of the Missouri Botanical Garden* 66 (3): 277. https://doi.org/10.2307/2398833.

Michener, C. D. 1999. "The Corbiculae of Bees." *Apidologie* 30 (1): 67–74. https://doi.org/10.1051/apido:19990108.

Michener, C. D. 2000. *The Bees of the World*, Baltimore, MD: Johns Hopkins University Press.

Miguel, I., M. Iriondo, L. Garnery, W. S. Sheppard, and A. Estonba. 2007. "Gene Flow Within the M Evolutionary Lineage of *Apis mellifera*: Role of the Pyrenees, Isolation by Distance and Post-Glacial Re-Colonization Routes in the Western Europe." *Apidologie* 38 (2): 141–155. https://doi.org/10.1051/apido:2007007.

Mikheyev, A. S. 2003. "Evidence for Mating Plugs in the Fire Ant *Solenopsis invicta*." *Insectes Sociaux* 50 (4): 401–402. https://doi.org/10.1007/s00040-003-0697.

Misof, B., S. Liu, K. Meusemann, R. S. Peters, A. Donath, C. Mayer, P. B. Frandsen, et al. 2014. "Phylogenomics Resolves the Timing and Pattern of Insect Evolution. *Science* 346 (6210): 763–767. https://doi.org/10.1126/science.1257570.

Mix, A. C., É. Bard, and R. R. Schneider. 2001. "Environmental Processes of the Ice Age: Land, Oceans, Glaciers (EPILOG)." *Quaternary Science Reviews* 20 (4): 627–657. https://doi.org/10.1016/s0277-3791(00)00145-1.

Miyakawa, M. O., and A. S. Mikheyev. 2015. "QTL Mapping of Sex Determination Loci Supports an Ancient Pathway in Ants and Honey Bees." *PLOS Genetics* 11 (11): e1005656. https://doi.org/10.1371/journal.pgen.1005656.

Moeller, F. E. 1961. *The Relationship between Colony Populations and Honey Production as Affected by Honey Bee Stock Lines*. Washington, DC: US Department of Agriculture, Agricultural Research Service.

Monnin, T., and F. L. W. Ratnieks. 2001. "Policing in Queenless Ponerine Ants." *Behavioral Ecology and Sociobiology* 50 (2): 97–108. https://doi.org/10.1007/s002650100351.

Montovan, K. J., N. J. Karst, L. Jones, and T. D. Seeley. 2013. "Local Behavioral Rules Sustain the Cell Allocation Pattern in the Combs of Honey Bee Colonies (*Apis mellifera*)." *Journal of Theoretical Biology* 336 (November): 75–86. https://doi.org/10.1016/j.jtbi.2013.07.010.

Moore, A. J., M. D. Breed, and M. J. Moor. 1987. "The Guard Honey Bee: Ontogeny and Behavioural Variability of Workers Performing a Specialized Task." *Animal Behaviour* 35 (4): 1159–1167. https://doi.org/10.1016/s0003-3472(87)80172-0.

Moritz, R. F. 1985a. "Two Parthenogenetical Strategies of Laying Workers in Populations of the Honeybee, *Apis mellifera* (Hymenoptera: Apidae)." *Entomologia Generalis* 11 (3–4): 159–164. https://doi.org/10.1127/entom.gen/11/1986/159.

Moritz, R. F. A. 1985b. "The Effects of Multiple Mating on the Worker-Queen Conflict in *Apis mellifera* L." *Behavioral Ecology and Sociobiology* 16 (4): 375–377. https://doi.org/10.1007/bf00295551.

Moritz, R. F. A. 1986. "Intracolonial Worker Relationship and Sperm Competition in the Honeybee (*Apis mellifera* L.)." *Experientia* 42 (4): 445–448. https://doi.org/10.1007/bf02118652.

Moritz, R., and Crewe, R. 2018. *The Dark Side of the Hive: The Evolution of the Imperfect Honeybee*. New York: Oxford University Press.

Moritz, R. F. A., and E. Hillesheim. 1985. "Inheritance of Dominance in Honeybees (*Apis mellifera capensis* Esch.)." *Behavioral Ecology and Sociobiology* 17 (1): 87–89. https://doi.org/10.1007/bf00299434.

Moritz, R., and E. Southwick. 1992. *Bees as Superorganisms: An Evolutionary Reality*. Berlin, Germany: Springer.

Moritz, R. F., P. Kryger, G. Koeniger, N. Kœniger, A. Estoup, and S. Tingek. 1995. "High Degree of Polyandry in *Apis dorsata* Queens Detected by DNA Microsatellite Variability." *Behavioral Ecology and Sociobiology* 37 (5): 357–363. https://doi.org/10.1007/bf00174141.

Moritz, R. F., H. M. G. Lattorff, P. Neumann, F. B. Kraus, S. E. Radloff, and H. R. Hepburn. 2005. "Rare Royal Families in Honeybees, *Apis mellifera*." *Science of Nature* 92 (10): 488–491. https://doi.org/10.1007/s00114-005-0025-6.

Moritz, R. F. A., and D. Kauhausen. 1984. "Hybridization between *Apis mellifera capensis* and Adjacent Races of *Apis mellifera*." *Apidologie* 15 (2): 211–222. https://doi.org/10.1051/apido:19840209.

Morse, R. A. 1965. "The Effect of Light on Comb Construction by Honeybees." *Journal of Apicultural Research* 4 (1): 23–29. https://doi.org/10.1080/00218839.1965.11100098.

Morse, R., and G. Eickwort. 1990. "*Acarapis woodi*, A Recently Evolved Species? Recent Research on Bee Pathology." Gent, Belgium: International Symposium of the International Federation of Beekeepers Associations.

Mountford, M. D. 1968. "The Significance of Litter-Size." *Journal of Animal Ecology* 37 (2): 363. https://doi.org/10.2307/2953.

Mueller, L. D., and V. F. Sweet. 1986. "Density-Dependent Natural Selection in Drosophila: Evolution of Pupation Height." *Evolution* 40 (6): 1354. https://doi.org/10.2307/2408963.

Müller, U., S. Klotz, M. A. Geyh, J. Pross, and G. C. Bond. 2005. "Cyclic Climate Fluctuations During the Last Interglacial in Central Europe." *Geology* 33 (6): 449. https://doi.org/10.1130/g21321.1.

Naug, D. 2009. "Nutritional Stress due to Habitat Loss May Explain Recent Honeybee Colony Collapses." *Biological Conservation* 142: 2369–2372.

Naug, D., and B. H. Smith. 2006. "Experimentally Induced Change in Infectious Period Affects Transmission Dynamics in a Social Group." *Proceedings of the Royal Society B: Biological Sciences* 274 (1606): 61–65. https://doi.org/10.1098/rspb.2006.3695.

Naumann, K., M. L. Winston, and K. N. Slessor. 1993. "Movement of Honey Bee (*Apis mellifera* L.) Queen Mandibular Gland Pheromone in Populous and

Unpopulous Colonies." *Journal of Insect Behavior* 6 (2): 211–223. https://doi.org/10.1007/bf01051505.

Naumann, Ken, Mark L. Winston, K. N. Slessor, Glenn D. Prestwich, and Bachir Latli. 1992. "Intra-Nest Transmission of Aromatic Honey Bee Queen Mandibular Gland Pheromone Components: Movement as a Unit." *Canadian Entomologist* 124 (5): 917–934. https://doi.org/10.4039/ent124917-5.

Nazzi, F. 2016. "The Hexagonal Shape of the Honeycomb Cells Depends on the Construction Behavior of Bees." *Scientific Reports* 6 (1): 28431. https://doi.org/10.1038/srep28341.

Nel, A., P. Roques, P. Nel, P., Prokin, T. Bourgoin, J. Prokop, J. Szwedo, et al. 2013. "The Earliest Known Holometabolous Insects." *Nature* 503 (7475): 257–261. https://doi.org/10.1038/nature12629.

Neumann, P., and R. Hepburn. 2002. "Behavioural Basis for Social Parasitism of Cape Honeybees (*Apis mellifera capensis*)." *Apidologie* 33 (2): 165–192. https://doi.org/10.1051/apido:2002008.

Neumann, P., and R. F. A. Moritz. 2002. "The Cape Honeybee Phenomenon: The Sympatric Evolution of a Social Parasite in Real Time?" *Behavioral Ecology and Sociobiology* 52 (4): 271–281. https://doi.org/10.1007/s00265-002-0518-7.

Neumann, P., C. W. W. Pirk, H. R. Hepburn, A. J. Solbrig, F. L. W. Ratnieks, P. J. Elzen, and J. R. Baxter. 2001. "Social Encapsulation of Beetle Parasites by Cape Honeybee Colonies (*Apis mellifera capensis* Esch.)." *Science of Nature* 88 (5): 214–216. https://doi.org/10.1007/s001140100224.

Nielsen, R., D. R. Tarpy, and H. K. Reeve. 2003. "Estimating Effective Paternity Number in Social Insects and the Effective Number of Alleles in a Population." *Molecular Ecology* 12 (11): 3157–3164. https://doi.org/10.1046/j.1365-294x.2003.01994.x.

Nijhout, H. F., and D. E. Wheeler. 1982. "Juvenile Hormone and the Physiological Basis of Insect Polymorphisms." *Quarterly Review of Biology* 57 (2): 109–133. https://doi.org/10.1086/412671.

Noirot, C. 1969. "Formation of Castes in Higher Termites." *Biology of Termites*, edited by K. Krishna and F. Weesner. New York: Academic Press.

Nolan, M. P., and K. S. Delaplane. 2016. "Distance Between Honey Bee *Apis mellifera* Colonies Regulates Populations of *Varroa destructor* at a Landscape Scale." *Apidologie* 48 (1): 8–16. https://doi.org/10.1007/s13592-016-0443-9.

Nonacs, P., and N. F. Carlin. 1990. "When Can Ants Discriminate the Sex of Brood? A New Aspect of Queen-Worker Conflict." *Proceedings of the National Academy of Sciences of the United States of America* 87 (24): 9670–9673. https://doi.org/10.1073/pnas.87.24.9670.

Nonacs, P., and J. E. Tobin. 1992. "Selfish Larvae: Development and the Evolution of Parasitic Behavior in the Hymenoptera." *Evolution* 46 (6): 1605. https://doi.org/10.2307/2410019.

Nowak, M. A., C. E. Tarnita, and E. O. Wilson. 2010. "The Evolution of Eusociality." *Nature* 466 (7310): 1057–1062. https://doi.org/10.1038/nature09205.

O'Brien, C. L, M. Huber, E. Thomas, M. Pagani, J. R. Super, L. E. Elder, and P. M. Hull. 2020. "The Enigma of Oligocene Climate and Global Surface Temperature Evolution." *Proceedings of the National Academy of Sciences of the United States of America* 117 (41): 25302–25309. https://doi.org/10.1073/pnas.2003914117.

Ohno, S. 1979. *Major Sex-Determining Genes*. Berlin, Germany: Springer.

Oldroyd, B. P., M. J. Clifton, K. Parker, S. Wongsiri, T. E. Rinderer, and R. H. Crozier. 1998. "Evolution of Mating Behavior in the Genus *Apis* and an Estimate of Mating Frequency in *Apis cerana* (Hymenoptera: Apidae)." *Annals of the Entomological Society of America* 91 (5): 700–709. https://doi.org/10.1093/aesa/91.5.700.

Oldroyd, B. P., M. J. Clifton, S. Wongsiri, T. E. Rinderer, H. Allen Sylvester, and R. H. Crozier. 1997. "Polyandry in the Genus *Apis* , Particularly *Apis andreniformis*." *Behavioral Ecology and Sociobiology* 40 (1): 17–26. https://doi.org/10.1007/s002650050311.

Oldroyd, B. P., and F. L. W. Ratnieks. 2000. "Evolution of Worker Sterility in Honey-bees (*Apis mellifera*): How Anarchistic Workers Evade Policing by Laying Eggs That Have Low Removal Rates." *Behavioral Ecology and Sociobiology* 47 (4): 268–273. https://doi.org/10.1007/s002650050665.

Oldroyd, B. P., T. E. Rinderer, J. R. Harbo, and S. M. Buco. 1992. "Effects of Intracolonial Genetic Diversity on Honey Bee (Hymenoptera: Apidae) Colony Performance." *Annals of the Entomological Society of America* 85 (3): 335–343. https://doi.org/10.1093/aesa/85.3.335.

Oldroyd, B. P., A. J. Smolenski, J.-M. Cornuet, S. Wongsiri, A. Estoup, T. E. Rinderer, and R. H. Crozier. 1995. "Levels of Polyandry and Intracolonial Genetic Relationships in *Apis florea*." *Behavioral Ecology and Sociobiology* 37 (5): 329–335. https://doi.org/10.1007/bf00174137.

Oldroyd, B. P., A. J. Smolenski, J.-M. Cornuet, S. Wongsiri, A. Estoup, T. E. Rinderer, and R. H. Crozier. 1996. "Levels of Polyandry and Intracolonial Genetic Relationships in *Apis dorsata* (Hymenoptera: Apidae)." *Annals of the Entomological Society of America* 89 (2): 276–283. https://doi.org/10.1093/aesa/89.2.276.

Oldroyd, B. P., T. C. Wossler, and F. L. W. Ratnieks. 2001. "Regulation of Ovary Activation in Worker Honey-Bees (*Apis mellifera*): Larval Signal Production and

Adult Response Thresholds Differ Between Anarchistic and Wild-Type Bees." *Behavioral Ecology and Sociobiology* 50 (4): 366–370. https://doi.org/10.1007/s002650100369.

Omholt, S. W., and G. V. Amdam. 2004. "Epigenetic Regulation of Aging in Honeybee Workers." *Science of Aging Knowledge Environment* 2004 (26). https://doi.org/10.1126/sageke.2004.26.pe28.

Onions, G. 1912. "South African Fertile Worker-Bees." *Agricultural Journal of the Union of South Africa* 3: 720. https://journals.co.za/doi/pdf/10.10520/AJA0000021_1356.

Oster, G. F., and E. O. Wilson. 1979. *Caste and Ecology in the Social Insects.* Princeton, NJ: Princeton University Press.

Oxley, P. R., M. Spivak, and B. P. Oldroyd. 2010. "Six Quantitative Trait Loci Influence Task Thresholds for Hygienic Behaviour in Honeybees (*Apis mellifera*)." *Molecular Ecology* 19 (7): 1452–1461. https://doi.org/10.1111/j.1365-294x.2010.04569.x.

Oxley, P. R., G. Thompson, and B. P. Oldroyd. 2008. "Four Quantitative Trait Loci That Influence Worker Sterility in the Honeybee (*Apis mellifera*)." *Genetics* 179 (3): 1337–1343. https://doi.org/10.1534/genetics.108.087270.3.

Pabalan, N., K. G. Davey, and L. Packer. 1996. "Comparative Morphology of Spermathecae in Solitary and Primitively Eusocial Bees (Hymenoptera; Apoidea)." *Canadian Journal of Zoology* 74 (5): 802–808. https://doi.org/10.1139/z96-092.

Packer, Laurence. 1992. "The Social Organisation of *Lasioglossum* (*Dialictus*) *Laevissimum* (Smith) in Southern Alberta." *Canadian Journal of Zoology* 70 (9): 1767–1774. https://doi.org/10.1139/z92-244..

Page, R. E. 1980. "The Evolution Of Multiple Mating Behavior by Honey Bee Queens (*Apis mellifera* L.)." *Genetics* 96 (1): 263–273. https://doi.org/10.1093/genetics/96.1.263.

Page, R. E. 1997. "The Evolution of Insect Societies." *Endeavour* 21 (3): 114–120. https://doi.org/10.1016/s0160-9327(97)80220-7.

Page, R. E., and E. H. Erickson. 1988. "Reproduction by Worker Honey Bees (*Apis mellifera* L.)." *Behavioral Ecology and Sociobiology* 23 (2): 117–126. https://doi.org/10.1007/bf00299895.

Page, R. E., and M. K. Fondrk. 1995. "The Effects of Colony-level Selection on the Social Organization of Honey Bee (*Apis mellifera* L.) Colonies: Colony-level Components of Pollen Hoarding." *Behavioral Ecology and Sociobiology* 36 (2): 135–144. https://doi.org/10.1007/bf00170718.

Page, R. E., and R. A. Metcalf. 1982. "Multiple Mating, Sperm Utilization, and Social Evolution." *American Naturalist* 119 (2): 263–281. https://doi.org/10.1086/283907.

Page, R. E., and R. A. Metcalf. 1984. "A Population Investment Sex Ratio for the Honey Bee (*Apis mellifera* L.)." *American Naturalist* 124 (5): 680–702. https://doi.org/10.1086/284306.

Page, R. E., and S. D. Mitchell. 1998. "Self-Organization and the Evolution of Division of Labor." *Apidologie* 29 (1–2): 171–190. https://doi.org/10.1051/apido:19980110.

Page, R. E., G. E. Robinson, and M. K. Fondrk. 1989. "Genetic Specialists, Kin Recognition and Nepotism in Honey-bee Colonies." *Nature* 338 (6216): 576–579. https://doi.org/10.1038/338576a0.

Page, R. E., G. E. Robinson, M. K. Fondrk, and M. E. Nasr. 1995. "Effects of Worker Genotypic Diversity on Honey Bee Colony Development and Behavior (*Apis mellifera* L.)." *Behavioral Ecology and Sociobiology* 36 (6): 387–396. https://doi.org/10.1007/s002650050161.

Page, R. E., R. Scheiner, J. Erber, and G. V. Amdam. 2006. "The Development and Evolution of Division of Labor and Foraging Specialization in a Social Insect (*Apis mellifera* L.)." *Current Topics in Developmental Biology* 74: 253–286.

Pain, J., and J. Maugenet. 1966. "Recherches biochimiques et physiologiques sur le pollen emmagasiné par les abeilles." *Les Annales de l'Abeille* 9: 209–236.

Paleolog, J. 2009. "Behavioural Characteristics of Honey Bee (*Apis mellifera*) Colonies Containing Mix of Workers of Divergent Behavioural Traits." *Animal Science Papers and Reports* 27 (3): 237–248. https://www.cabdirect.org/abstracts/20093296447.html.

Palmer, K. A., and B. P. Oldroyd. 2000. "Evolution of Multiple Mating in the Genus *Apis*." *Apidologie* 31 (2): 235–248. https://doi.org/10.1051/apido:2000119.

Paxton, R. J., P Thorén, J. Tengö, A. Estoup, and P. Pamilo. 1996. "Mating Structure and Nestmate Relatedness in a Communal Bee, *Andrena jacobi* (Hymenoptera, Andrenidae), Using Microsatellites." *Molecular Ecology* 5 (4): 511–519. https://doi.org/10.1111/j.1365-294x.1996.tb00343.x.

Perepelova, L. I. 1928. "Laying Workers, the Egg-Laying Activity of the Queen, and Swarming." *Bee World* 10: 69–71.

Pesante, D., T. E. Rinderer, and A. M. Collins. 1987a. "Differential Nectar Foraging by Africanized and European Honeybees in the Neotropics." *Journal of Apicultural Research* 26 (4): 210–216. https://doi.org/10.1080/00218839.1987.11100762.

Pesante, D., T. E. Rinderer, and A. M. Collins. 1987b. "Differential Pollen Collection by Africanized and European Honeybees in Venezuela." *Journal of Apicultural Research* 26 (1): 24–29. https://doi.org/10.1080/00218839.1987.11100730.

Pfeiffer, K., and K. Crailsheim. 1998. "Drifting of Honeybees." *Insectes Sociaux* 45: 151–167.

Phillips, E. F., and G. S. Demuth. 1914. “The Temperature of the Honeybee Cluster in Winter.” *Bulletin of the USDA*.

Pillay, P., M. S. Allen, and J. Littleton. 2022. “Canine Companions or Competitors? A Multi-Proxy Analysis of Dog-Human Competition.” *Journal of Archaeological Science* 139 (March): 105556. https://doi.org/10.1016/j.jas.2022.105556..

Pinker, S. 1994. *The Language Instinct: How the Mind Creates Language*. New York: Harper Collins.

Pinto, L. Z., M. M. G. Bitondi, and Z. L. P. Simões. 2000. “Inhibition of Vitellogenin Synthesis in *Apis mellifera* Workers by a Juvenile Hormone Analogue, Pyriproxyfen.” *Journal of Insect Physiology* 46 (2): 153–160. https://doi.org/10.1016/s0022-1910(99)00111-0.

Ponge, J.-F.. 2005. “Emergent Properties from Organisms to Ecosystems: Towards a Realistic Approach.” *Biological Reviews* 80 (3): 403–411. https://doi.org/10.1017/s146479310500672x.

Pratt, S. C. 2000. “Gravity-Independent Orientation of Honeycomb Cells.” *Science of Nature* 87 (1): 33–35. https://doi.org/10.1007/s001140050005.

Pratt, S. C. 2004. “Collective Control of the Timing and Type of Comb Construction by Honey Bees (*Apis mellifera*).” *Apidologie* 35 (2): 193–205. https://doi.org/10.1051/apido:2004005.

Price, R. I’A., and C. Grüter. 2015. “Why, When and Where Did Honey Bee Dance Communication Evolve?” *Frontiers in Ecology and Evolution* 3 (November). https://doi.org/10.3389/fevo.2015.00125.

Qin, M., H. Wang, Z. Liu, Y. Wang, W. Zhang, and B. Xu. 2019. “Changes in Cold Tolerance During the Overwintering Period in *Apis mellifera ligustica*.” *Journal of Apicultural Research* 58 (5): 702–709. https://doi.org/10.1080/00218839.2019.1634461.

Queller, D. C. 1997. *Cooperators Since Life Began*. Chicago: University of Chicago Press.

Queller, D. C. 2016. “Kin Selection and Its Discontents.” *Philosophy of Science* 83 (5): 861–872. https://doi.org/10.1086/687870.

Queller, D. C., and J. E. Strassmann. 1998. “Kin Selection and Social Insects.” *BioScience* 48 (3): 165–175. https://doi.org/10.2307/1313262.

Queller, D. C., and J. E. Strassmann. 2002. “The Many Selves of Social Insects.” *Science* 296 (5566): 311–313. https://doi.org/10.1126/science.1070671.

Queller, D. C., and J. E. Strassmann 2009. “Beyond Society: The Evolution of Organismality.” *Philosophical Transactions of the Royal Society B* 364 (1533): 3143–3155. https://doi.org/10.1098/rstb.2009.0095.

Rachinsky, A., C. Strambi, A. Strambi, and K. Hartfelder. 1990. “Caste and Metamorphosis: Hemolymph Titers of Juvenile Hormone and Ecdysteroids in Last

Instar Honeybee Larvae." *General and Comparative Endocrinology* 79 (1): 31–38. https://doi.org/10.1016/0016-6480(90)90085-z.

Raffiudin, R., and R. H. Crozier. 2007. "Phylogenetic Analysis of Honey Bee Behavioral Evolution." *Molecular Phylogenetics and Evolution* 43 (2): 543–552. https://doi.org/10.1016/j.ympev.2006.10.013.

Ramírez-Barahona, S., H. Sauquet, and S. Magallón. 2020. "The Delayed and Geographically Heterogeneous Diversification of Flowering Plant Families." *Nature Ecology and Evolution* 4 (9): 1232–1238. https://doi.org/10.1038/s41559-020-1241-3.

Ramírez-Delgado, V. H., S. Sanabria-Urbán, M. A. Serrano-Meneses, and R. C. Del Castillo. 2016. "The Converse to Bergmann's Rule in Bumblebees, a Phylogenetic Approach." *Ecology and Evolution* 6 (17): 6160–6169. https://doi.org/10.1002/ece3.2321

Rangel, J., S. R. Griffin, and T. D. Seeley. 2010. "An Oligarchy of Nest-site Scouts Triggers a Honeybee Swarm's Departure From the Hive." *Behavioral Ecology and Sociobiology* 64 (6): 979–987. https://doi.org/10.1007/s00265-010-0913-4.

Rangel, J., and T. D. Seeley. 2008. "The Signals Initiating the Mass Exodus of a Honeybee Swarm From Its Nest." *Animal Behaviour* 76 (6): 1943–1952. https://doi.org/10.1016/j.anbehav.2008.09.004.

Ransome, H. 1937. *The Sacred Bee in Ancient Times and Folklore*. London: George Allen & Unwin.

Rasmussen, C., and J. M. F. De Camargo. 2008. "A Molecular Phylogeny and the Evolution of Nest Architecture and Behavior in *Trigona* s.s.(Hymenoptera: Apidae: Meliponini)." *Apidologie* 39 (1): 102–118. https://doi.org/10.1051/apido:2007051.

Ratcliff, W. C., R. F. Denison, M. E. Borrello, and M. Travisano. 2012. "Experimental Evolution of Multicellularity." *Proceedings of the National Academy of Sciences of the United States of America* 109 (5): 1595–1600. https://doi.org/10.1073/pnas.1115323109.

Ratnieks, F L. 1990. "The Evolution of Polyandry by Queens in Social Hymenoptera: The Significance of the Timing of Removal of Diploid Males." *Behavioral Ecology and Sociobiology* 26 (5). https://doi.org/10.1007/bf00171100.

Ratnieks, F. L. 1993. "Egg-Laying, Egg-Removal, and Ovary Development by Workers in Queenright Honey Bee Colonies." *Behavioral Ecology and Sociobiology* 32 (3). https://doi.org/10.1007/bf00173777.

Ratnieks, F. L. 1995. "Evidence for a Queen-Produced Egg-Marking Pheromone and Its Use in Worker Policing in the Honey Bee." *Journal of Apicultural Research* 34 (1): 31–37. https://doi.org/10.1080/00218839.1995.11100883.

Ratnieks, F. L., and Heikki Helanterä. 2009. "The Evolution of Extreme Altruism and Inequality in Insect Societies."*Philosophical Transactions of the Royal Society B* 364 (1533): 3169–3179. https://doi.org/10.1098/rstb.2009.0129.

Ratnieks, F. L., and H. K. Reeve. 1992. "Conflict in Single-Queen Hymenopteran Societies: The Structure of Conflict and Processes That Reduce Conflict in Advanced Eusocial Species." *Journal of Theoretical Biology* 158 (1): 33–65. https://doi.org/10.1016/s0022-5193(05)80647-2.

Ratnieks, F. L., and T. Wenseleers. 2008. "Altruism in Insect Societies and Beyond: Voluntary or Enforced?" *Trends in Ecology and Evolution* 23 (1): 45–52. https://doi.org/10.1016/j.tree.2007.09.013.

Rehan, S. M., R. Leys, and M. P. Schwarz. 2012. "A Mid-Cretaceous Origin of Sociality in Xylocopine Bees With Only Two Origins of True Worker Castes Indicates Severe Barriers to Eusociality." *PLOS ONE* 7 (4): e34690. https://doi.org/10.1371/journal.pone.0034690.

Rehan, S. M., and A. L. Toth. 2015. "Climbing the Social Ladder: The Molecular Evolution of Sociality." *Trends in Ecology and Evolution* 30 (7): 426–433. https://doi.org/10.1016/j.tree.2015.05.004.

Rembold, H. 1966. "Biologically Active Substances in Royal Jelly." In *Vitamins and Hormones*, 359–382. Amsterdam, Netherlands: Elsevier.

Reznick, D. N., M. J. Bryant, and F. Bashey. 2002. "*r*- and *K*-Selection Revisited: The Role of Population Regulation in Life-History Evolution." *Ecology* 83 (6): 1509. https://doi.org/10.2307/3071970.

Ribbands, C. R. 1953. *The Behaviour and Social Life of Honeybees*, London: International Bee Research Association.

Rinderer, T. E. 1988. "Evolutionary Aspects of the Africanization of Honey-Bee Populations in the Americas." In *Africanized Honey Bees and Bee Mites*. Chichester, UK: Ellis Horwood.

Rinderer, T. E., A. M. Collins, D. Pesante, R. F. Daniel, V. Lancaster, and J. S. Baxter. 1985. "A Comparison of Africanized and European Drones: Weights, Mucus Gland and Seminal Vesicle Weights, and Counts of Spermatozoa." *Apidologie* 16 (4): 407–412. https://doi.org/10.1051/apido:19850405.

Robeau, R., and S. Vinson. 1976. "Effects of Juvenile Hormone Analogues on Caste Differentiation in the Imported Fire Ant, *Solenopsis invicta*." *Journal Georgia Entomological Society* 11: 198–203.

Roberts, K., S. E. F. Evison, B. Baer, and W. O. Hughes. 2015. "The Cost of Promiscuity: Sexual Transmission of Nosema Microsporidian Parasites in Polyandrous Honey Bees." *Scientific Reports* 5 (1). https://doi.org/10.1038/srep10982.

Robinson, G. E. 1992. "Regulation of Division of Labor in Insect Societies." *Annual Review of Entomology* 37 (1): 637–665. https://doi.org/10.1146/annurev.ento.37.1.637.

Robinson, G. E., and R. E. Page. 1988. "Genetic Determination of Guarding and Undertaking in Honey-Bee Colonies." *Nature* 333 (6171): 356–358. https://doi.org/10.1038/333356a0.

Robinson, G. E., and R. E. Page.1989. "Genetic Determination of Nectar Foraging, Pollen Foraging, and Nest-Site Scouting in Honey Bee Colonies." *Behavioral Ecology and Sociobiology* 24 (5): 317–323. https://doi.org/10.1007/bf00290908.

Robinson, G. E., and E. L. Vargo. 1997. "Juvenile Hormone in Adult Eusocial Hymenoptera: Gonadotropin and Behavioral Pacemaker." *Archives of Insect Biochemistry and Physiology* 35 (4): 559–583. https://doi.org/10.1002/(sici)1520-6327(1997)35:4.

Roebroeks, W., and P. Villa. 2011. "On the Earliest Evidence for Habitual Use of Fire in Europe." *Proceedings of the National Academy of Sciences of the United States of America* 108 (13): 5209–5214. https://doi.org/10.1073/pnas.1018116108.

Romiguier, J., S. A. Cameron, S. H. Woodard, B. J. Fischman, L. Keller, and C. Praz. 2015. "Phylogenomics Controlling for Base Compositional Bias Reveals a Single Origin of Eusociality in Corbiculate Bees." *Molecular Biology and Evolution* 33 (3): 670–678. https://doi.org/10.1093/molbev/msv258.

Ronai, I., B. P. Oldroyd, and V. Vergoz. 2016. "Queen Pheromone Regulates Programmed Cell Death in the Honey Bee Worker Ovary." *Insect Molecular Biology* 25 (5): 646–652. https://doi.org/10.1111/imb.12250.

Ronquist, F., S. Klopfstein, L. Vilhelmsen, S. Schulmeister, D. Murray, and A. P. Rasnitsyn. 2012. "A Total-Evidence Approach to Dating With Fossils, Applied to the Early Radiation of the Hymenoptera." *Systematic Biology* 61 (6): 973–999. https://doi.org/10.1093/sysbio/sys058.

Röseler, P.-F. 1975. "Die Kasten der sozialen Bienen." In *Informationsaufnahme und Informationsverarbeitung im lebenden Organismus.* Mainz, Germany: Academie der Wissenschaften und der Literatur.

Rosenkranz, P., P. Aumeier, and B. Ziegelmann. 2010. "Biology and Control of *Varroa destructor*." *Journal of Invertebrate Pathology* 103 (January): S96–S119. https://doi.org/10.1016/j.jip.2009.07.016.

Roubik, D., T. Heard, and P. Kwapong. 2018. "Stingless bee colonies and pollination." In *The Pollination of Cultivated Plants: A Compendium for Practitioners*, edited by D. W. Roubik. Rome, Italy: FAO.

Rovelli, C. 2016. *Seven Brief Lessons on Physics*. New York: Riverhead Books.

Royce, L. A., P. A. Rossignol, D. M. Burgett, and B. A. Stringer. 1991. "Reduction of Tracheal Mite Parasitism of Honey Bees by Swarming." *Philosophical Transactions of the Royal Society B* 331 (1260): 123–129. https://doi.org/10.1098/rstb.1991.0003.

Rueppell, O., C. Bachelier, M. K. Fondrk, and R. E. Page. 2007. "Regulation of Life History Determines Lifespan of Worker Honey Bees (*Apis mellifera* L.)." *Experimental Gerontology* 42 (10): 1020–1032. https://doi.org/10.1016/j.exger.2007.06.002.

Rueppell, O., N. G. Johnson, and J. Rychtář. 2008. "Variance-based Selection May Explain General Mating Patterns in Social Insects." *Biology Letters* 4 (3): 270–273. https://doi.org/10.1098/rsbl.2008.0065.

Russell, S. C., A. B. Barron, and D. Harris. 2013. "Dynamic Modelling of Honey Bee (*Apis mellifera*) Colony Growth and Failure." *Ecological Modelling* 265 (September): 158–169. https://doi.org/10.1016/j.ecolmodel.2013.06.005.

Ruttner, F. 1977. "The Problem of the Cape Bee (*Apis mellifera capensis* Escholtz): Parthenogenesis—Size of Population—Evolution." *Apidologie* 8: 281–294.

Ruttner, F. 1988. *Biogeography and Taxonomy of Honeybees*. Berlin, Germany: Springer.

Ruttner, F. 1992. *Naturgeschichte der Honigbienen*. Munich, Germany: Ehrenwirth.

Ruttner, F., M. Elmi, and S. Fuchs. 2000. "Ecoclines in the Near East Along 36° N Latitude in *Apis mellifera* L." *Apidologie* 31 (2): 157–165. https://doi.org/10.1051/apido:2000113.

Ruttner, H., and F. Ruttner. 1972. "Untersuchungen über die Flugaktivität und das Paarungsverhalten der Drohnen. V.-Drohnensammelplätze und Paarungsdistanz." *Apidologie* 3: 203–232.

Ruttner, F., J. Woyke, and N. Kœniger. 1972. "Reproduction in *Apis cerana* 1. Mating Behaviour." *Journal of Apicultural Research* 11 (3): 141–146. https://doi.org/10.1080/00218839.1972.11099714.

Ruttner, F., J. Woyke, and N. Kœniger. 1973. "Reproduction in *Apis cerana* 2. Reproductive Organs and Natural Insemination." *Journal of Apicultural Research* 12 (1): 21–34. https://doi.org/10.1080/00218839.1973.11099727.

Sakagami, S. 1966. "Comparative Ethology of Apidae." *Society of Systematic Zoology Circ.* 35: 1–6.

Sakagami, S.F., and H. Fukuda. 1968. "Life Tables for Worker Honeybees." *Population Ecology* 10 (2): 127–139. https://doi.org/10.1007/bf02510869.

Sakagami, S. F., and Y. Maeta. 1987. In *Animal Societies: Theories and Facts*, edited by Y. Ito, J. J. Brown, and J. Kikkawa. Tokyo, Japan: Scientific Society Press.

Sarzetti, L. C., J. F. Genise, and M. V. López Sánchez. 2014. "Nest Architecture of *Oxaea austera* (Andrenidae, Oxaeinae) and Its Significance for the Interpretation of Uruguayan Fossil Bee Cells." *Journal of Hymenoptera Research* 39 (September): 59–70. https://doi.org/10.3897/jhr.39.8201.

Sasaki, T., and S. C. Pratt. 2018. "The Psychology of Superorganisms: Collective Decision Making by Insect Societies." *Annual Review of Entomology* 63 (1): 259–275. https://doi.org/10.1146/annurev-ento-020117-043249.

Sauter, A., M. J. F. Brown, B. Baer, and P. Schmid-Hempel. 2001. "Males of Social Insects Can Prevent Queens From Multiple Mating." *Proceedings of the Royal Society B: Biological Sciences* 268 (1475): 1449–1454. https://doi.org/10.1098/rspb.2001.1680.

Scannapieco, A C, M. C. Mannino, G. Soto, M. A. Palacio, J. L. Cladera, and S. B. Lanzavecchia. 2017. "Expression Analysis of Genes Putatively Associated with

Hygienic Behavior in Selected Stocks of *Apis mellifera* L. from Argentina." *Insectes Sociaux* 64 (4): 485–494. https://doi.org/10.1007/s00040-017-0567-6.

Scheiner, R., R. E. Page, and J. Erber. 2004. "Sucrose Responsiveness and Behavioral Plasticity in Honey Bees (*Apis mellifera*)." *Apidologie* 35 (2): 133–142. https://doi.org/10.1051/apido:2004001.

Schlüns, H., R. F. A. Moritz, P. Neumann, P. Kryger, and G. Koeniger. 2005. "Multiple Nuptial Flights, Sperm Transfer and the Evolution of Extreme Polyandry in Honeybee Queens." *Animal Behaviour* 70 (1): 125–131. https://doi.org/10.1016/j.anbehav.2004.11.005.

Schlüns, H., E. A. Schlüns, J. Van Praagh, and R. F. A. Moritz. 2003. "Sperm Numbers in Drone Honeybees (*Apis mellifera*) Depend on Body Size." *Apidologie* 34 (6): 577–584. https://doi.org/10.1051/apido:2003051.

Schmickl, T., and K. Crailsheim. 2001. "Cannibalism and Early Capping: Strategy of Honeybee Colonies in Times of Experimental Pollen Shortages." *Journal of Comparative Physiology A-Neuroethology Sensory Neural and Behavioral Physiology* 187 (7): 541–547. https://doi.org/10.1007/s003590100226.

Schmid-Hempel, P. 1987. "Efficient Nectar-Collecting by Honeybees I. Economic Models." *Journal of Animal Ecology* 56 (1): 209. https://doi.org/10.2307/4810.

Schmid-Hempel, P., and T. Wolf. 1988. "Foraging Effort and Life Span of Workers in a Social Insect." *Journal of Animal Ecology* 57 (2): 509. https://doi.org/10.2307/4921.

Schmidt Capella, I. C., and K. Hartfelder. 1998. "Juvenile Hormone Effect on DNA Synthesis and Apoptosis in Caste-specific Differentiation of the Larval Honey Bee (*Apis mellifera* L.) Ovary." *Journal of Insect Physiology* 44 (5–6): 385–391. https://doi.org/10.1016/s0022-1910(98)00027-4.

Schmidt Capella, I. C., and K. Hartfelder. 2002. "Juvenile-Hormone-Dependent Interaction of Actin and Spectrin Is Crucial for Polymorphic Differentiation of the Larval Honey Bee Ovary." *Cell and Tissue Research* 307 (2): 265–272. https://doi.org/10.1007/s00441-001-0490-y.

Schmieder, S., D. Colinet, and M. Poirié. 2012. "Tracing Back the Nascence of a New Sex-Determination Pathway to the Ancestor of Bees and Ants." *Nature Communications* 3 (1). https://doi.org/10.1038/ncomms1898.

Schmitzová, J., J. Klaudiny, Š. Albert, W. Schröder, W. E. Schreckengost, J. Hanes, J. Júdová, and J. Šimúth. 1998. "A Family of Major Royal Jelly Proteins of the Honeybee *Apis mellifera* L." *Cellular and Molecular Life Sciences* 54 (9): 1020–1030. https://doi.org/10.1007/s000180050229.

Schulte, P., L. Alegret, I. Arenillas, J. A. Arz, P. J. Barton, P. R. Bown, T. J. Bralower, et al. 2010. "The Chicxulub Asteroid Impact and Mass Extinction at the Cretaceous-Paleogene Boundary." *Science* 327 (5970): 1214–1218. https://doi.org/10.1126/science.1177265.

Scotese, C. R., and J. Golonka. 1997. *Paleogeographic Atlas*. PALEOMAP Project. Arlington, TX: University of Texas at Arlington.

Seeley, T. D. 1977. "Measurement of Nest Cavity Volume by the Honey Bee (*Apis mellifera*)." *Behavioral Ecology and Sociobiology* 2 (2): 201–227. https://doi.org/10.1007/bf00361902.

Seeley, T. D. 1978. "Life History Strategy of the Honey Bee, *Apis mellifera*." *Oecologia* 32 (1): 109–118. https://doi.org/10.1007/bf00344695.

Seeley, T. D. 1982. "Adaptive Significance of the Age Polyethism Schedule in Honeybee Colonies." *Behavioral Ecology and Sociobiology* 11 (4): 287–293. https://doi.org/10.1007/bf00299306.

Seeley, T. D. 1985. *Honeybee Ecology: A Study of Adapatation in Social Life*. Princeton, NJ: Princeton University Press.

Seeley, T. D. 1997. "Honey Bee Colonies Are Group-Level Adaptive Units." *American Naturalist* 150 (S1): S22–S41. https://doi.org/10.1086/286048.

Seeley, T. D. 2003. "Consensus Building During Nest-Site Selection in Honey Bee Swarms: The Expiration of Dissent." *Behavioral Ecology and Sociobiology* 53 (6): 417–424. https://doi.org/10.1007/s00265-003-0598-z.

Seeley, T. D. 2010. *Honeybee Democracy*. Princeton, NJ: Princeton University Press.

Seeley, T. D. 2017. "Life-History Traits of Wild Honey Bee Colonies Living in Forests Around Ithaca, NY, USA." *Apidologie* 48 (6): 743–754. https://doi.org/10.1007/s13592-017-0519-1.

Seeley, T. D., and S. A. Kolmes. 1991. "Age Polyethism for Hive Duties in Honey Bees—Illusion or Reality?" *Ethology* 87 (3–4): 284–297. https://doi.org/10.1111/j.1439-0310.1991.tb00253.x.

Seeley, T. D., and R. A. Morse. 1976. "The Nest of the Honey Bee (*Apis mellifera* L.)." *Insectes Sociaux* 23 (4): 495–512. https://doi.org/10.1007/bf02223477.

Seeley, T. D., and R. A. Morse. 1978. "Nest Site Selection by the Honey Bee, *Apis mellifera*." *Insectes Sociaux* 25: 323–337. https://doi.org/10.1007/BF02224297.

Seeley, T. D., and D. R. Tarpy. 2006. "Queen Promiscuity Lowers Disease Within Honeybee Colonies." *Proceedings of the Royal Society B: Biological Sciences* 274 (1606): 67–72. https://doi.org/10.1098/rspb.2006.3702.

Seeley, T. D., and P. K. Visscher. 1985. "Survival of Honeybees in Cold Climates: The Critical Timing of Colony Growth and Reproduction." *Ecological Entomology* 10 (1): 81–88. https://doi.org/10.1111/j.1365-2311.1985.tb00537.x.

Seeley, T. D., and P. K. Visscher. 2004. "Quorum Sensing During Nest-Site Selection by Honeybee Swarms." *Behavioral Ecology and Sociobiology* 56 (6): 594–601. https://doi.org/10.1007/s00265-004-0814-5.

Seeley, T. D., P. K. Visscher, T. Schlegel, P. M. Hogan, N. R. Franks, and J. A. R. Marshall. 2012. "Stop Signals Provide Cross Inhibition in Collective Decision-

Making by Honeybee Swarms." *Science* 335 (6064): 108–111. https://doi.org/10.1126/science.1210361.

Shanley, D. P., and T. B. L. Kirkwood. 2001. "Evolution of the Human Menopause." *BioEssays* 23 (3): 282–287. https://doi.org/10.1002/1521-1878(200103)23:3.

Sharma, P. 1960. "Observations on the Swarming and Mating Habits of the Indian Honeybee." *Bee World* 41: 121–125.

Shelomi, M.. 2012. "Where Are We Now? Bergmann's Rule Sensu Lato in Insects." *American Naturalist* 180 (4): 511–519. https://doi.org/10.1086/667595.

Sheppard, W. S., and M. D. Meixner. 2003. "*Apis mellifera pomonella*, a New Honey Bee Subspecies From Central Asia." *Apidologie* 34 (4): 367–375. https://doi.org/10.1051/apido:2003037.

Sherman, G., and P. K. Visscher. 2002. "Honeybee Colonies Achieve Fitness Through Dancing." *Nature* 419 (6910): 920–922. https://doi.org/10.1038/nature01127.

Sherman, P. W., E A. Lacey, H. K. Reeve, and L. Keller. 1995. "The Eusociality Continuum." *Behavioral Ecology* 6 (1): 102–108. http://dx.doi.org/10.1093/beheco/6.1.102.

Sherman, P. W., T. D. Seeley, and H. K. Reeve. 1988. "Parasites, Pathogens, and Polyandry in Social Hymenoptera." *American Naturalist* 131 (4): 602–610. https://doi.org/10.1086/284809.

Shriner, D., F. Tekola-Ayele, A. Adeyemo, and C. N. Rotimi. 2014. "Genome-Wide Genotype and Sequence-Based Reconstruction of the 140,000 Year History of Modern Human Ancestry." *Scientific Reports* 4 (1). https://doi.org/10.1038/srep06055.

Shuel, R., and S. Dixon. 1960. "The Early Establishment of Dimorphism in the Female Honeybee, *Apis mellifera* L." *Insectes Sociaux* 7: 265–282.

Shykoff, J. A., and P. Schmid-Hempel. 1991. "Parasites and the Advantage of Genetic Variability Within Social Insect Colonies." *Proceedings of the Royal Society B: Biological Sciences* 243 (1306): 55–58. https://doi.org/10.1098/rspb.1991.0009.

Silberrad, R. E. 1976. *Bee-Keeping in Zambia,* Bucharest, Romania: Apimondia.

Simmons, L. W. 2002. *Sperm Competition and Its Evolutionary Consequences in the Insects.* Princeton, NJ: Princeton University Press.

Simone-Finstrom, M., B. Foo, D. R. Tarpy, and P. T. Starks. 2014. "Impact of Food Availability, Pathogen Exposure, and Genetic Diversity on Thermoregulation in Honey Bees (*Apis mellifera*)." *Journal of Insect Behavior* 27 (4): 527–539. https://doi.org/10.1007/s10905-014-9447-3.

Simone-Finstrom, M., and M. Spivak. 2010. "Propolis and Bee Health: The Natural History and Significance of Resin Use by Honey Bees." *Apidologie* 41 (3): 295–311. https://doi.org/10.1051/apido/2010016.

Simone-Finstrom, M., and D. R. Tarpy. 2018. "Honey Bee Queens Do Not Count Mates to Assess Their Mating Success." *Journal of Insect Behavior* 31 (2): 200–209. https://doi.org/10.1007/s10905-018-9671-3.

Simpson, J. 1957. "Observations on Colonies of Honey-Bees Subjected to Treatmets Designed to Induce Swarming." *Proceedings of the Royal Entomological Society of London A* 32: 185–192.

Skinner, M. 1991. "Bee Brood Consumption: An Alternative Explanation for Hypervitaminosis a in KNM-ER 1808 *Homo erectus* From Koobi Fora, Kenya." *Journal of Human Evolution* 20 (6): 493–503. https://doi.org/10.1016/0047-2484(91)90022-n.

Slaa, E. J. 2006. "Population Dynamics of a Stingless Bee Community in the Seasonal Dry Lowlands of Costa Rica." *Insectes Sociaux* 53: 70–79.

Smith, C. R., K. E. Anderson, C. V. Tillberg, J. Gadau, and A. V. Suarez. 2008. "Caste Determination in a Polymorphic Social Insect: Nutritional, Social, and Genetic Factors." *American Naturalist* 172 (4): 497–507. https://doi.org/10.1086/590961.

Smith, D. 2021. "Biogeography of Honey Bees." In *Encyclopedia of Social Insects*, edited by C. Starr. Cham, Switzerland: Springer Nature.

Smith, F. G. 1961. "The Races of Honeybees in Africa." *Bee World* 42: 255–260.

Smith, M. 1959. "A Note on the Capping Activities of an Individual Honeybee." *Bee World* 40: 153–154.

Soman, A., and S. Chawda. 1996. "A Contribution to the Biology and Behaviour of the Dwarf Bee, *Apis florea* F. and its Economic Importance in Kutch, Gujarat, India." *Indian Bee Journal* 58: 81–88.

Southwick, E. E., and Robin F. A. Moritz. 1987. "Social Control of Air Ventilation in Colonies of Honey Bees, *Apis mellifera*." *Journal of Insect Physiology* 33 (9): 623–626. https://doi.org/10.1016/0022-1910(87)90130-2.

Spivak, M., and R. G. Danka. 2020. "Perspectives on Hygienic Behavior in *Apis mellifera* and Other Social Insects." *Apidologie* 52 (1): 1–16. https://doi.org/10.1007/s13592-020-00784-z

Spivak, M., and M. Gilliam. 1998. "Hygienic Behaviour of Honey Bees and its Application for Control of Brood Diseases and Varroa: Part II. Studies on Hygienic Behaviour since the Rothenbuhler Era." *Bee World* 79: 169–186.

Spivak, M., and G. S. Reuter. 1998. "Performance of Hygienic Honey Bee Colonies in a Commercial Apiary." *Apidologie* 29 (3): 291–302. https://doi.org/10.1051/apido:19980308.

Spivak, M., and G. S. Reuter. 2001. "Resistance to American Foulbrood Disease by Honey Bee Colonies *Apis mellifera* Bred for Hygienic Behavior." *Apidologie* 32 (6): 555–565. https://doi.org/10.1051/apido:2001103.

Spötter, A., P. Gupta, M. Mayer, N. Reinsch, and K. Bienefeld. 2016. "Genome-Wide Association Study of a Varroa-Specific Defense Behavior in Honeybees (*Apis mellifera*)." *Journal of Heredity* 107 (3): 220–227. https://doi.org/10.1093/jhered/esw005.

Stabentheiner, A., H. Kovac, and R. Brodschneider. 2010. "Honeybee Colony Thermoregulation—Regulatory Mechanisms and Contribution of Individuals in Dependence on Age, Location and Thermal Stress." *PLOS ONE* 5 (1): e8967. https://doi.org/10.1371/journal.pone.0008967.

Stabentheiner, A., H. Kovac, M. Mandl, and H. Käfer. 2021. "Coping With the Cold and Fighting the Heat: Thermal Homeostasis of a Superorganism, the Honeybee Colony." *Journal of Comparative Physiology A-Neuroethology Sensory Neural and Behavioral Physiology* 207 (3): 337–351. https://doi.org/10.1007/s00359-021-01464-8.

Starks, P. T., and D. C. Gilley. 1999. "Heat shielding: A Novel Method of Colonial Thermoregulation in Honey Bees." *Naturwissenschaften* 86: 438–440.

Starr, C. K. 1985. "Sperm competition, kinship, and sociality in the aculeate Hymenoptera." In *Sperm Competition and the Evolution of Animal Mating Systems*, edited by R. L. Smith. Orlando, FL: Academic Press.

Steinthorsdottir, M., H. Coxall, A. M. De Boer, M. Huber, N. Barbolini, C. Bradshaw, Na. J. Burls, et al. 2021. "The Miocene: The Future of the Past." *Paleoceanography and Paleoclimatology* 36 (4). https://doi.org/10.1029/2020pa004037.

Stephen, W. P., and G. E. Bohart. 1969. *The Biology and External Morphology of Bees with a Synopsis of the Genera of Northwestern America*. Corvallis, OR: Oregon State University Agricultural Experiment Station.

Stork, N. E. 2018. "How Many Species of Insects and Other Terrestrial Arthropods Are There on Earth?" *Annual Review of Entomology* 63 (1): 31–45. https://doi.org/10.1146/annurev-ento-020117-043348.

Strang, G. E. 1970. "A Study of Honey Bee Drone Attraction in the Mating Response." *Journal of Economic Entomology* 63: 641–645.

Strassmann, J. 2001. "The Rarity of Multiple Mating by Females in the Social Hymenoptera." *Insectes Sociaux* 48: 1–13.

Strassmann, J. E., and D. C. Queller. 2007. "Insect Societies as Divided Organisms: The Complexities of Purpose and Cross-Purpose." *Proceedings of the National Academy of Sciences of the United States of America* 104 (Suppl_1): 8619–8626. https://doi.org/10.1073/pnas.0701285104.

Strassmann, J. E., and D. C. Queller. 2010. "The Social Organism: Congresses, Parties, and Committees." *Evolution* 64 (3): 605–616. https://doi.org/10.1111/j.1558-5646.2009.00929.x.

Strauss, K., H. Scharpenberg, R. M. Crewe, F. Glahn, H. Foth, and R. F. A. Moritz. 2008. "The Role of the Queen Mandibular Gland Pheromone in Honeybees (*Apis mellifera*): Honest Signal or Suppressive Agent?" *Behavioral Ecology and Sociobiology* 62 (9): 1523–1531. https://doi.org/10.1007/s00265-008-0581-9.

Strauss, U., C. W. W. Pirk, V. Dietemann, R. M. Crewe, and H. Human. 2014. "Infestation Rates of *Varroa destructor* and *Braula coeca* in the Savannah Honey Bee (*Apis mellifera scutellata*)." *Journal of Apicultural Research* 53 (4): 475–477. https://doi.org/10.3896/ibra.1.53.4.10.

Stürup, M., B. Baer-Imhoof, D. R. Nash, J. J. Boomsma, and B. Baer. 2013. "When Every Sperm Counts: Factors Affecting Male Fertility in the Honeybee *Apis mellifera*." *Behavioral Ecology* 24 (5): 1192–1198. https://doi.org/10.1093/beheco/art049.

Su, S., Š. Albert, S. W. Zhang, S. Maier, S. Chen, H.-H. Du, and J. Tautz. 2007. "Non-Destructive Genotyping and Genetic Variation of Fanning in a Honey Bee Colony." *Journal of Insect Physiology* 53 (5): 411–417. https://doi.org/10.1016/j.jinsphys.2007.01.002.

Svendsen, J. I., H. Alexanderson, V. Astakhov, I. Demidov, J. A. Dowdeswell, S. Funder, V. Gataullin, et al. 2004. "Late Quaternary Ice Sheet History of Northern Eurasia." *Quaternary Science Reviews* 23 (11–13): 1229–1271. https://doi.org/10.1016/j.quascirev.2003.12.008.

Swenson, W. M., D. Sloan Wilson, and R. Elias. 2000. "Artificial Ecosystem Selection." *Proceedings of the National Academy of Sciences of the United States of America* 97 (16): 9110–9114. https://doi.org/10.1073/pnas.150237597.

Szabo, T. I. 1983. "Effect of Various Combs on the Development and Weight Gain of Honeybee Colonies." *Journal of Apicultural Research* 22: 45–48.

Tân, N. Q., M. Mardan, P. H. Thái, and P. H. Chinh. 1999. "Observations on Multiple Mating Flights of *Apis dorsata* Queens." *Apidologie* 30 (4): 339–346. https://doi.org/10.1051/apido:19990410.

Tarpy, D. R. 2003. "Genetic Diversity Within Honeybee Colonies Prevents Severe Infections and Promotes Colony Growth." *Proceedings of the Royal Society B: Biological Sciences* 270 (1510): 99–103. https://doi.org/10.1098/rspb.2002.2199.

Tarpy, D. R., D. C. Gilley, and T. D. Seeley. 2004a. "Levels of Selection in a Social Insect: A Review of Conflict and Cooperation During Honey Bee (*Apis mellifera*) Queen Replacement." *Behavioral Ecology and Sociobiology* 55 (6): 513–523. https://doi.org/10.1007/s00265-003-0738-5.

Tarpy, D. R., R. Nielsen, and D. I. Nielsen. 2004b. "A Scientific Note on the Revised Estimates of Effective Paternity Frequency in *Apis*." *Insectes Sociaux* 51 (2): 203–204. https://doi.org/10.1007/s00040-004-0734-4.

Tarpy, D. R., and R. E. Page. 2000. "No Behavioral Control Over Mating Frequency in Queen Honey Bees (*Apis mellifera* L.): Implications for the Evolution of Extreme Polyandry." *American Naturalist* 155 (6): 820–827. https://doi.org/10.1086/303358.

Tarpy, D. R., and R. E. Page. 2001. "The Curious Promiscuity of Queen Honey Bees (*Apis mellifera*): Evolutionary and Behavioral Mechanisms." *Annales Zoologici Fennici* 38(3): 255–265. https://www.jstor.org/stable/23735844.

Tarpy, D. R., D. vanEngelsdorp, and J. S. Pettis. 2013. "Genetic Diversity Affects Colony Survivorship in Commercial Honey Bee Colonies." *Naturwissenschaften* 100 (8): 723–728. https://doi.org/10.1007/s00114-013-1065-y.

Tarpy, D. R., and T. D. Seeley. 2006. "Lower Disease Infections in Honeybee (*Apis mellifera*) Colonies Headed by Polyandrous Vs Monandrous Queens." *Naturwissenschaften* 93 (4): 195–199. https://doi.org/10.1007/s00114-006-0091-4.

Taylor, P. 1981. "Sex Ratio Compensation in Ant Populations." *Evolution* 35 (6): 1250. https://doi.org/10.2307/2408138.

Thackray, G. D. 1994. "Fossil Nest of Sweat Bees (Halictinae) From a Miocene Paleosol, Rusinga Island, Western Kenya." *Journal of Paleontology* 68 (4): 795–800. https://doi.org/10.1017/s0022336000026238.

Thompson, C., J. C. Biesmeijer, T. R. Allnutt, S. Pietravalle, and G. E. Budge. 2014. "Parasite Pressures on Feral Honey Bees (*Apis mellifera* Sp.)." *PLOS ONE* 9 (8): e105164. https://doi.org/10.1371/journal.pone.0105164.

Tierney, S. M., J. A. Smith, L. B. Chenoweth, and M. P. Schwarz. 2008. "Phylogenetics of Allodapine Bees: A Review of Social Evolution, Parasitism and Biogeography." *Apidologie* 39 (1): 3–15. https://doi.org/10.1051/apido:2007045.

Tihelka, E., C. Cai, D. Pisani, and P. C. J. Donoghue. 2020. "Mitochondrial Genomes Illuminate the Evolutionary History of the Western Honey Bee (*Apis mellifera*)." *Scientific Reports* 10 (1). https://doi.org/10.1038/s41598-020-71393-0.

Tindale, R. S., and T. Kameda. 2017. "Group Decision-Making From an Evolutionary/Adaptationist Perspective." *Group Processes & Intergroup Relations* 20 (5): 669–680. https://doi.org/10.1177/1368430217708863.

Tokarev, Y. S., W.-F. Huang, L. F. Solter, J. M. Malysh, J. J. Becnel, and C. R. Vossbrinck. 2020. "A Formal Redefinition of the Genera *Nosema* and *Vairimorpha* (Microsporidia: Nosematidae) and Reassignment of Species Based on Molecular Phylogenetics." *Journal of Invertebrate Pathology* 169 (January): 107279. https://doi.org/10.1016/j.jip.2019.107279.

Tollis, M., J. D. Schiffman, and A. M. Boddy. 2017. "Evolution of Cancer Suppression as Revealed by Mammalian Comparative Genomics." *Current Opinion in Genetics & Development* 42: 40–47.

Toth, A. L., and S. M. Rehan. 2017. "Molecular Evolution of Insect Sociality: An Eco-Evo-Devo Perspective." *Annual Review of Entomology* 62 (1): 419–442. https://doi.org/10.1146/annurev-ento-031616-035601.

Toth, A. L., and G. E. Robinson. 2007. "Evo-Devo and the Evolution of Social Behavior." *Trends in Genetics* 23 (7): 334–341. https://doi.org/10.1016/j.tig.2007.05.001.

Trivers, R., and H. L. Hare. 1976. "Haploidploidy and the Evolution of the Social Insect." *Science* 191 (4224): 249–263. https://doi.org/10.1126/science.1108197.

Tschinkel, W. 2006. *The Fire Ants*. Cambridge, MA: Harvard University Press.

Tsvetkov, N., S. Bahia, B. Calla, M. R. Berenbaum, and A. Zayed. 2023. "Genetics of Tolerance in Honeybees to the Neonicotinoid Clothianidin." *iScience* 26 (3): 106084. https://doi.org/10.1016/j.isci.2023.106084.

Urban, M. C. 2015. "Accelerating Extinction Risk From Climate Change." *Science* 348 (6234): 571–573. https://doi.org/10.1126/science.aaa4984.

Urquhart, F. A., and N. R. Urquhart. 1978. "Autumnal Migration Routes of the Eastern Population of the Monarch Butterfly (*Danaus P. plexippus* L.; Danaidae; Lepidoptera) in North America to the Overwintering Site in the Neovolcanic Plateau of Mexico." *Canadian Journal of Zoology* 56 (8): 1759–1764. https://doi.org/10.1139/z78-240.

Van Baalen, M. 1998. "Coevolution of Recovery Ability and Virulence." *Proceedings of the Royal Society B: Biological Sciences* 265 (1393): 317–325. https://doi.org/10.1098/rspb.1998.0298.

Van Doorn, A., and J. Heringa. 1986. "The Ontogeny of a Dominance Hierarchy in Colonies of the Bumblebee *Bombus terrestris* (Hymenoptera, Apidae)." *Insectes Sociaux* 33 (1): 3–25. https://doi.org/10.1007/bf02224031.

Van Zweden, J. S. 2010. "The Evolution of Honest Queen Pheromones in Insect Societies." *Communicative & Integrative Biology* 3 (1): 50–52. https://doi.org/10.4161/cib.3.1.9655.

Vandenberg, John D., Diane Redfield Massie, H. Shimanuki, J. R. Peterson, and D. M. Poskevich. 1985. "Survival, Behavior and Comb Construction by Honey Bees, *Apis mellifera*, in Zero Gravity Aboard NASA Shuttle Mission STS-13." *Apidologie* 16 (4): 369–384. https://doi.org/10.1051/apido:19850402.

Vásquez, A., and T. Olofsson. 2009. "The Lactic Acid Bacteria Involved in the Production of Bee Pollen and Bee Bread." *Journal of Apicultural Research* 48 (3): 189–195. https://doi.org/10.3896/ibra.1.48.3.07

Velthuis, H. H., M. C. Laurino, Z. Pereboom, and V. L. Imperatriz-Fonseca. 2003. "The Conservative Egg of the Genus *Melipona* and its Consequences for Speciation." *Apoidea Neotropica* 209–216. https://www.researchgate.net

/publication/237720383_The_conservative_egg_of_the_genus_Melipona_and_its_consequences_for_speciation.

Viljakainen, L., J. D. Evans, M. Hasselmann, O. Rueppell, S. Tingek, and P. Pamilo. 2009. "Rapid Evolution of Immune Proteins in Social Insects." *Molecular Biology and Evolution* 26 (8): 1791–1801. https://doi.org/10.1093/molbev/msp086.

Visick, O. D., and F. L. W. Ratnieks. 2022. "Density of Wild-Living Honey Bee, *Apis mellifera*, Colonies Worldwide: A Review." *Authorea (Authorea)* May 24. https://doi.org/10.22541/au.165342071.17994585/v1.

Visscher, P. K. 1996. "Reproductive Conflict in Honey Bees: A Stalemate of Worker Egg-Laying and Policing." *Behavioral Ecology and Sociobiology* 39 (4): 237–244. https://doi.org/10.1007/s002650050286.

Visscher, P. K. 2007. "Group Decision Making in Nest-Site Selection among Social Insects." *Annual Review of Entomology* 52 (1): 255–275. https://doi.org/10.1146/annurev.ento.51.110104.151025.

Visscher, P. K., R. S. Vetter, and G. E. Robinson. 1995. "Alarm Pheromone Perception in Honey Bees Is Decreased by Smoke (Hymenoptera: Apidae)." *Journal of Insect Behavior* 8 (1): 11–18. https://doi.org/10.1007/bf01990966.

Von Frisch, K., and L. E. Chadwick. 1967. *The Dance Language and Orientation of Bees*. Cambridge, MA: Belknap Press of Harvard University Press.

Wager, B. R., and M. D. Breed. 2000. "Does Honey Bee Sting Alarm Pheromone Give Orientation Information to Defensive Bees?" *Annals of the Entomological Society of America* 93 (6): 1329–1332. https://doi.org/10.1603/0013-8746(2000)093.

Wållberg, A., H. Fan, G. Wellhagen, B. Dahle, M. Kawata, N. Haddad, Z. L. P Simões, et al. 2014. "A Worldwide Survey of Genome Sequence Variation Provides Insight into the Evolutionary History of the Honeybee *Apis mellifera*." *Nature Genetics* 46 (10): 1081–1088. https://doi.org/10.1038/ng.3077.

Wang, D.-I, and F. E. Mofller. 1970. "The Division of Labor and Queen Attendance Behavior of Nosema-Infected Worker Honey Bees." *Journal of Economic Entomology* 63 (5): 1539–1541. https://doi.org/10.1093/jee/63.5.1539.

Ware, J. L., D. A. Grimaldi, and M. S. Engel. 2010. "The Effects of Fossil Placement and Calibration on Divergence Times and Rates: An Example From the Termites (Insecta: Isoptera)." *Arthropod Structure & Development* 39 (2–3): 204–219. https://doi.org/10.1016/j.asd.2009.11.003.

Wcislo, W. T., and B. N. Danforth. 1997. "Secondarily Solitary: The Evolutionary Loss of Social Behavior." *Trends in Ecology and Evolution* 12 (12): 468–474. https://doi.org/10.1016/s0169-5347(97)01198-1.

Weinstock, G. M., G. E. Robinson, R. A. Gibbs, K. C. Worley, J. D. Evans, R. Maleszka, H. M. Robertson, et al. 2006. "Insights Into Social Insects From the

Genome of the Honeybee *Apis mellifera*." *Nature* 443 (7114): 931–949. https://doi.org/10.1038/nature05260.

Weismann, A. 1889. *Essays upon Heredity*, London: Oxford University Press.

Weiss, K. 1965. "Über den Zuckerverbrauch und die Beanspruchung der Bienen bei der Wachserzeugung." *Z. Bienenforsch* 8: 106–124.

Weiss, K. 1984. "Regulierung des proteinhaushaltes im bienenvolk (*Apis mellifica* L.) durch brutkannibalismus." *Apidologie* 15: 339–354.

Wenseleers, T., H. Helanterä, A. G. Hart, and F. L. W. Ratnieks. 2004. "Worker Reproduction and Policing in Insect Societies: An ESS Analysis." *Journal of Evolutionary Biology* 17 (5): 1035–1047. https://doi.org/10.1111/j.1420-9101.2004.00751.x.

Wenseleers, T., and F. L. W. Ratnieks. 2006. "Comparative Analysis of Worker Reproduction and Policing in Eusocial Hymenoptera Supports Relatedness Theory." *American Naturalist* 168 (6): E163–E179. https://doi.org/10.1086/508619.

Wenzel, John W. 1990. "A Social Wasp's Nest From the Cretaceous Period, Utah, USA, and Its Biogeographical Significance." *Psyche* 97 (1–2): 21–29. https://doi.org/10.1155/1990/24696.

West, S. A. 2009. *Sex Allocation*. Princeton, NJ: Princeton University Press.

West, S. A., R. M. Fisher, A. Gardner, and E. T. Kiers. 2015. "Major Evolutionary Transitions in Individuality." *Proceedings of the National Academy of Sciences of the United States of America* 112 (33): 10112–10119. https://doi.org/10.1073/pnas.1421402112.

West, S. A., and A. Gardner. 2010. "Altruism, Spite, and Greenbeards." *Science* 327 (5971): 1341–1344. https://doi.org/10.1126/science.1178332.

West, S. A., A. S. Griffin, and A. Gardner. 2006. "Social Semantics: Altruism, Cooperation, Mutualism, Strong Reciprocity and Group Selection." *Journal of Evolutionary Biology* 20 (2): 415–432. https://doi.org/10.1111/j.1420-9101.2006.01258.x.

West-Eberhard, M. 1969. *The Social Biology of Polistine Wasps*. Ann Arbor, MI: University of Michigan.

West-Eberhard, M. J. 1996. "Wasp Societies as Microcosms for the Study of Development and Evolution." Chapter 17 in *Natural History and Evolution of Paper Wasps*, edited by S. Turillazz and M. J. West-Eberhard. Oxford, UK: Oxford University Press. https://doi.org/10.1093/oso/9780198549475.003.0017.

Westerkamp, C. 1991. "Honeybees Are Poor Pollinators? Why?" *Plant Systematics and Evolution* 177 (1–2): 71–75. https://doi.org/10.1007/bf00937827.

Wetterer, J. K. 1999. "The Ecology and Evolution of Worker Size-Distribution in Leaf-Cutting Ants (Hymenoptera: Formicidae)." *Sociobiology* 34: 119–144.

Wheeler, D. E. 1986. "Developmental and Physiological Determinants of Caste in Social Hymenoptera: Evolutionary Implications." *American Naturalist* 128: 13–34.

Wheeler, W. M. 1911. "The Ant-Colony as an Organism." *Journal of Morphology* 22: 307–325.

Wheeler, W. M. 1928. *The Social Insects: Their Origin and Evolution*. New York: Harcourt Brace. https://doi.org/10.5962/bhl.title.140774.

Whitfield, C. W., S. K. Behura, S. H. Berlocher, A. G. Clark, J. S. Johnston, W. S. Sheppard, D. R. Smith, A. V. Suarez, D. Weaver, and N. D. Tsutsui. 2006. "Thrice Out of Africa: Ancient and Recent Expansions of the Honey Bee, *Apis mellifera*." *Science* 314 (5799): 642–645. https://doi.org/10.1126/science.1132772.

Whitham, T. G., G. J. Allan, H. F. Cooper, and S. M. Shuster. 2020. "Intraspecific Genetic Variation and Species Interactions Contribute to Community Evolution." *Annual Review of Ecology, Evolution, and Systematics* 51 (1): 587–612. https://doi.org/10.1146/annurev-ecolsys-011720-123655.

Wille, A., and C. D. Michener. 1973. "The Nest Architecture of Stingless Bees with Special Reference to Those of Costa Rica (Hymenoptera, Apidae)." *Revista de Biologia Tropical* 21.

Williams, G. C. 1966. "Natural Selection, the Costs of Reproduction, and a Refinement of Lack's Principle." *American Naturalist,* 100: 687–690.

Williams, T. A., C. J. Cox, P. G. Foster, G. J. Szöllősi, and T. M. Embley. 2019. "Phylogenomics Provides Robust Support for a Two-Domains Tree of Life." *Nature Ecology and Evolution* 4 (1): 138–147. https://doi.org/10.1038/s41559-019-1040-x.

Wilson, D., and E. Sober. 1989. "Reviving the Superorganism." *Journal of Theoretical Biology* 136: 337–356.

Wilson, E. O. 1971. *The Insect Societies*. Cambridge, MA: Belknap Press, Harvard Universtiy Press.

Wilson, E. O. 1985. "The Sociogenesis of Insect Colonies." *Science* 28: 1489–1495.

Wilson, E. O. 1987. "Causes of Ecological Success: The Case of the Ants." *Journal of Animal Ecology* 56: 1–9.

Wilson, E. O. 2012. *The Social Conquest of Earth*. New York: WW Norton & Company.

Wilson, E. O. 2014. *The Meaning of Human Existence*. New York: WW Norton & Company.

Wilson, E. O., and B. Hölldobler. 2005. "Eusociality: Origin and Consequences." *Proceedings of the National Academy of Sciences of the United States of America,* 102: 13367–13371.

Wilson, E. O., and O. Kinne. 1990. *Success and Dominance in Ecosystems: The Case of the Social Insects*. Oldendorf/Luhe, Germany: Ecology Institute.

Wilson-Rich, N., D. R. Tarpy, and P. T. Starks. 2012. "Within- and Across-colony Effects of Hyperpolyandry on Immune Function and Body Condition in Honey Bees (*Apis mellifera*)." *Journal of Insect Physiology* 58 (3): 402–407. https://doi.org/10.1016/j.jinsphys.2011.12.020.

Winfree, R., N. M. Williams, H. R. Gaines, J. S. Ascher, and C. Kremen. 2007. "Wild Bee Pollinators Provide the Majority of Crop Visitation Across Land-Use Gradients in New Jersey and Pennsylvania, USA." *Journal of Applied Ecology* 45 (3): 793–802. https://doi.org/10.1111/j.1365-2664.2007.01418.x.

Winston, M. 1980a. "Swarming, Afterswarming, and Reproductive Rate of Unmanaged Honeybee Colonies (*Apis mellifera*)." *Insectes Sociaux* 27: 391–398.

Winston, M. L. 1980b. "Seasonal Patterns of Brood Rearing and Worker Longevity in Colonies of the Africanized Honey Bee (Hymenoptera: Apidae) in South America." *Journal of the Kansas Entomological Society* 157–165.

Winston, M. 1987. *The Biology of the Honey Bee.* Cambridge, MA: Harvard University Press.

Winston, M., and O. Taylor. 1980. "Factors Preceding Queen Rearing in the Africanized Honeybee (*Apis mellifera*) in South America." *Insectes Sociaux* 27: 289–304.

Winston, M. L., J. A. Dropkin,and O. R. Taylor. 1981. "Demography and Life History Characteristics of Two Honey Bee Races (*Apis mellifera*)." *Oecologia* 407–413.

Winston, M. L., and L. A. Fergusson. 1985. "The Effect of Worker Loss on Temporal Caste Structure in Colonies of the Honeybee (*Apis mellifera* L.)." *Canadian Journal of Zoology* 63: 777–780.

Winston, M. L., H. Higo, S. J. Colley, T. Pankiw, and K. N. Slessor. 1991. "The Role of Queen Mandibular Pheromone and Colony Congestion in Honey Bee (*Apis mellifera* L.) Reproductive Swarming (Hymenoptera: Apidae)." *Journal of Insect Behavior* 4 (5): 649–660. https://doi.org/10.1007/bf01048076.

Winston, M. L., and S. J. Katz. 1981. "Longevity of Cross-Fostered Honey Bee Workers (*Apis mellifera*) of European and Africanized Races." *Canadian Journal of Zoology* 59 (8): 1571–1575. https://doi.org/10.1139/z81-214.

Withrow, J. M, and D. R. Tarpy. 2018. "Cryptic 'Royal' Subfamilies in Honey Bee (*Apis mellifera*) Colonies." *PLOS ONE* 13 (7): e0199124. https://doi.org/10.1371/journal.pone.0199124.

Woese, C. R., O. Kandler, and M. L. Wheelis. 1990. "Towards a Natural System of Organisms: Proposal for the Domains Archaea, Bacteria, and Eucarya." *Proceedings of the National Academy of Sciences of the United States of America* 87 (12): 4576–4579. https://doi.org/10.1073/pnas.87.12.4576.

Woese, C. R., L. J. Magrum, and G. E. Fox. 1978. "Archaebacteria." *Journal of Molecular Evolution* 11: 245–252.

Wongsiri, S., C. Lekprayoon, R. B. Thapa, K. Thirakupt, T. E. Rinderer, H. A. Sylvester, B. P. Oldroyd, and U. Booncham. 1997. "Comparative Biology of *Apis andreniformis* and *Apis florea* in Thailand." *Bee World* 78 (1): 23–35. https://doi.org/10.1080/0005772x.1997.11099328.

Woodgate, J. L., J. C. Makinson, N. Rossi, K. S. Lim, A. Reynolds, C. J. Rawlings, and L. Chittka. 2021. "Harmonic Radar Tracking Reveals That Honeybee Drones Navigate Between Multiple Aerial Leks." *iScience* 24 (6): 102499. https://doi.org/10.1016/j.isci.2021.102499.

Wossler, T. C. 2002. "Pheromone Mimicry by *Apis mellifera capensis* Social Parasites Leads to Reproductive Anarchy in Host *Apis mellifera scutellata* Colonies." *Apidologie* 33 (2): 139–163. https://doi.org/10.1051/apido:2002006.

Woyciechowski, M., and D. Moroń. 2009. "Life Expectancy and Onset of Foraging in the Honeybee (*Apis mellifera*)." *Insectes Sociaux* 56 (2): 193–201. https://doi.org/10.1007/s00040-009-0012-6.

Woyke, J. 1960. "Naturaine i sztuczne unasienianie matek pszczelich [Natural and Artificial Insemination of Honey Bee Queens]." *Pszczeinicze Zeszyty Naukowe* 4: 183–275.

Woyke, J. 1963. "What Happens to Diploid Drone Larvae in a Honeybee Colony." *Journal of Apicultural Research* 2: 73–75.

Woyke, J. 1966. "Wovon hängt die Zahl der Spermien in der Samenblase der auf natürlichem Wege begatteten Königinnen ab." *Z. Bienenforsch* 8: 236–247.

Woyke, J. 1975. "Natural and Instrumental Insemination of *Apis cerana indica* in India." *Journal of Apicultural Research* 14: 153–159.

Woyke, J. 1988. "A Mathematical Model for the Dynamics of Spermatozoa Entry Into the Spermathecae of Instrumentally Inseminated Queen Honeybees." *Journal of Apicultural Research* 27 (2): 122–125. https://doi.org/10.1080/00218839.1988.11100790.

Woyke, J. 2011. "The Mating Sign of Queen Bees Originates From Two Drones and the Process of Multiple Mating in Honey Bees." *Journal of Apicultural Research* 50 (4): 272–283. https://doi.org/10.3896/ibra.1.50.4.04.

Yang, S., Q. Meng, W. Zhao, J. Wang, Y. Liu, X. Gong, and K. Dong. 2022. "Cell Orientation Characteristics of the Natural Combs of Honey Bee Colonies." *PLOS ONE* 17 (2): e0263249. https://doi.org/10.1371/journal.pone.0263249.

Yasui, Y. 2001. "Female Multiple Mating as a Genetic Bet-Hedging Strategy When Mate Choice Criteria Are Unreliable." *Ecological Research* 16 (4): 605–616. https://doi.org/10.1046/j.1440-1703.2001.00423.x.

Yoshida, T., J. Saito, and N. Kajigaya. 1994. "The Mating Flight Times of Native *Apis cerana japonica* Radoszkowski and Introduced *Apis mellifera* L in Sympatric Conditions." *Apidologie* 25 (4): 353–60. https://doi.org/10.1051/apido:19940401.

Zagwijn, W.H. 1989. "Vegetation and Climate During Warmer Intervals in the Late Pleistocene of Western and Central Europe." *Quaternary International* 3–4 (January): 57–67. https://doi.org/10.1016/1040-6182(89)90074-8.

Zareba, J., P. Błażej, A. Łaszkiewicz, Ł. Śnieżewski, M. Majkowski, S. Janik, and M. Cebrat. 2017. "Uneven Distribution of Complementary Sex Determiner (Csd) Alleles in *Apis mellifera* Population." *Scientific Reports* 7 (1). https://doi.org/10.1038/s41598-017-02629-9.

Zhukovskaya, M. I., A. Yanagawa, and B. T. Forschler. 2013. "Grooming Behavior as a Mechanism of Insect Disease Defense." *Insects* 4 (4): 609–630. https://doi.org/10.3390/insects4040609.

INDEX